WITHDRAWN

GLOBAL WARMING: PHYSICS AND FACTS

AIP CONFERENCE PROCEEDINGS 247

GLOBAL WARMING: PHYSICS AND FACTS

WASHINGTON, D.C. 1991

EDITORS:
BARBARA GOSS LEVI
PHYSICS TODAY

DAVID HAFEMEISTER
SENATE FOREIGN RELATIONS COMMITTEE

RICHARD SCRIBNER
GEORGETOWN UNIVERSITY

American Institute of Physics   New York

Authorization to photocopy items for internal or personal use, beyond the free copying permitted under the 1978 U.S. Copyright Law (see statement below), is granted by the American Institute of Physics for users registered with the Copyright Clearance Center (CCC) Transactional Reporting Service, provided that the base fee of $2.00 per copy is paid directly to CCC, 27 Congress St., Salem, MA 01970. For those organizations that have been granted a photocopy license by CCC, a separate system of payment has been arranged. The fee code for users of the Transactional Reporting Service is: 0094-243X/87 $2.00.

© 1992 American Institute of Physics.

Individual readers of this volume and nonprofit libraries, acting for them, are permitted to make fair use of the material in it, such as copying an article for use in teaching or research. Permission is granted to quote from this volume in scientific work with the customary acknowledgment of the source. To reprint a figure, table, or other excerpt requires the consent of one of the original authors and notification to AIP. Republication or systematic or multiple reproduction of any material in this volume is permitted only under license from AIP. Address inquiries to Series Editor, AIP Conference Proceedings, AIP, 335 East 45th Street, New York, NY 10017-3483.

L.C. Catalog Card No. 91-78423
ISBN 0-88318-932-1
DOE CONF-9104324

Printed in the United States of America.

CONTENTS

Preface vii

Introduction ix

Biographical Notes on Contributors xiii

A Tutorial on Global Atmospheric Energetics
and the Greenhouse Effect 1
    T. P. Ackerman
Global Climate Models: What and How 24
    D. A. Randall
Comparison of General Circulation Models 46
    R. D. Cess
Climate and the Earth's Radiation Budget 55
    V. Ramanathan, B. R. Barkstrom, and E. F. Harrison
Temperature and Sea Level Change 78
    G. A. Maul
Short-Term Climate Variability and Predictions 113
    J. Shukla
The Great Ocean Conveyor 129
    W. S. Broecker
Trace Gases in the Atmosphere: Temporal and Spatial Trends 162
    D. R. Blake
The Geochemical Carbon Cycle and the Uptake of Fossil Fuel CO_2 175
    J. F. Kasting and J. C. G. Walker
Forestry and Global Warming: The Physical and Policy Linkages 201
    M. C. Trexler
Policy Implications of Greenhouse Warming 222
    R. Coppock
Options for Lowering U.S. Carbon Dioxide Emissions 237
    R. M. Bierbaum, R. M. Friedman, H. Levenson, R. D. Rapoport, and N. Sundt
Options for Reducing Carbon Dioxide Emissions 261
    A. H. Rosenfeld and L. Price
Science and Diplomacy: A New Partnership to Protect the Environment 292
    R. E. Benedick

Author Index 311

Preface

What should be done about global warming? The issue is being debated at the highest governmental levels, where decision makers are asking: Are the projections valid? Will the Earth's climate grow warmer? Should we take any actions now? What should those actions be? Governmental leaders as well as ordinary citizens are looking to the scientists for guidance but there are no easy answers. The phenomenon of global warming is not a simple one: The deeper one delves into the basic physics of global warming, the more layers of complex issues and unanswered questions one finds. Thus this book is not a clarion call to action but rather an invitation to learn. It is specifically aimed at both scientists and policy makers who need to understand this critical issue on the deepest possible level so that they may more wisely function either as advisors to governmental, industrial, or community bodies, or as teachers to citizens and students. In the various chapters, experts in the field present the physics principles underlying global warming and tell which facts are scientifically established, which await further study, and what policy actions might be appropriate.

The book is based on a short course held on April 19–21, 1991 at Georgetown University. The sponsors were the Forum on Physics and Society and the Panel on Public Affairs of the American Physical Society, and the School of Foreign Service and the Department of Physics of Georgetown University. The 5 sessions drew participants of between 50 and 150 people. We gratefully acknowledge the support of Edward Finn, Chair of the Georgetown physics department, Peter Krogh, Dean of Georgetown's School of Foreign Service, and many students and faculty from both the Physics Deparment and the School of Foreign Service. They all provided generous assistance and kind hospitality in a most pleasant environment.

Barbara G. Levi
PHYSICS TODAY
New York, NY

David Hafemeister*
Senate Foreign Relations Committee
Washington, DC

Richard Scribner
Georgetown University
Washington, DC

*On leave from California Polytechnic State University, San Luis Obispo, CA.

INTRODUCTION

This book is about the possible global warming of our planet due to effluents dumped into our atmosphere as byproducts of development and industrialization. Global warming is not yet a factual certainty: Empirical evidence for any temperature rise is so far inconclusive. However, elaborate and successively refined efforts in computer modeling of the global climate, calibrated where possible by observations, strongly suggest that a doubling of the preindustrial atmospheric concentration of carbon dioxide, which is expected by the middle of the next century, could raise global average temperatures by several degrees. Given this possibility of climate change and the global environmental and social disruptions that might attend it, many have concluded that the prudent course is to take some mitigation measures now. These measures could have enormous impact on the direction and pace of development in both the industrialized and developing countries. But, without them, global warming could have a far more disastrous impact. The problem is that policy makers must choose a course now, while the scientific evaluation is far from complete.

The chapters in this book are intended to summarize the current state of knowledge about the phenomenon of global warming as well as recent deliberation of policy bodies about appropriate measures to ameliorate its effect. This introduction surveys some of the highlights.

Model Predictions

The lead chapter, by Thomas Ackerman, illustrates the basic mechanisms of radiation and convection with a simple one-dimensional model that treats the atmosphere as a single layer. The temperature of the Earth's upper atmosphere is fixed, to first order, by the balance that must be kept between the rate of energy absorbed from the Sun and the rate of energy radiated from the Earth. The temperature determined by this energy balance, applied at the outer edge of the atmosphere, is 255 K ($-18\,°C$). However, Earth is blanketed by an atmosphere that contains some so-called "greenhouse gases," such as carbon dioxide. These gases tend to transmit the downward solar radiation yet block or trap Earth's outward thermal radiation. Thus the global average temperature near the surface of the Earth (below this blanket) is 15 °C, i.e., 33 °C warmer than the globe would be without an atmosphere. This natural "greenhouse effect" is welcome, as it has made our planet more comfortable and habitable. However, the concern is that additions of more CO_2 or other gases with similar properties (such as methane, nitrous oxide, or chlorofluorocarbons) may cause the planet to grow still warmer.

Ackerman goes on to refine the one-layer model by adding multiple layers, accounting for the deposition of solar energy throughout the atmosphere, recognizing the variation in thermal absorption with altitude, and discussing the feedbacks from water vapor and from clouds.

While the basic mechanisms of convection and radiation are readily demonstrated by such a one-dimensional model, the real world, of course, is far more complex than can be represented by such a simple approach. Generating realistic simulations of the climate requires sophisticated three-dimensional global circulation models (GCMs).

In Chap. 2, David Randall presents the physics of the climate system and describes the numerical methods and physical parameterizations that have been developed for climate modeling. The current GCMs are able to reproduce many aspects of the observed climate, but most have fixed the sea surface temperatures at their observed values. The sea surface temperatures must be allowed to change if one is to predict the response of Earth's climate to a change in greenhouse-gas concentration. That in turn demands that the models feature more realistic representations of the oceans. The other major deficiency of GCMs is their inability to parameterize the effects of clouds. Scientists need to develop a successful theory of cloud formation and dissipation and they must then parameterize the interactions of the predicted clouds with solar and terrestrial radiation.

In Chap. 3, Robert Cess describes a recent comparison of a number of the GCMs currently being run. In the comparison the models all tried to predict the temperature rise that would result from a doubling of carbon dioxide. They were in close agreement when clouds were excluded from the simulations. However, the predicted temperature rises varied by as much as a factor of 3 (from 1.9 to 5.2 °C) when clouds were included. These results underscore the biggest uncertainty in these models—the feedback that will result from the cloud cover. Clouds tend to cool because they reflect the incoming sunlight, but they also tend to warm because they absorb the outgoing thermal radiation. High clouds tend to warm the system while low clouds tend to cool the system. In the current climate, the cooling effect dominates but no one knows whether a change in CO_2 will produce a positive or negative feedback from the clouds.

Although the feedback mechanism from clouds is not well understood, there is good evidence that the feedback from water vapor is positive: More carbon dioxide raises the temperature, hence evaporating more water vapor, which enhances the trapping of infrared radiation. In Chap. 4, V. Ramanathan, Bruce Barkstrom, and Edwin Harrison discuss measurements by the satellites in the Earth Radiation Budget Experiment. These satellites measured the flux of long-wave and short-wave radiation from the Earth. They helped to confirm that the net effect of the clouds in today's atmosphere is to cool the Earth, although no one yet knows whether an increase in greenhouse gases will increase or decrease that cooling effect. The measurements also determined that the water-vapor feedback is positive, and that its quantitative effect agrees closely with what the computer models predict.

Empirical Data

Because most of the current GCMs predict that Earth's temperature will rise with the levels of atmospheric carbon dioxide, it is only natural to wonder if the increase of carbon dioxide concentrations over the last 100 years has yet produced any observable warming. Several groups of scientists have concluded that the temperature may have risen somewhere between 0.2 and 0.5 °C in the past century. In Chap. 5, George Maul reviews the data on which such assessments of temperature trends are based, and describes the difficulties in discerning a statistically significant effect, especially in view of the large natural temperature and sea-level fluctuations. For example, most measurements of temperature were made on land whereas the bulk of Earth's surface area is oceans. The land-based temperature data must be corrected for the effects of "urban heat islands," while the data on sea surface temperatures suffer from changes in methods of measurement. Maul conducts a similar analysis of evidence for changes in the sea level.

In Chap. 6, Jagadish Shukla describes the nature of the short-term variability of climate systems, i.e., changes that occur on time scales from 10 to 1000 days. He discusses evidence that most of these fluctuations—such as monsoons or tropical droughts—are consequences of interactions among the different components of the climate system—atmosphere, ocean, and biosphere. Better simulation and prediction of short-term climate variability will increase confidence in models used to predict climate change.

To further complicate the picture, assumptions that warming trends may occur gradually may not be valid. Temperature changes may occur abruptly, rather than slowly, because the climate system is nonlinear. For example, in Chap. 7, Wally Broecker discusses his theory that the deep ocean current, which now operates as a kind of "conveyor belt," may flip-flop quite quickly between two extreme states, with consequent changes in land temperatures. Broecker presents evidence that such a sudden change of state has occurred in the past.

Greenhouse Gases

Water vapor is the atmospheric gas that accounts for the largest fraction of the natural greenhouse effect, and carbon dioxide comes next. But carbon dioxide has been the focus of most public concern because anthropogenic emissions have increased its concentrations from a preindustrial value of 280 ppmV to its present concentration of about 350 ppmV. As discussed in Chap. 8 by Donald Blake, other greenhouse gases—CFCs, methane, and nitrous oxide—are also increasing at a rapid rate. Although the concentrations of these other gases are smaller than CO_2 by factors of one thousand or one million, they can be more effective absorbers of thermal radiation. For example, a single molecule of, say, CFC-12, can be thousands of times more effective as an absorber than a given molecule of CO_2. At this time, CO_2 accounts for 55% of the increase in atmospheric greenhouse gases above the preindustrial atmosphere, with CFCs responsible for 24%, methane 15%, and nitrous oxide 6%.

In Chap. 9, James Kasting and James Walker describe the geochemical carbon cycle, putting the anthropogenic additions of carbon dioxide into a striking perspective. The major sink for atmospheric CO_2 is the weathering of silicate rocks and the major source of CO_2 is the emission from volcanoes. The present annual infusion of about 6–7 billion tonnes of carbon into the atmosphere swamps the natural source from volcanoes. Of the man-made CO_2 emissions, about 5–6 billion tonnes comes from the burning of fossil fuels and 0.3 to 2.0 billion tonnes comes from deforestation. According to a model by Kasting and Walker, if society eventually burns all presently calculated fossil fuel reserves, mitigation measures will not prevent a warming but only elongate the time scale and reduce the peak temperature.

In Chap. 10, Mark Trexler discusses the contribution of deforestation to effective increases in carbon dioxide and assesses the potential of reforestation to ameliorate the problem. Reforestation can remove carbon in two different scenarios: One is to reforst and keep the vegetation in a steady state. This procedure will remove modest amounts of carbon dioxide. A second scenario is to reforest, use the carbon sequestered by the vegetation (as in biomass energy), and reforest continually. This second procedure can have a much larger effect, which increases with the years. Most of the evaluations of the potential of reforestation done to date, Trexler warns, use only physical criteria in estimating how much land might be available for reforestation. Accounting for social and economic constraints is much more difficult, resulting in daunting uncertainty about what could actually be accomplished.

Prevention and Mitigation

If global warming is a reality, what should be done? A study of global warming released in the spring of 1991 by the National Academy of Sciences concluded that the "greenhouse warming is a potential threat sufficient to justify action now." In Chap. 11, Rob Coppock summarizes the findings of the NAS panel, whose report is entitled "Policy Implications of Greenhouse Warming." The report includes recommendations to change the U.S. energy policy, to promote forest offsets, and to support the proposed ban on CFCs by the end of this century. Mitigation options might reduce U.S. emissions by an estimated 10% to 40% at modest cost. Such actions could serve as insurance against the possibility of dramatic surprises. The NAS panel also looked at adaptation and international considerations.

At the same time that the NAS study appeared, the Office of Technology Assessment also released a report on global warming, called "Changing by Degrees: Steps to Reduce Greenhouse Gases." The OTA report concentrates on what remedies might be applied and at what cost. As summarized in Chap. 12 by Rosina Bierbaum, Robert Freidman, Howard Levenson, Richard Rapoport, and Nick Sundt, OTA arrived at two sets of mitigation measures: One that might hold U.S. carbon emissions (currently 20% of the world's output) to about a 15% increase over 1987 levels by 2015, and a second set of tougher options that could drop U.S. emissions to more than 20% below 1987 levels by 2015. How much might such efforts cost? Somewhat surprisingly, depending on a number of as-yet undetermined factors, the first set of measures might yield a net savings. The second set of measures might also yield a net savings, but it might also cost as much as $150 billion.

In Chap. 13, Art Rosenfeld and Lynn Price give some of the details about the potential savings from conservation measures. The measures they discuss illustrate the possibility of simultaneously saving both carbon dioxide and energy. For example, they calculate that the cost of conserved energy in an efficient refrigerator is about 2.5 cents per kilowatt hour, far below the average price of electricity of 6.3 cents per kilowatt hour.

If greenhouses gases indeed pose a threat, then action will be required at many levels, worldwide as well as nationally. In Chap. 14, Richard Benedick discusses the path-breaking Montreal Protocol to protect the ozone, which calls for a worldwide ban on CFC production by the year 2000, and compares the ozone issue to the problem of global warming: The Montreal Protocol was negotiated when the empirical data confirming linkage between CFCs and ozone depletion were still uncertain, and it set a precedent for a possible international agreement to limit the greenhouse gases. Benedick draws many lessons from the ozone experience that apply not only to international action on global warming but also more generally to the increasing overlap between science and diplomacy and the role of scientists in such negotiations.

Next Steps

With the wisdom of hindsight we can be glad that the prevailing scientific uncertainties did not delay the Montreal Protocol. Before the pact the problem seemed staggeringly complex, and formidable forces were arrayed against a limit on CFCs. Global warming is a more complex scientific problem by far than the ozone question, and the proof of a link between greenhouse gases and climate change is at least as tenuous as was the connection between CFCs and ozone depletion before the Montreal Protocol. Moreover, the mitigation measures for carbon dioxide are far more disruptive of our economies and lifestyles, and hence are meeting even greater resistance. Perhaps the greatest legacy of the Montreal Protocol may have been to point the way to cooperative action on the basis of strong, yet inconclusive, scientific evidence: It also set an example of paying an economically painful short-term cost to avoid incurring a possibly greater long-term price. This is precisely the problem the U.S. and the world now face with the issue of global warming.

BIOGRAPHICAL NOTES ON CONTRIBUTORS

Thomas Ackerman is an associate professor in the department of meteorology at Pennsylvania State University. After earning a Ph.D. in atmospheric science at the University of Washington in 1976, Ackerman worked at the Australian Numerical Meteorology Research Center before going to NASA Ames Research Center in 1979 and then to Penn State in 1988. Ackerman was one of five scientists known as TTAPS (from the first initials of their last names) who did the first "nuclear winter" calculations.

Bruce Barkstrom is a senior research scientist in the Radiation Sciences Branch of the Atmospheric Sciences Division at NASA's Langley Research Center. During the past 12 years he has been the Experiment Scientist for the Earth Radiation Budget Experiment (ERBE) and leader of the ERBE science team. He is currently principal investigator of the Earth Observing System (EOS) investigation of Clouds and the Earth's Radiant Energy System. Before going to Langley in 1979, Barkstrom was on the faculty of George Washington University. Barkstrom holds a Ph.D. in astronomy from Northwestern University.

Ambassador Richard Benedick, a career diplomat, is currently on detail from the U.S. Department of State to the World Wildlife Fund in Washington, DC. After several overseas assignments, he was appointed Coordinator of Population Affairs and, more recently, Deputy Assistant Secretary for Environment, Health, and Natural Resources. He was the chief U.S. negotiator of the 1987 Montral Protocol on protecting the ozone layer and is author of *Ozone Diplomacy: New Directions in Safeguarding the Planet* (Harvard University, Cambridge, MA, 1991).

Rosina Bierbaum is a senior associate and project director for climate change in the Oceans and Environment Program of the U.S. Congress' Office of Technology Assessment. In that capacity she was the primary author of the OTA report, *Changing by Degrees: Steps to Reduce Greenhouse Gases*. After receiving a Ph.D. in ecology and evolution from the State University of New York at Stony Brook, Bierbaum first came to OTA in 1980 as a congressional science fellow. At OTA she has worked on studies of acid rain, ozone depletion, marine resources, and global warming.

Donald Blake is a research specialist in the chemistry department at the University of California at Irvine, which is the school at which he earned his Ph.D. in chemistry in 1984. He has appeared as an expert witness both for the state and federal governments. Much of his research has concerned atmospheric methane and the possible contribution of that gas—and other gases—to global warming.

Wally Broecker is the Newberry Professor of Geology at Columbia University, where he is associated with the University's Lamont-Doherty Geological Observatory. Broecker earned his Ph.D. from Columbia in 1958 and joined its faculty a year later. He is the auther of four textbooks, the most recent of which is *How to Build a Habitable Planet* (Eldigio, New York, 1987). Broecker is a member of the National Academy of Sciences.

Robert Cess is a Distinguished Service Professor in the Institute for Terrestrial and Planetary Atmospheres at the State University of New York at Stony Brook. His research interests involve climate change, climate feedback mechanisms, and global warming. He has been the recipient of the Heat Transfer Memorial Award of the American Society of Mechanical Engineers (1977), the NASA Exceptional Scientific Achievement Medal (1989), and the Oregon State University Distinguished Alumnus Award (1991).

Rob Coppock has a background both in science (a Bachelor's Degree in physics science from Washington State) and in economics (a Ph.D. from the University of Wuppertal, West Germany). He worked in Germany for about 13 years, first in the Battelle Lab in Frankfurt and then at the International Institute for Environment and Society of the Science Center in Berlin. He has been on the staff of the National Research Council since 1986. Most recently he has been the staff director of the Panel on Policy Implications of Greenhouse Warming of the National Academy of Sciences. Coppock has written about the regulation of chemicals, social constraints on technological progress, and the perceptions of technological risk. He serves as chairman of the global risk analysis section of the Society for Risk Analysis.

Robert M. Friedman is a senior associate in the Oceans and Environment Program of the Office of Technology Assessment. He came to OTA in 1979 as a congressional fellow after earning a Ph.D. in environmental science from the University of Wisconsin at Madison. He has directed OTA assessments concerned with acid rain and with ozone, both undertaken in support of the reauthorization of the Clean Air Act. He most recently worked on the OTA report on climate change, *Changing by Degrees*.

David Hafemeister is a professor of physics at the California Polytechnic State University. He served as a science advisor in the U.S. Senate from 1975–77 and a special assistant to an Under Secretary of State in 1977–79. He was a visiting scientist at MIT in 1983–84, Lawrence Berkeley Laboratory in 1985, and the U.S. Department of State in 1987. From 1989–91, while this book was being produced, he served on the staff of the U.S. Senate Foreign Relations Committee.

Edwin F. Harrison is a senior research scientist and head of the Radiation Sciences Branch of the Atmospheric Sciences Division at NASA's Langley Research Center. He has played a significant role in the development of the Earth Radiation Budget Experiment beginning in the mid-1970s. He is currently a principal investigator on the ERBE science team, and a co-investigator on both an experiment on the Earth Observing System and the First International Cloud Climatology Project Regional Experiment. Harrison earned a B.S. in mathematics from East Carolina University in 1957, and started working at Langley immediately after graduation.

James Kasting holds an M.S. in physics and a Ph.D. in atmospheric sciences from the University of Michigan. After a postdoctoral appointment at the National Center for Atmospheric Research, Kasting worked for seven years in the Space Science Division at NASA Ames. He currently holds a joint position in geosciences and meteorology at Pennsylvania State University. His research interests are in planetary atmospheres, such as the runaway greenhouse effect on Venus, and in the long-term evolution of Earth's atmosphere. He is also interested in the problem of the uptake of carbon dioxide from fossil fuel combustion.

Howard Levenson is a senior associate in the Oceans and Environment Program at the Office of Technology Assessment. He has worked on a range of environmental issues, including marine pollution, solid waste management, endangered species, and the relationships between natural resource policies and global climate change. Levenson has been at OTA since 1983, has directed two studies, contributed to several others, including *Changing by Degrees*, and is now co-authoring another assessment. He received a Ph.D. in biology from the University of Kansas.

Barbara Goss Levi is a senior associate editor at *Physics Today*, a magazine with which she has been associated since earning her Ph.D. from Stanford in 1971. She has also taught physics at Fairleigh Dickinson University, Georgia Tech, and Rutgers University. From 1980–1987, Levi was on the research staff of Princeton University's Center for Energy and Environmental Studies.

George Maul is a supervisory oceanographer in the physical oceanography division of the National Oceanic and Atmospheric Administration in Miami, Florida, where he has worked since 1969. He holds a Ph.D. in physical oceanography from the University of Miami. From 1960 to 1969 he served as an officer in the U.S. Coast and Geodetic Survey. His research interests include reestablishing a Caribbean Sea and Adjacent Regions sea-level network for climate studies, determining absolute sea-level change, interannual variability in the circulation and climate of the ocean, and satellite altimetry research. He won an award for a 1984 book, *Introduction to Satellite Oceanography*, and has served on the editorial boards of "Marine Geodesy Journal" and "Remote Sensing of the Environment."

Lynn Price is a science writer for the Center for Building Science at Lawrence Berkeley Laboratory. She holds an M.S. in environmental science from the University of Wisconsin at Madison. Prior to assuming her present position at LBL, she provided litigation support and expert witness testimony in utility power-plant rate case proceeedings.

Veerabhadran Ramanathan is an Alderson Professor of Ocean Sciences at Scripps Institution of Oceanography of the University of California at San Diego. His research is concerned with the greenhouse effect of trace gases, Earth's radiation budget, general circulation of the atmosphere, and climate change. In 1973, Ramanathan earned a Ph.D. in planetary atmospheres from the State University of New York at Stony Brook. Before coming to Scripps in 1990, Ramanathan worked at the National Center for Atmospheric Research, Colorado State University, and the University of Chicago. He is the principal investigator for the science team of NASA's Earth Radiation Budget Experiment, and he is a fellow of the American Meteorological Society and of the American Association for the Advancement of Science.

David Randall holds a Ph.D. in atmospheric sciences from UCLA. He was an assistant professor in meteorology at MIT for three years before moving to NASA's Goddard Space Flight Center, where he worked for nine years. He has been on the faculty of the department of atmospheric sciences at Colorado State University since 1988. His main interests are general circulation modeling, cloud-climate studies, and cloud parametrization. He is a fellow of the American Meteorological Society.

Richard Rapoport is a senior research scientist at Battelle Pacific Northwest Laboratories in Washington, DC. He holds a Ph.D. in environmental sciences and engineering from the University of Califonia at Los Angeles. Until recently he was an analyst at the Office of Technology Assessment, where he participated in the study of climate change, *Changing by Degrees*.

Arthur Rosenfeld began his professional life as a particle physicist (he was Enrico Fermi's last graduate student) before switching to work on energy conservation. He is now a professor of physics at the University of California at Berkeley and the director of the Center for Building Science at the Lawrence Berkeley Laboratory. He has written four books, the most recent of which is *Supplying Energy Through Greater Efficiency*. In 1986 he was the recipient of the Forum's Szilard Award for doing physics in the public interest. He recently served on the Mitigation Panel for the National Academy of Sciences' "Report on Policy Implications of Greenhouse Warming."

Richard Scribner is an associate professor and director of the science, technology, and international affairs program in Georgetown University's School of Foreign Service. He was originally trained as a low-temperature, solid-state physicist, having earned a Ph.D. from the University of Florida in 1968. He has served in senior staff and programmatic roles in the U.S. State Department and with the American Association for the Advancement of Science. In both organizations, he was involved with global and transborder environmental issues including, e.g., acid rain, as well as a variety of arms control, international technology, and science policy matters.

Jagadish Shukla earned a Ph.D. in geophysics from Banaras Hindu University in India and a Sc.D. in meteorology from MIT. From 1979 to 1983 he headed the climate modeling group at NASA's Goddard Space Flight Center. Since 1984 Shukla has been on the faculty of the department of meteorology at the University of Maryland, where he also directs the Center for Ocean-Land-Atmosphere Interactions. He has served on numerous national and international panels dealing with meteorology and climate and is a fellow of the American Meteorological Society.

Nick Sundt is an independent consultant in Washington, DC, specializing in energy, natural resources, and climate. From 1982 through 1990 he was an analyst with the Office of Technology Assessment, where he worked among others on the report, *Changing by Degrees*. He earned an M.A. in energy and resources from the University of California at Berkeley.

Mark Trexler directed work on global warming mitigation through forestry for the World Resources Institute until May 1990. Now living in Portland, Oregon, he continues to work with the Institute on forestry, energy policy, and global warming issues. Before joining WRI Trexler did extensive evaluations of the implementation of international environmental agreements for the Economic Commission for Europe and the World Wildlife Fund. Trexler formerly served on the staff of the California Energy Commissions, has been a research associate of the International Union for the Conservation of Nature, and has been a Fulbright Fellow. He holds a Ph.D. in public policy from the University of California at Berkeley.

James Callan Gray Walker holds two positions at the University of Michigan: He is a professor of geochemical dynamics in the college of literature, science, and the arts, and a professor of atmospheric and oceanic science in the college of engineering. Walker earned a Ph.D. in geophysics from Columbia University in 1964. He was a professor of geology and geophysics at Yale University from 1967 to 1974, an ionosphere group leader and senior research associate at the Arecibo Observatory in Chile (1974–80), and an associate director of the Space Physics Research Laboratory at the University of Michigan (1980–84), before assuming his current position in the engineering school. He has served numerous publications in an editorial capacity, and he is a fellow of the American Geophysical Society.

A TUTORIAL ON GLOBAL ATMOSPHERIC ENERGETICS AND THE GREENHOUSE EFFECT

Thomas P. Ackerman
Department of Meteorology, The Pennsylvania State University
University Park, PA 16802

INTRODUCTION

Although the climate of the earth is complicated and highly non-linear, many of its properties can be illustrated with rather simple models. In this paper, we focus our attention on a few of these simple models and use them to demonstrate the physics of global warming and some of the important feedback processes. In some cases, we even can use the models to come up with back-of-the-envelope estimates of the magnitudes of the warming produced by different physical mechanisms. As we proceed, however, it should be remembered that reality is far more complex than the simple models we use here.

PLANETARY RADIATIVE EQUILIBRIUM

If we consider our planetary atmosphere as a single thermodynamic system, we can apply to it the first law of thermodynamics, expressed as:

$$\frac{dU}{dt} = Q - W \qquad (1)$$

where U is the internal energy, Q the time rate at which heat is added to the system, and W the rate at which work is done by the system. Note that both Q and W are powers or energy per unit time.

For an ideal gas, the internal energy may be expressed as

$$\frac{dU}{dt} = Mc_V \frac{dT}{dt}$$

2 Global Atmospheric Energetics

where M is the mass of the atmosphere, c_V the specific heat capacity at constant volume of air, and T the temperature. Here, T represents some average temperature of our system since the entire atmosphere need not be isothermal (and indeed is unlikely to be isothermal). Recognizing that the work performed by an ideal gas is given by $V\,dp$, where V is the volume and p the pressure, we can transform Eq. 1 into the form

$$Mc_P \frac{dT}{dt} = Q + MRT\frac{d}{dt}\ln(p) \qquad (2)$$

where c_P is the specific heat capacity at constant pressure and R is the gas constant. In this expression, the left side of the equation again simply represents the time rate of change in internal energy, while the right side contains two terms. The first of these is the heat added to the system and, in general, is referred to as the diabatic term. Although Q is often equated only with radiative processes, heating associated with phase changes is a very important component of Q in the atmosphere. The second term represents temperature changes due to adiabatic processes or fluid motions. Since we are concerned with integrating over the entire atmosphere, this term must integrate to zero. Adiabatic compression of some column of air must be balanced on average by adiabatic expansion of another column, leaving no net temperature change.

If the planet is observed from outside the atmosphere, then the diabatic processes can consist only of radiation absorbed and emitted. For convenience, we break Q into two components: Q_{SOL}, representing the absorbed solar radiation, and Q_{IR}, representing the thermal radiation emitted by the planet. For the atmosphere as a whole, we can rewrite (2) as:

$$Mc_p\frac{dT}{dt} = Q_{SOL} + Q_{IR}. \qquad (3)$$

If we allow sufficient time for equilibrium to occur, the left side becomes zero and the equation becomes

$$Q_{SOL} + Q_{IR} = 0, \qquad (4)$$

which is usually referred to as *radiative equilibrium*.

We can use (4) to compute the global mean temperature by balancing the energy absorbed and emitted. The rate of solar absorption is simply

$$Q_{SOL} = S_0(1 - A_P)\pi R_e^2$$

where A_P is the planetary albedo (or reflectivity) and S_0 is the solar irradiance at the radius of the earth. For the earth A_P has a value of about 0.3, due to scattering by molecules (also called Rayleigh scattering), particles, clouds and the earth's surface. For computational purposes in this paper, the solar irradiance is taken to be 1365

W/m². It is actually not constant, but varies slightly with solar activity and possibly with some long-term trend as well. R_e is the radius of the earth. It appears in the equation as the cross-sectional area of the planet because we are interested in the flux absorbed, which is the component of the solar radiation perpendicular to the earth's surface. Thus for the planet as a whole, the intercepting area is its cross-section.

The rate of thermal emission can be written as

$$-Q_{IR} = \sigma T_e^4 (4\pi R_e^2)$$

where σ is the Stefan-Boltzmann constant, and T_e is the radiative equilibrium temperature, i.e., the average temperature at which the planet must radiate in order to balance the absorbed solar radiation. Note that we have written Q_{IR} with a negative sign because it is defined to be the energy absorbed *by* the system. The geometrical factor in this case is the surface area of the planet, since the exiting flux is emitted perpendicular to the planet everywhere. Note that we do not distinguish between the radius of the planetary surface and the radius of the emitting surface since for the earth they vary by less then 10 km on average. Or, to put it more simply, the thickness of the atmosphere (which is not a well-defined quantity) is very small relative to the radius of the earth.

Equating these two expressions results in the equilibrium condition:

$$\frac{1}{4}(1 - A_P)S_0 = \sigma T_e^4. \tag{5}$$

As discussed, the factor of 1/4 in this equation arises from considering the planetary geometry. If we consider a single average column on the earth, we can argue for the same factor by noting that the sun shines only half the day and the cosine of the incident angle is on average about one half. From (5), we compute a value of the radiative equilibrium temperature of 255 K or -18°C.

From observations we can show that the average surface air temperature of the earth is about 286 K, which is significantly higher than the radiative equilibrium temperature. Does this mean that our energy balance expression is incorrect? Actually, no. In fact, the computed value of T_e is very close to the average value observed by satellite. This temperature of 255 K represents the average temperature of the thermal radiation emitted by the surface plus atmosphere. In the earth's atmosphere, this temperature occurs on average at an altitude of about 5 km. Thus, the effective radiating level of the combined system is not at the surface but at approximately the middle (by mass) of the atmosphere. (We will return to this point in Section VII.)

THE ATMOSPHERIC GREENHOUSE EFFECT

As we saw above, the surface air temperature is higher than the radiative equilibrium temperature of the planet. This is presumably due to the thin sheath of gas that surrounds the planet. What is it about the atmosphere that produces this warming of the surface?

Over 99% of the atmosphere is composed of diatomic and triatomic molecules. These molecules can absorb radiative energy over a wide range of the electromagnetic spectrum. However, they are in general much more effective at absorbing energy at infrared wavelengths (from about 1 to 100 μm) than at visible and near-infrared wavelengths (from about 0.3 to 1 μm). The reasons for this difference are complex and beyond the scope of this paper. As a consequence, the bulk of the solar radiation passes through the atmosphere and is absorbed at the surface. This warm surface emits thermal radiation, essentially as a black-body. The thermal radiation is partially absorbed in the atmosphere by a variety of atmospheric constituents such as H_2O, CO_2, and O_3. Although this absorption varies with wavelength, we will for the moment assume that it is spectrally invariant. This assumption is equivalent to stating that the atmosphere absorbs as a grey-body with an emissivity, ϵ_a, that is independent of wavelength (at thermal infrared wavelengths) and less than unity.

Based on this simple interpretation of the physics of the atmosphere, we can construct a one-layer model of the earth-atmosphere system as shown in Figure 1. The atmosphere is represented by a single layer with a uniform temperature, and the earth by another layer with a uniform temperature. It is assumed that no solar radiation is absorbed in the atmosphere. The validity of this assumption will be addressed in the following section. At equilibrium, the energy balance in this simple system can be represented by two equations in two unknowns, the temperature of the surface (T_s) and the temperature of the atmosphere (T_a). These two equations are:

$$\frac{1}{4}(1 - A_P)S_0 + \epsilon_a \sigma T_a^4 = \sigma T_s^4 \tag{6}$$

$$2\epsilon_a \sigma T_a^4 = \epsilon_a \sigma T_s^4 \tag{7}$$

The first of this pair represents the energy balance at the surface, and the second the energy balance of the one-layer atmosphere. The factor of 2 arises because a layer radiates in two directions, both up and down. Solving these two equations, we find that

$$T_a = \frac{1}{\sqrt[4]{2}} T_s \tag{8}$$

$$\sigma T_s^4 = \frac{\frac{1}{4}S_0(1 - A_P)}{(1 - \frac{1}{2}\epsilon_a)} \tag{9}$$

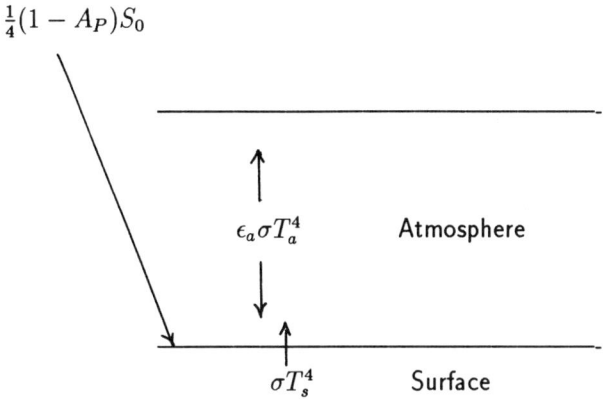

Figure 1: A schematic diagram of a one-layer atmospheric model.

Note that in this simple model, the atmospheric temperature depends only on the surface temperature. This result arises from Kirchhoff's law, which states that the absorptivity and emissivity of a black (or grey) body must be equal. The surface temperature is both a function of the solar energy absorbed and the emissivity of the atmosphere because the surface receives energy both from the sun and the thermal emission of the atmosphere. This latter term is the source of what we commonly call the "greenhouse effect". Although a greenhouse operates on quite another principle (suppressing heat loss by preventing convection and turbulent mixing), the name has become ingrained in the popular press and perhaps conveys a useful image. It might be more useful to describe this process as the "insulation effect" since the atmosphere behaves somewhat as insulation, preventing the loss of heat from the surface. Of course, the effects of home insulation are largely produced by preventing convective, rather than radiative, losses so this analogy is also not scientifically accurate.

Th surface and atmospheric temperatures computed from (8) and (9) are presented in Figure 2 as a function of atmospheric emissivity. It is perhaps easier to understand the curves shown in Figure 2 if we rewrite the temperatures making use of the relationship $\sigma T_e^4 = \frac{1}{4}(1 - A_P)S_0$. Here T_e is the radiative equilibrium temperature found in the previous section. In this case, we have

$$T_s = T_e/(1 - \tfrac{1}{2}\epsilon_a)^{1/4} \geq T_e \qquad (10)$$
$$T_a = T_e/(2 - \epsilon_a)^{1/4} \leq T_e \qquad (11)$$

The inequalities arise because the emissivity of the atmosphere in our simple model is constrained to lie between 0 and 1. The lower value represents the surface with

6 Global Atmospheric Energetics

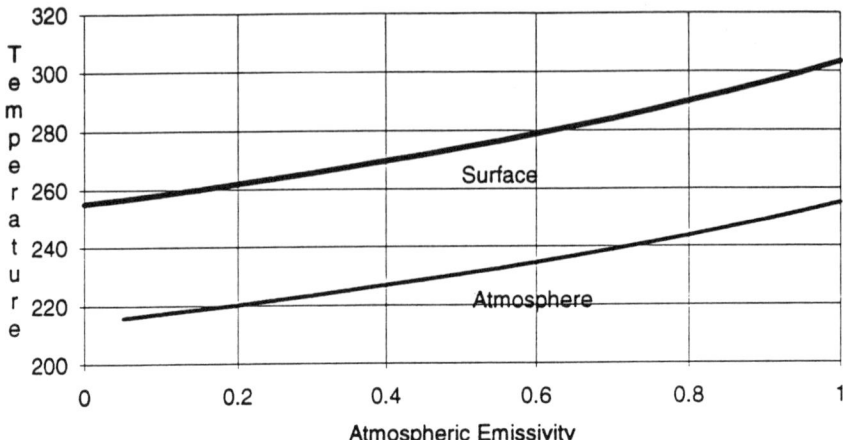

Figure 2: One-layer model temperatures as a function of atmospheric emissivity.

no atmosphere and the upper value represents the theoretical limit of a black-body atmosphere. As the emissivity approaches 0, $T_s \to T_e$, since there is no atmosphere to radiate energy down to the surface. We can derive an atmospheric temperature for this case, but it is physically meaningless. An examination of (7) shows that we are solving an equation in which both sides are 0; hence there is no solution for T_a. As the emissivity approaches 1, we find that $T_a = T_e$. This is due to the fact that no thermal emission emitted by the surface is transmitted by the atmosphere. Thus the energy emitted from the atmosphere must be equal to the radiation absorbed by the planet.

The spectrally-averaged emissivity of the earth's atmosphere is about 0.8. This leads to a surface temperature of about 288 K and an atmosphere temperature of about 244 K. The former is higher than the radiative equilibrium temperature, the latter is lower. In short, the atmosphere warms the surface by about 33 K over the radiative equilibrium temperature. Without this effect, the surface of our planet would be a substantially colder and less hospitable environment.

Before concluding this section, a word of warning should be given. It is often stated that if there were no atmosphere, the surface temperature of the earth would be lower by about 33 K. This is, however, a meaningless statement because the solar absorption would also change if the atmosphere were absent. The plantary reflectivity is determined largely by the presence of clouds and molecules, both of which would be absent if the atmosphere were absent. In addition, the surface of the planet would be quite different in the absence of oceans and vegetation. The earth would then resemble the moon or Mars. For both these bodies, the surface temperature can be predicted quite accurately from (9) with $\epsilon_a = 0$.

CHANGING THE ATMOSPHERIC EMISSIVITY

The emissivity of the earth's atmosphere is determined primarily by the concentrations of H_2O and CO_2, and secondarily by the concentrations of other trace gases that absorb infrared radation. Thus, altering the concentration of any of these gases will alter the emissivity of the atmosphere, which in turn changes the surface temperature via the relationship in (9). It is of interest to develop a relationship between the change in surface temperature in our simple model for a given change in atmospheric emissivity. This is essentially the slope of the curve shown in Figure 2.

A rearrangment of (9) gives

$$(1 - \frac{1}{2}\epsilon_a)\sigma T_s^4 = \frac{1}{4}S_0(1 - A_P) \tag{12}$$

We now differentiate (12) with respect to emissivity. The right side of the equation is independent of emissivity so the derivitive is zero. Approximating derivatives as differences and rearranging, we find

$$\frac{\Delta T_s}{T_s} = \frac{1}{4} \frac{\frac{1}{2}\Delta\epsilon_a(\sigma T_s^4)}{(1 - \frac{1}{2}\epsilon_a)(\sigma T_s^4)} \tag{13}$$

which has an straightforward physical interpretation. The denominator on the right (not including the factor of 4) is the solar energy absorbed in our system. The numerator is the change in energy radiated to the surface due to the change in emissivity ($\Delta\epsilon_a$). (Note that, from (4), $T_a^4 = T_s^4/2$.) Thus the ratio of the change in surface temperature to the surface temperature itself is equal to one-fourth the ratio of the change in emitted energy by the atmosphere to the energy absorbed in the system. The factor of one-fourth arises from the fourth power dependence of black-body radiation.

Let us now apply this expression to an actual atmospheric problem. From more sophisticated radiative transfer calculations (see for example, Ramanathan and Coakley, 1979), we can determine that doubling the atmospheric CO_2 concentration will increase the downward infrared radiative flux at the surface (the numerator in our expression above) by roughly 4 W/m². For current conditions, the absorbed solar energy amounts to some 240 W/m² and $T_s \approx 288K$. Plugging these values into (13), we find $\Delta T_s = 1.2 K$. In other words, for each doubling of CO_2, we expect the surface temperature to increase by about 1.2 K.

At this point, we might conclude that we have solved the greenhouse problem. Given a radiative transfer algorithm sufficiently powerful to calculate the change in downward infrared flux for a particular change in the concentration of some gaseous absorber, we can use our simple model to evaluate the corresponding change in surface temperature.

Is this all there is to the greenhouse problem? Well, not quite! In the course of developing our simple model and solution, we introduced several simplifying assumptions and neglected some other important processes. Five questions come to mind:

1. Is a one-layer atmosphere model adequate?

2. Is solar energy actually deposited only at the surface? If not, what impact does this have on our results?

3. Is the atmosphere grey?

4. What about the response of that other important absorber, water vapor?

5. What about clouds?

In the succeeding sections, each of these questions will be addressed and an attempt made to relate the implications of the answers to our problem of greenhouse warming.

A MULTI-LAYER MODEL

The answer to our first question is an emphatic "no". A simple examination of any observed vertical profile of atmospheric temperature shows that the temperature decreases typically by 60 to 80 °C from the surface to the tropopause, defined as the boundary between the relatively well-mixed lower 8 to 12 km of the atmosphere (the troposphere) and the stably stratified stratosphere. Above the tropopause, the temperature increases with height to the top of the stratosphere. This increase is caused by the absorption of solar energy by ozone and oxygen. Although this increase in temperature has a number of important consequences, it has little effect on surface temperature and atmospheric energetics, so we will not discuss if further. The fact that temperature decreases by such a large amount across the tropopause means that our assumption of a single atmospheric layer with a uniform temperature is incorrect.

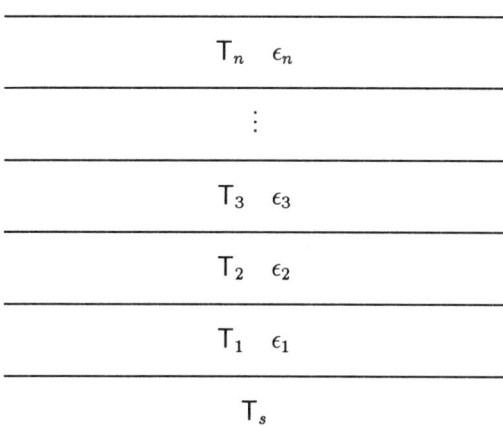

Figure 3: A schematic diagram of a multi-layer atmospheric model.

The shape and magnitude of the tropospheric temperature profile is discussed in a later section. However, it is fairly easy to demonstrate from radiative considerations why the temperature must decrease with altitude in an atmosphere with more than one layer. A schematic diagram of a multi-layer model is shown in Figure 3. Here each layer is labeled with a temperature T_i and an emissivity ϵ_i. If we again assume that solar absorption occurs only at the surface, we find at equilibrium that

$$T_s > T_1 > T_2 > T_3 > \cdots > T_n \tag{14}$$

While the actual values of the layer temperatures depend on the layer emissivities, the inequalities hold regardless of the emissivity values. The relationships in (14) may not be intuitively obvious, but can be rationalized from simple energetics arguments. The surface is the warmest, just as in our one-layer model, because it receives both solar radiation and downward thermal radiation from each of the layers above. The temperature of any atmospheric layer is determined by balancing the partial absorption of radiation emitted by all the layers above and below it against the radiation emitted by the layer itself. The top layer can only absorb radiation emitted from the layers below, although it still radiates in both directions. We can view this problem as each of the layers exchanging thermal radiation with each other, with the most important exchange occurring between adjacent layers. However, the closer the layer is to the upper boundary, the fewer layers there are between that layer and cold space. Thus, the higher the layer, the more energy it radiates to space without receiving any compensating exchange in return. Hence, the layer temperatures decrease with height.

It is a relatively straight-forward problem to consider a two-layer atmosphere. The result is a set of three coupled equations in three unknowns. Interested readers are encouraged to write out this set and solve it as a function of layer emissivities to convince themselves that indeed the temperature decreases with height. It is also instructive to consider a system of n black layers. In this case, the equations simplify considerably because the layers are nontransmitting. This simple model also shows that the temperature decreases with altitude.

Given this behavior of our atmospheric system, it seems quite likely that we need to expand our one-layer model to a multi-layer model in order to do justice to the impact of increasing concentrations of absorbing gases. The implications of this change are considered below.

ATMOSPHERIC RADIATIVE TRANSFER

Answering the second and third questions from our list requires us to first take a small sidetrip and discuss the transfer of solar and infrared radiation in the atmosphere. In considering the interaction between a beam of radiation and a volume of atmosphere, three processes are of interest. First of all, photons may be absorbed by either molecules or particles found in that volume. Absorption by molecules is of course highly dependent on the wavelength of the incident radiation, while absorption by particles tends to be a much weaker function of wavelength. In either case, the absorption is strongly dependent on the specific molecules and particle composition. Thus, we need to be able to specify the composition of the atmosphere in some detail to be able to compute absorption of radiation.

The second process that may occur is scattering, which may be thought of in classical terms as a simple redirection of some part of the beam energy due to interaction with a molecule or particle. Scattering is a complex phenomenon and an adequate description is beyond the scope of this short paper. Excellent treatments of the various aspects of the problem can be found in books such as van de Hulst (1975) and Bohren and Huffman (1983). For our purpose it is sufficient to note that the effect of scattering is to remove energy from the incident beam of radiation and direct in into other directions. For particles with dimensions near and larger than the wavelength, the preferential direction of scatter is in the forward cone, i. e., in directions not very far from the original direction of the beam. This is a consequence of the nature of the scattering of electromagnetic waves and may not be intuitively understandable in terms of typical, larger-scale scattering events such as balls on a billiard table. As a result of this predominant forward scatter, thin layers of large particles (for example, a thin cirrus cloud) tend to simply diffuse an incoming beam of radiation slightly about its original direction without causing substantial change to the fraction of the beam that traverses the layer.

In a optically thick medium, i.e., one in which virtually none of the incident beam traverses the medium without being scattered, the incident beam is converted by scattering into a diffuse, transmitted beam and a diffuse, reflected beam. This is the effect of a layer of stratus clouds on solar radiation. Although the cloud is sufficiently thick that relatively little of the unscattered, direct beam from the sun penetrates the layer (the solar disc cannot be observed), much of the incident beam does penetrate the layer as a diffuse beam (hence, the white light of layer clouds). Of course, much of the incident beam is also reflected, leading to the bright appearance of clouds on visible wavelength satellite imagery. The fact that so much of the incident beam actually is transmitted through the cloud is due to the tendency for forward scattering. If this were not so, our world would be considerably darker whenever we found ourselves under any substantial cloud cover.

The third process of interest is emission. Since the volume of atmosphere must be in thermal equilibrium, any absorption of energy must be compensated for by emission. This emission is of course a function of temperature as specified by Planck's law, and is also uniform in direction.

A consideration of these fundamental processes leads us to make a distinction between solar and thermal radiative transfer. The bulk of the solar radiation is emitted between 0.25 and 4.0 μm, with a peak emission around 0.5 μm. Because atmospheric temperatures are much lower than the temperature of the sun, peak atmospheric emission is around 15 μm, and very little emission occurs at wavelengths less than about 4 μm. Furthermore, much of the solar radiation occurs at wavelengths that are not absorbed by atmospheric gaseous constituents because the energy is too high to be effective in exciting rotational or vibrational transitions, and not sufficiently high to cause electronic transitions or dissociation. Hence, the solar radiative transfer problem can be posed largely as the transfer of an external beam through a principally scattering medium. There are important exceptions to this generalization such as the absorption of ultraviolet radiation by ozone and near-infrared radiation by water vapor. The thermal radiative transfer problem is in some ways much more complicated since it is an internal source problem, i.e., each volume is a source as well as a sink of energy, with highly variable absorption as a function of wavelength. For a clear atmosphere, scattering may be largely neglected, which simplifies our problem enormously. However, in a cloudy atmosphere, we may be forced to consider all three processes simultaneously.

12 Global Atmospheric Energetics

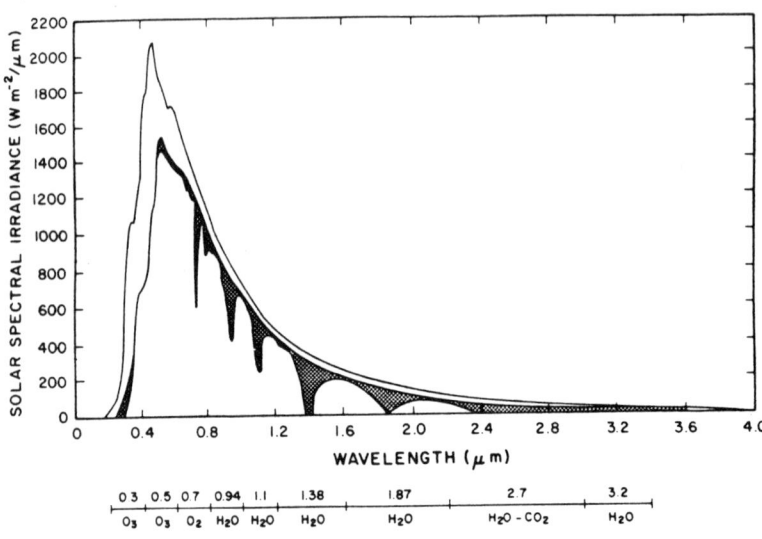

Figure 4: Spectral distribution of solar irradiance at the top of the atmosphere (upper curve), and at sea level (lower curve). The shaded areas represent absorption due to various gases (indicated on bar below the horizontal axis); the upper envelope of the shaded areas indicates the reduction of solar radiation due to molecular scattering. (taken from Liou, 1980)

Considering first the transfer of solar radiation through a cloudfree atmosphere, we obtain the result shown in Figure 4 (taken from Liou, 1980).

The upper curve depicts the spectrally-dependent solar irradiance at the top of the atmosphere. This curve is closely approximated by a 5500 °C black body curve. The curve just beneath the upper curve represents the amount of solar radiation that would reach the surface if only molecular scattering by the atmosphere were considered. Because the molecular scattering coefficient is a function of λ^{-4}, the effect of scattering is much more pronounced at short than at long wavelengths. (This dependence gives rise to the blue color of the clear sky.) The cross-hatched areas indicate areas of gaseous absorption and the particular molecular species responsible for each absorption band is shown on the bar on the horizontal axis. Since most of the solar radiation under the lower curve is absorbed and since the hatched area is only a small fraction of that total area, the major share of the solar absorption occurs at the surface.

Furthermore, neglecting for the moment the small but very important role played by ozone at wavelengths shorter than 0.4 μm, most of the solar absorption in the atmosphere is due to water vapor. If we note that about 70% of the water vapor in an atmospheric column is found within 2 km of the ground, it is apparent that most of the atmospheric absorption indicated in the figure occurs near the ground. As we shall see a bit later, the surface and lower atmosphere are coupled together by convection. Thus, from the point of view of atmospheric energetics, there is relatively little difference between absorption at the ground or in the lowest layers of the atmosphere. So, our intial assumption that absorption occurred only at the ground is fairly reasonable.

The one notable exception to this assumption is absorption of ultraviolet radation by ozone and molecular oxygen. Although only about 5% of the solar irradiance lies in the ultraviolet, the absorption of this energy has two important consequences. It shields plants and animals from potential DNA damage, and it produces the stratosphere, which we shall discuss in more detail below.

Thermal radiative transfer in clear sky conditions is dominated by gaseous absorption and emission. A useful way of visualizing this is shown in Figure 5 (Liou, 1980). The smooth curves on the diagram represent the Planck function at the indicated temperature, while the jagged curve is an interferometric measurement made at nadir view from a satellite over the Pacific Ocean. This measurement is expressed in terms of equivalent radiative temperature, which is simply the temperature that a black-body would have if it were emitting the measured radiance. Several features of interest are immediately apparent. In the region between 800 and 1200 cm^{-1}, the observational curve nearly parallels the Planck curves at a temperature of about 295 K. In this spectral region, often called the atmospheric window, gaseous absorption is very small. There is a weak continuum absorption due to water vapor in this window region, but the absorption is of importance energetically only in tropical atmospheres with large water concentrations. Thus, the observed radiation, corresponding to a relatively high temperature of about 295 K, is being emitted from the surface or boundary layer and transmitted to space, thereby cooling the surface and lower atmosphere.

Several regions of the spectrum show radiative temperatures much lower than that of the window. In particular, the absorption feature at 667 cm^{-1} (15 μm) due to the fundamental vibrational band of CO_2 stands out. This very low temperatures associated with this band indicate that emission is occurring high in the atmosphere. Similar low temperatures are associated with the ozone feature at 1050 cm^{-1} and the water vapor vibrational feature shortward of about 1450 cm^{-1}. Low emission temperatures would also be seen below 400 cm^{-1} due to water vapor rotational features if the interferometer had been capable of measuring at these wavelengths. While it is again beyond the scope of this paper to discuss these processes in detail (the

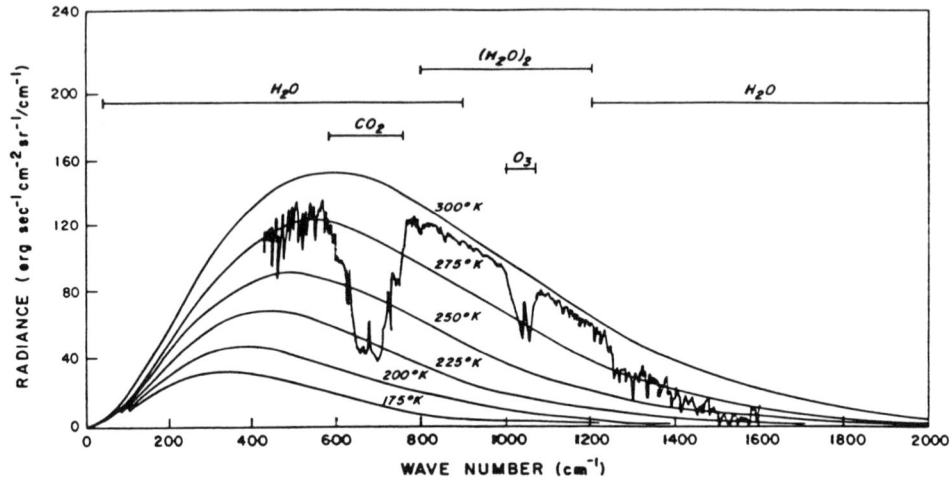

Figure 5: The terrestrial infrared spectrum observed at nadir view by the Nimbus IV IRIS instrument near Guam (15.1 °N, 215.5 °W) on April 27,1970. The horizontal bars indicate the atmospheric gases responsible for the observed absorption features. The smooth curves are the Planck function calculated at the indicated temperature. (taken from Liou, 1980).

interested reader is referred to the excellant book by Goody and Yung, 1989), the important inference to draw from this figure is that emission occurs at a variety of levels in the earth system because the opacity of the cloud-free atmosphere varies enormously with wavelength.

Returning now to our two questions about radiative transfer, we can provide some answers based on the preceding discussion.

- Absorption of solar radiation occurs principally at the surface or in the lowest kilometer or two or the atmosphere. The most notable exception is absorption of UV radiation in the upper atmosphere. Therefore, our assumption that absorption occurs only at the surface is moderately good, but will not give a correct stratospheric temperature profile.

- Molecular absorption in the infrared is highly spectrally-dependent. Coupling this with the fact that the distribution of absorbers varies with altitude, it becomes apparent that the opacity of the atmosphere varies strongly with frequency and altitude. Consequently, infrared absorption and emission are strong functions of altitude, as is the associated radiative cooling of the atmosphere. Therefore, our assumption that the atmosphere is a grey absorber is poor.

So now what do we do to patch up our model?

RADIATIVE EQUILIBRIUM IN MULTIPLE LAYERS

The straightforward solution to fixing our model is to rewrite the equation for radiative equilibrium (Eq. 4) in terms of a vertical coordinate:

$$\rho c_p \frac{dT(z)}{dt} = Q_{SOL}(z) + Q_{IR}(z) = 0. \tag{15}$$

The solution to this equation is considerably more complicated than that outlined above for the multi-layer model. Not only must solar absorption in the atmosphere be computed, but infrared absorption and emission cannot be treated with a simple layer-dependent emissivity. However, without going into elaborate detail, the solution steps can be described.

We begin by subdividing the atmosphere into discrete layers (Fig. 3). Concentrations of H_2O, CO_2, and O_3 are specified for each layer from a given atmospheric profile. The criteria used for the layer divisions are not rigorous, but basically are that the spectrally-averaged gaseous emissivity for each layer should be considerably less than one, and that, when the equilibrium temperature is found, the temperature change from layer to layer should not be too large. Then, clouds and surface properties are specified. An initial temperature profile, T(z), is assumed and Q_{SOL} and Q_{IR} are computed as functions of z. Using (16), a new T(z) profile is computed from the initial T(z) profile plus the calculated heating rates. This latter procedure is iterated until the condition $dT(z)/dt = 0$ is met everywhere.

The results of a radiative equilibrium calculation for the earth's atmosphere are shown in Figure 6. Note the extremely high surface temperature of about 345 K or 60°C and the rapid decrease of temperature with height. This is the result of the strong solar heating at the surface and in the lowest layers of the atmosphere. In order to balance this heating with infrared emission, the surface must get very warm. The temperatures in the region from 5 to 15 km are extremely cold because the greenhouse gases emit infrared radiation in this region but there is very little solar absorption. Finally, warming occurs due to solar absorption by O_3 and molecular oxygen in the region above 15 km.

16 Global Atmospheric Energetics

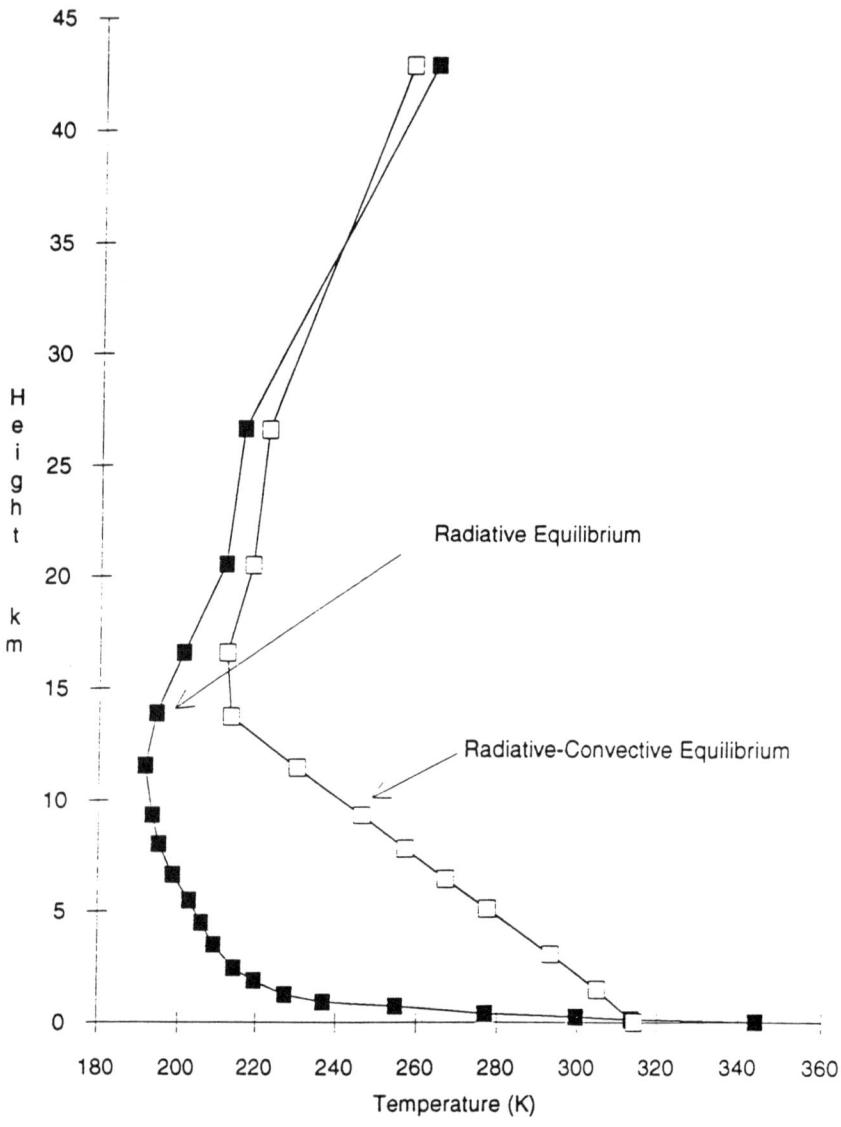

Figure 6: Radiative equilibrium (solid markers) and radiative-convective equilibrium (open markers) temperature profile for the earth atmosphere.

The radiative-equilibrium temperature profile shown in Figure 6 is obviously incorrect. The surface temperature is far too high and the lower atmosphere too low because we have neglected convection. The atmosphere is a compressible fluid. Thus, as a parcel of air rises in the atmosphere, it expands and does work against its environment, thereby cooling. We define this rate of temperature decrease as the lapse rate $= \Gamma = -dT(z)/dz$. For a parcel in which no phase changes are taking place, the rate of temperature change during an adiabatic ascent or descent is given by the *dry adiabatic lapse rate* defined as

$$\Gamma_D = g/c_P = 9.8 K/km.$$

For a saturated parcel, condensation of water occurs during the lifting process and latent heat is released, thereby reducing the parcel cooling. For this process, we define the *moist adiabatic lapse rate*, Γ_M, which is not constant but is a function of parcel temperature. For our purpose here, we simply note that $\Gamma_M \leq \Gamma_D$ always.

For a dry parcel convection will occur if $\Gamma > \Gamma_D$, while for a saturated parcel convection will occur if $\Gamma > \Gamma_M$. Looking again at our radiative equilibrium profile in Figure 6, we see that the lapse rate in the lower km or so of the atmosphere exceeds these conditions. Simply put, if we heat the surface air to these high temperatures, it will become unstable and rise. The effect of this process is to move heat upward from the surface into the atmosphere. The average atmospheric lapse rate is a result of some complex mixture of dry and saturated processes and, from observations, we find that it is on the order of 6.5 K/km. Thus, to fix up our model, we make one simple additional assumption: the lapse rate between any two layers is constrained such that $\Gamma \leq 6.5 K/km$. The resulting equilibrium temperature profile produced by the solution of (16) subject to this constraint is shown in Figure 6. The solution is generally achieved by successive iterations, where each iteration consists of a calculation of radiative heating rates and temperature changes, followed by a convective adjustment of temperature lapse rates that exceed the imposed constraint. This atmospheric temperature profile resulting from this process is called *radiative-convective equilibrium* (RCE).

Comparing the two profiles in Figure 6, we see that the RCE surface temperature has dropped relative to radiative equilibrium, while the atmospheric temperatures have risen. A comparison with observations shows that this type of model does a good job of representing the average temperature profile of the atmosphere (as well it might, since we chose our lapse rate based on observations). More important, however, we have now developed what might be called the poor man's Global Climate Model. The free parameters in the model that adjust to changes in model forcing are the surface temperature, the tropopause depth, and stratospheric temperatures. Since the tropospheric lapse rate is constrained, the temperature profile in the troposphere is fixed by the combination of calculated surface temperature and fixed lapse rate. These simple RCE models can be (and have been) used to investigate the impact of a variety of interesting climate-related processes such as changes in greenhouse gas concentrations on temperature profiles. Thus, in trying to answer our questions about the nature of radiative transfer in the atmosphere, we have been led to the development of a far more sophisticated and useful model than our simple one-layer emissivity model.

WATER VAPOR

Our fourth question dealt with the response of water vapor concentration in the atmosphere to changes in temperature. The concentration of water vapor in an atmospheric parcel is limited by the saturation vapor pressure, $e_s(T)$, which depends on temperature according to the Clausius-Clapeyron equation. If we assume that the latent heat of vaporization is independent of temperature, we can integrate the Clausius-Clapeyron equation to give

$$e_s(T) = e_{s0} \exp[(L_v/R_W)(1/T_0 - 1/T)].$$

Here, e_{s0} is the saturation vapor pressure at a reference temperature T_0, L_v is the latent heat of vaporization for water, and R_W is the gas constant for water. A plot of this equation is shown in Figure 7. Note the exponential rise in the vapor pressure with increasing temperature.

Air parcels are in general not saturated with water vapor. If they were, fog and cloud would be ubiquitous. Meteorologists typically define the water vapor concentration of a parcel in terms of the relative humidity (RH), which is the ratio of the parcel water vapor concentration to the saturation vapor pressure (i.e., $RH = 100\%(e/e_s(T))$). Thus, RH is simply an indicator of how close a parcel is to saturation. On both theoretical and observational grounds we are led to suggest that, averaged over large regions of the globe, RH is approximately constant with time in the atmosphere. Although the reasons behind this fact are fairly complex, we can intuitively understand it by noting that most of our planet is covered by water. Hence, there is plenty of water availability at the surface. The limiting process with regard to atmospheric water vapor concentrations then becomes the rate at which water vapor can evaporate[1] into the near-surface atmospheric layer and be transported upwards.

---

[1] Some confusion can occur regarding the use of this term. Meteorologists generally mean *net* evaporation, the difference between condensation and evaporation at the surface. In meteorology we are predominantly concerned with the net transfer of water from the surface to the atmosphere and hence we simply equate the concept of evaporation with this net transfer.

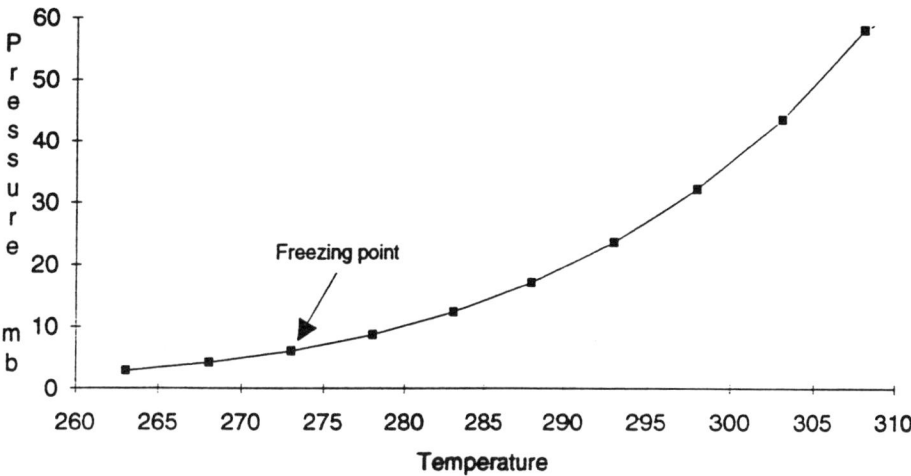

Figure 7: Saturation vapor pressure for water as a function of temperature. The freezing point of water is indicated for reference.

The evaporation rate is controlled by two factors: the wind speed and RH of the near-surface layer. As the RH of the layer approaches 100%, the evaporation rate becomes smaller and smaller. Thus, for very dry air, the evaporation rate is rapid and the RH rises but for moist air the evaporation rate slows to near zero. This tends to maintain a high and approximately constant RH in the atmosphere when averaged over large areas and times.

We now consider the issue of water vapor feedback. From the preceeding, we see that if the average temperature of an air parcel should increase (say due to an increasing CO_2 concentration), then the saturation vapor pressure of that parcel would also increase (Fig. 7). But if, as we have just argued, the RH is to remain approximately constant, the water vapor concentration (e) in the parcel must also increase. But, because water vapor is a greenhouse gas, increases in water vapor concentration lead to increases in temperature. In short, we have a positive feedback. As T increases, e increases and as e increases, T increases.

We can use our simple one-layer model to quantify this feedback. We stated earlier that doubling the CO_2 concentration will increase the downward atmospheric radiation by about 4 W/m² and raise the temperature by about 1.2 K. If we then compute the additional water vapor that would be put into the atmosphere by this temperature rise (using the Clausius-Clapeyron Equation) and the additional rise in downward atmospheric radiation, we find the increase in downward radiation to be about a factor of four relative to that produced by the increased CO_2 concentration. This factor of four increase in forcing translates directly into a factor of four increase in temperature (Eqn. 14). Thus for a doubling of CO_2, we expect an increase in temperature on the order of 4.8 K, which is in fact very near that predicted by 3-dimensional climate models.

CLOUDS

Our final question dealt with the role of clouds in the greenhouse problem. This is, without a doubt, the most complicated component of the problem and cannot be treated comprehensively in this short article. However, it is possible to summarize briefly the important radiative properties of clouds and their impact on climate. From this, in turn, we can deduce their role in climate change.

We begin by considering the bulk optical properties of a cloud. The most important of these properties is the combined effect of cloud optical thickness and absorption. To a first approximation, absorption of solar radiation by clouds can be neglected. Clouds simply increase the planetary albedo by reflecting incident solar radiation. The amount of sunlight reflected by the cloud increases monotonically with increasing optical thickness to some asymptotic limit. At thermal wavelengths, clouds are absorbers and increasing optical thickness increases the absorption. The absorption process varies approximately exponentially with optical thickness. Thus, an optical thickness of 1 or 3 causes the cloud to be nearly opaque. (These phenomena, and others of interest, are discussed in a lucid and elegant treatment of multiple scattering by Bohren, 1987.)

Because low clouds have relatively high liquid water concentrations, they have large optical thicknesses and are essentially black (non-transmitting) in the thermal infrared region. Thus increasing their optical thickness has little impact on infrared radiation, but does increase their solar reflectivity. High clouds, conversely, have relatively small optical depths, so changing their optical depths has a significant impact on the thermal radiation field.

The second important optical property of clouds has to do with the probability of scattering radiation either forward or backward, i.e., scattered into either the forward or backward hemisphere relative to the original direction of propagation. This particular quantity is a property of the size of the particles that make up the cloud. In general, for a fixed wavelength, the larger the particle, the higher the probability of forward scattering. Because ice particles in cirrus clouds are typically much larger than water droplets, cirrus clouds are more strongly forward scattering. As a consequence, they have a relatively small impact on the planetary albedo. The reflectivity of stratus clouds, however, can potentially be affected by changing the average particle size. Some evidence suggests that the number of drops in a stratus cloud can be increased by adding small particles to the environment. These particles may serve as additional nucleation sites and, for a given amount of condensation, this results in more and smaller drops. This increases the optical depth of the clouds and, since the drops are smaller, also increases the probability of back-scatter. Both effects cause the cloud to be more reflective.

The impact of clouds on climate is complicated. Apart from the difficulty of simply calculating the cloud optical properties, we find that, for any particular cloud, the solar and infrared effects are opposite in sign. Because low clouds primarily affect the solar radiation budget through increased reflectivity, they have an overall tendency to cool the earth. Conversely, high clouds tend to warm the system because they primarily act to reduce the outgoing infrared radiation. Thus, we also need to know the vertical location of any particular cloud and the vertical distribution of clouds in general.

Rather than trying to quantify the locations and properties of all clouds and then compute the impact on climate, an alternative approach to the problem is to try to observe directly the impact of clouds on the earth's climate. At this point, we do not have a completely adequate answer to this question. Satellite observations provide us with a means of measuring the impact of clouds on the planetary radiation budget, but this is by no means a trivial task. The Earth Radiation Budget Experiment (described by V. Ramanathan in a later chapter in this volume) is an ongoing effort to quantify from space cloud impacts on the radation budget. This, however, still does not provide the entire answer because we need to know the impact of clouds on the surface energy budget as well. Surface observation systems are too coarse spatially to provide adequate coverage of the globe and we have not yet been able to devise sufficiently accurate algorithms to measure the surface budget from satellites.

Estimates of the effect of clouds on climate suggest that about a half to two-thirds of the planetary albedo is due to clouds, i.e., the planet would absorb an additional 15 to 20% of the incident solar radiation if no clouds were present. Presumably, most of this energy would be absorbed at or very near the surface. It is more difficult to estimate the greenhouse impact of clouds, but clouds probably reduce the outgoing longwave radiation by a factor of 20 to 40% over what would be emitted by a cloud-free planet with the same thermal structure. Clouds increase the downward thermal radiation at the surface by more than a factor of two.

Given our inability to measure cloud effects accurately, we might choose to try to compute them instead. Unfortunately, our understanding of the detailed cloud microphysical processes that are important in cloud droplet formation and maintenance is also limited. Furthermore, these calculations need to be carried out on very small spatial grids and very short timescales that are completely incompatible with climate models. The best that we can do in climate models is use empirical relationships that attempt to relate cloud properties to large-scale variables (such as average relative humidity and temperature) predicted by the model. The relationships are not well grounded observationally and are inadequate for the task at hand. The improvement of our understanding of cloud processes and the quantification of this understanding in climate models is the current focus of considerable research in meteorology.

If we have so much uncertainty about the role of clouds in the current climate, then we must have a greater uncertainty about the role of clouds if that climate is changed. The issue of cloud feedback in climate change is currently unresolved and the subject of considerable debate, as will be discussed by other authors in later papers in this volume.

SUMMARY

All of the preceding discussion can be summarized in a few major points that cut to the heart of the global warming issue.

- The atmosphere, because in contains gasses such as CO_2 and H_2O that absorb infrared radiation, raises the average surface temperature from about 255 K to the observed 288 K. The primary greenhouse gas is water vapor, followed by carbon dioxide and ozone. Clouds also play an important role in maintaining the observed surface temperature.

- Model results susggest that doubling the CO_2 concentration of the atmosphere would raise the surface temperature by about 1.2 K, all other processes remaining constant.

- Because the saturation vapor pressure of water increases with increasing temperature, this increase in CO_2 will be accompanied by a rise in the water vapor concentration. This leads to a further increase in surface temperature. In RCE models, the effect of this water vapor feedback is to increase the CO_2 temperature increase by a factor of about 3 or 4.

- The troposphere is in radiative-convective equilibrium. Convection transports energy upwards from the surface and this energy is radiated to space throughout the atmospheric. Because of this convective linking, the temperature of the entire troposphere will increase if the surface temperature increases.

- The impact of clouds on the current climate is complicated and not well understood. It is possible that, in a changing climate, cloud feedbacks could damp out the temperature changes suggested above. It is also possible (and no less probable!) that cloud feedbacks could increase the temperature changes above.

The remainder of this collection of papers deals with these issues in more detail and with more sophisticated models. It is vital that our society come to grips with the nature of the greenhouse problem and the potential for climate change. We are already in the midst of an unprecedented, inadvertant global climate change experiment; our questions now are how far are we going to carry out this experiment and where will it take us?

REFERENCES

Bohren, C. F., 1987: Multiple scattering of light and some of its observable consequences. *Am. J. Phys.*, **55**, 524-533.

Bohren, C. F., and D. R. Huffman, 1983: *Absorption and Scattering of Light by Small Particles*. John Wiley & Sons, New York, NY, 530 pp.

Goody, R. M., and Y. L. Yung, 1989: *Atmospheric Radiation, Theoretical Basis*. Oxford University Press, New York, NY, 519 pp.

Liou, K. N., 1980: *An Introduction to Atmospheric Radiation*. Academic Press, New York, NY, 392pp.

Ramanathan, V., and J. A. Coakley, 1978: Climate modeling through radiative-convective models. *Rev. Geophys. and Space Phys.*, **16**, 465-490.

van de Hulst, H. C., 1957: *Light Scattering by Small Particles*. Dover Publications, New York, NY, 470 pp.

GLOBAL CLIMATE MODELS: WHAT AND HOW

David A. Randall

Department of Atmospheric Science, Colorado State University
Fort Collins, Colorado 80523

ABSTRACT

The physical basis of global climate models is reviewed, with emphasis on the atmospheric sub-model. An introductory discussion of the physics of the climate system is followed by a description of the numerical methods and physical parameterizations that have been developed for climate modeling. Current issues in model development are highlighted, and future directions are briefly discussed.

INTRODUCTION

Global climate models are in the news. Anyone who reads the newspaper is aware that climate models bring supercomputers to their knees, provide a basis for dire predictions of global warming and nuclear winter, and have serious weaknesses including a chronic inability to deal with the effects of clouds. Such predictions are influencing national and international policies on energy and the environment. Rumor has it that there is a simple climate model running somewhere in the White House.

There are approximately twenty-five climate modeling groups in the world, mainly based at national laboratories, but including a handful of university-based efforts. Most of these groups deal primarily with simulation of the atmospheric circulation, but out of necessity ocean circulation modeling is rapidly coming into its own.

Of course, not all climate models require supercomputers. "Pencil and paper" climate models, often called energy balance climate models (EBCMs), were developed and applied to a variety of climate change problems by Budyko[1], Sellers [2], North [3], and others. EBCMs typically resolve variations with latitude but not with longitude or height. They do not explicitly simulate the effects of atmospheric motions (the reasons for wanting to do so are discussed below), but include them very indirectly using mixing lengths or other comparably simple ideas. The strength of EBCMs lies in their simplicity and economy. As recently reviewed by Lindzen[4], EBCMs have produced a number of interesting results, including evidence that the Earth's climate may have more than one stable equilibrium under some conditions. Nevertheless, EBCMs are so drastically simplified that they can do little more than suggest possibilities for investigation with more complete models.

Radiative-convective models (RCMs) were pioneered by Manabe and Strickler[5] and Manabe and Wetherald[6], and have been extensively discussed by Ramanathan and Coakley[7]. They represent the vertical structure of the atmosphere but do not attempt to deal with horizontal variations. Because many important physical processes occur primarily in vertical columns, in particular radiative transfer and buoyancy-driven natural convection, RCMs can include fairly elaborate process models. In fact, in their most fully developed form the RCMs are essentially one-dimensional subsets of the complex three-dimensional climate models known as general circulation models (GCMs). Further discussion of RCMs is given by T. Ackerman elsewhere in this volume.

GCMs represent in most cases the atmosphere, or in an increasing number of cases the ocean, and in a few cases both. These behemoths typically require at least a few supercomputer cpu minutes (roughly 10^{10} floating-point operations) to simulate one day's "weather;" in fact, atmospheric GCMs are now routinely used for weather prediction and have led to major improvements in forecast skill[8,9].

For climate simulation, the GCM approach is purely brute force: simulate the hour-by-hour evolution of the atmosphere for as many simulated years as the available computer time will allow, then simply compute the statistics of the solution and interpret these in terms of the weather statistics that we call climate. The more elegant approach of computing the climate directly, without the tedious simulation and averaging of individual weather events, appears to be quite intractable.

This paper deals primarily with GCMs. We first briefly mention, however, that there is a fourth class of climate model: simplified GCMs that sacrifice realism for the sake of computational speed and ease of interpretation. Perhaps the best known of these is the Held-Suarez model[10]. Simplified GCMs are capable of simulating some aspects of the climate, including simplified representations of atmospheric dynamics, for many thousands of simulated years. They can thus be applied to such problems as ice-age transitions, which cannot yet be approached with GCMs.

WHAT ARE WE TRYING TO MODEL?

The climate system includes the atmosphere, the oceans, the cryosphere, and the biosphere; all are important players on sufficiently long time scales. Since I am a meteorologist, I have little choice but to concentrate on the atmosphere and especially its motion systems, but some discussion of the oceans and the land surface will also be included here. The cryosphere will not be discussed, except for very brief remarks on the role of sea ice.

The Earth-atmosphere system absorbs more radiation than it emits in low latitudes, and emits more than it absorbs in higher latitudes[11]. As discussed by T. Ackerman elsewhere in this volume, the distributions of absorption and emission depend on many factors, including the Earth's orbital parameters, the distributions of temperature, water vapor and clouds inside the atmosphere, and the temperature and optical properties of the ocean and the land surface.

Over one or more annual cycles, the radiative imbalances at the top of the atmosphere, at particular latitudes, must be compensated for by energy transports inside the system. One point of view often adopted is that the circulations of the ocean and atmosphere are essentially those required to carry out these energy transports. The transports are accomplished in roughly equal measure by the atmosphere and the oceans[12, 13]. The mechanisms involve large-scale mass circulations, with high-energy mass (air or water) flowing poleward, and low-energy mass returning to the tropics.

The atmospheric transports are often discussed in terms of "symmetric" circulations that are independent of longitude, and "eddies" that have varying degrees of longitudinal structure. The tropical atmosphere is home to a vast symmetric circulation called the Hadley cell, a giant "ferris wheel in the sky" that circulates on the order of 10^{14} grams of air per second (Fig. 1).

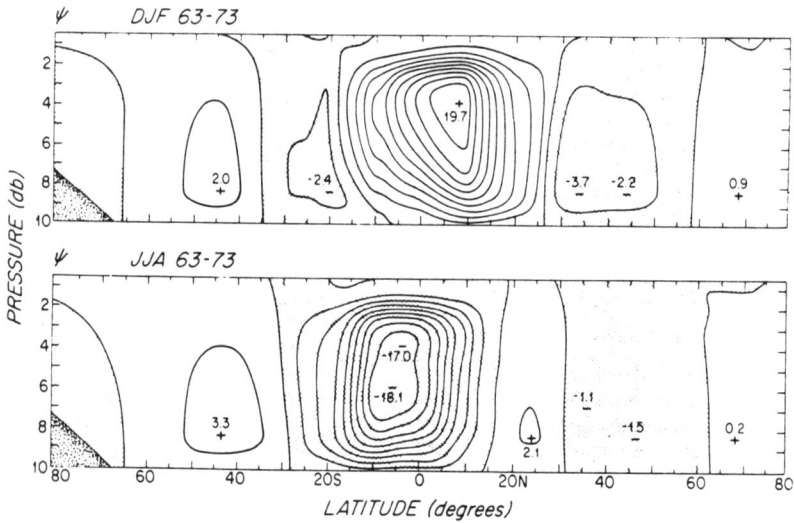

Figure 1: The observed stream function of the mean meridional circulation for December-January-February (top panel) and June-July-August (bottom panel). The horizontal axis is latitude, from 80 °S on the right to 80 °N on the left. The vertical axis is pressure, from 10 decabars (near the Earth's surface) to 1 decabar (near the tropopause). The isopleths of the stream function are parallel to the streamlines of the motion in the latitude-height plane. The units of the stream function are 10^{13} g s^{-1}. Positive values denote clockwise circulation, and negative values (shaded) counter-clockwise. The Hadley Cells are the large, intense features in the tropics. They have their rising branches in the summer hemisphere (in each season), but the bulk of each cell is found in the winter hemisphere. This figure is taken from a paper by Lindzen and Hou[14].

Eddy transports dominate in middle latitudes. The eddy population includes the familiar cyclones and anticyclones that appear on weather maps, but it also extends to enormous waves of planetary scale, as well as smaller features such as hurricanes. Eddy activity arises from hydrodynamical instabilities and from interactions of the wind with mountain ranges, and is most vigorous in the winter. The most energetic eddies in the atmosphere have scales of thousands of kilometers[15], while the those in the oceans have scales either comparable to the basin size (the "gyres") or considerably smaller, on the order of 100 kilometers. Most of the kinetic energy of the oceans actually resides in these small eddies. Because of the importance of relatively small eddies in the oceans, ocean models need much higher spatial resolution than atmosphere models[16].

Just as there is a meridional radiation imbalance that demands meridional mass circulations to transport energy poleward, there are also vertical radiation imbalances that demand vertical mass circulations to transport energy upward. The most important mechanism for upward energy transport in the atmosphere is cumulus convection[17]. The mass circulations associated with cumulus clouds are driven by buoyancy forces that are maintained by the release of latent heat as water vapor condenses. This is not a simple mixing process; the cumulus updrafts penetrate the entire troposphere, and produce transports that are independent of local gradients. Of course, the individual cumulus clouds are much too small to explicitly simulate in a climate model. The collective effects of cumulus clouds are included in climate models through a kind of statistical mechanics of cumulus convection[18]. A great deal of observational and theoretical work has established that cumulus convection is promoted by a wide variety of processes, including broad rising motion associated with large-scale atmospheric circulations, surface fluxes of moisture and sensible heat, and vertically non-uniform radiative heating or cooling.

Although the ocean is about 200 times more massive than the atmosphere, and contains about 10^3 times more enthalpy, the atmosphere actually carries about 30 times more kinetic energy than the oceans. This is of course due to the slowness of ocean currents (centimeters per second) relative to the winds (tens of meters per second).

There are many types of motion in the atmosphere and oceans, covering a wide range of spatial scales and kinetic energies. "Small scale" motions tend to be three dimensional, and are hardly affected by the Earth's rotation, while larger-scale motions are quasi-horizontal because the atmosphere is thin compared to the radius of the Earth, and are strongly constrained by the Earth's rotation. Both large and small-scale motions are strongly affected by density gradients and the resulting buoyancy forces[19]. The governing equations for the large-scale motions of the atmosphere are briefly discussed in the next section.

Of course, both the atmosphere and the oceans are highly nonlinear, chaotic systems that quickly "forget" their initial conditions[20]. The time required for nonlinear scrambling of the initial conditions is colloquially called the predictability time. For simulations long compared with the predictability time, we are essentially solving a boundary value problem, rather than an initial value problem, even though formally the equations are hyperbolic. The predictability time for the atmosphere is on the order of 20 days[21]. The corresponding time for the deep ocean circulation is not well known but is at least on the order of decades, and probably on the order of centuries. An atmospheric GCM that uses prescribed ocean temperatures as input thus "forgets" its initial conditions in a few weeks, while a coupled ocean-atmosphere model, of the type needed for climate studies, remembers important aspects of its initial conditions a thousand times longer.

THE BASIC GOVERNING EQUATIONS FOR ATMOSPHERIC GCMS

We now summarize the basic governing equations used to simulate the large-scale circulation of the global atmosphere. Similar equations are used to model the ocean[22]; they are omitted here for brevity.

The horizontal components of the winds are governed by the so-called "equation of motion," which is nothing more than Newton's second law, expressed in a coordinate frame that is rotating with the Earth's angular velocity Ω:

$$\rho \left\{ \frac{\partial V}{\partial t} + [(2\Omega + \nabla \times V) \cdot k] k \times V + \nabla \left[\frac{1}{2} (V \cdot V) \right] + w \frac{\partial V}{\partial z} \right\} + \nabla p = T_V \quad (1)$$

Here ρ is the density of the air, V is the horizontal wind vector, t is time, ∇ is the horizontal operator (following surfaces of constant height), w is the vertical velocity, z is the vertical coordinate (height, in this case), φ is latitude, k is a unit vector pointing upward, p is pressure, and T_V represents "friction" (per unit volume) which is mainly due to turbulent momentum exchange, primarily near the Earth's surface. Contributions to T_V can also come from convective and/or wave motions in the free atmosphere. Collectively, these frictional processes are referred to as "subgrid-scale" momentum exchanges, since they cannot be explicitly resolved in a climate model.

In addition to the continuum assumption, some further approximations[23] have been used in deriving (1). In particular, the atmosphere has been assumed to be thin compared with the radius of the Earth, and the centripetal acceleration associated with the Earth's rotation has been neglected, although the Coriolis acceleration has been kept.

The large-scale motions of the atmosphere are nearly in "geostrophic" balance. This means that the coriolis acceleration nearly balances the pressure gradient force in (1), so that the motion is strongly controlled by the Earth's rotation, i.e.

$$(2\Omega \cdot k) k \times V + \frac{\nabla p}{\rho} \approx 0. \quad (2)$$

From (2), it follows that geostrophic motion is parallel to lines of constant pressure, rather than from high pressure towards low pressure. Geostrophic motion is nondivergent, except for the fact that $\Omega \cdot k$ varies with latitude.

The total motion field can be regarded as the superposition of a geostrophic component and an ensemble of highly divergent buoyancy waves, called "gravity waves." Gravity waves with large horizontal scales are modified by the effects of rotation, and are called gravity-interia waves. They are of secondary imporannce, however; the kinetic energy of the atmosphere is strongly dominated by the geostrophic component of the motion.

An obvious question is: What about the vertical component of the motion? Why do we apply (1) only to the horizontal components? The reason is that the equation governing vertical motions, which looks very similar to (1), is very well approximated by a balance between the vertical component of the pressure gradient force and the weight of the air under gravity. The vertical equation of motion is thus replaced by

$$\frac{\partial p}{\partial z} = -\rho g, \qquad (3)$$

which is called the hydrostatic equation. Here g is the acceleration due to gravity, which can be treated as a constant since the atmosphere is thin compared with the radius of the Earth. An advantage of (3) is that it filters sound waves, which are meteorologically irrelevant, while doing little harm to the meteorologically important motions.

Conservation of mass is expressed by

$$\frac{\partial \rho}{\partial t} + \nabla \cdot (\rho \mathbf{V}) + \frac{\partial}{\partial z}(\rho w) = 0. \qquad (4)$$

Obviously one such equation can be written for each chemical species. Normally the mixture of gases that we call "dry air" is treated as a single species, and a separate conservation equation is introduced for water vapor. This has the form

$$\frac{\partial}{\partial t}(\rho q) + \nabla \cdot (\rho q \mathbf{V}) + \frac{\partial}{\partial z}(\rho q w) = \rho (E - C) + T_q. \qquad (5)$$

Here q is the "mixing ratio," i.e. the ratio of the water vapor density to the density of dry air, E is the rate of conversion of liquid (or ice) into vapor, and C is the rate of conversion of vapor into liquid. Finally, T_q represents the subgrid-scale transport of water vapor. Typically E, C, and T_q are dominated by their subgrid-scale components.

One could use additional equations to predict the concentrations of liquid and ice. This is just beginning to become standard practice. In the past, condensed water was typically assumed to fall out (or to re-evaporate) instantly.

A few existing models include additional conservation equations for ozone and other chemical species, but this is still quite rare because of the large additional computational burden. It is a fact of life that the present generation of climate models largely ignores atmospheric chemistry.

The conservation law for entropy is

$$\frac{\partial}{\partial t}(\rho s) + \nabla \cdot (\rho s \mathbf{V}) + \frac{\partial}{\partial z}(\rho s w) = \frac{Q}{T} + T_s, \qquad (6)$$

where s is the entropy per unit mass, Q is the heating rate due to various subgrid-scale processes described below, and T_s represents the subgrid-scale transport of entropy. The heating is discussed further below. The gain in entropy due to the dissipation of kinetic energy is included in only a few existing models; it amounts to a few Watts per square meter.

To close our system of equations, we need, "parameterizations" (parametric representations) for the terms on the right-hand sides of (1), (5), and (6); these are discussed below. In addition, we need the ideal gas law and a recipe to relate the entropy to the temperature, pressure, and moisture content. These are omitted here for brevity.

In practice, the equations are almost always expressed in terms of a vertical coordinate other than height. I have chosen to use height-coordinates here just to simplify the discussion. The various alternative vertical coordinates, such as pressure and entropy, each have their own advantages and disadvantages[24]. A discussion of this topic would take us too far afield.

If the turbulence and heating terms [those on the right-hand sides of (1), (5), and (6)] are considered to be known, then the equations given above allow us to prognostically determine ("time step") the horizontal velocity, the density of the air, the mixing ratio of water vapor, and the entropy. The additional thermodynamic variables are easily worked out. This leaves only the vertical velocity as an unknown; recall that we cannot simply time-step it because we use the hydrostatic approximation. This is a practical problem only in the z-coordinate system; with alternative vertical coordinates the vertical velocity is easily determined, and in fact this is part of the motivation for their use. An equation to determine the vertical velocity in the z-coordinate system can be derived by differentiating the hydrostatic equation with respect to time; further discussion is given by Kasahara[24] and Ooyama[25].

The real difficulties in using (1-6) arise from the source/sink terms on the right-hand sides of (1), (5), and (6). In a sense these terms are at the core of the climate modeling problem, because they represent the sources and sinks of energy and momentum that keep the system going and determine its long-term behavior, i.e. its climate.

The governing equations for the large-scale circulation of the oceans are essentially similar to those given above for the atmosphere. The main differences are: 1) sea water is nearly incompressible, so that a different equation of state is needed; and 2) heating and related processes are mainly confined to the top hundred meters of the ocean and are relatively simple for the ocean.

DISCRETIZATION

Because the governing equations are highly nonlinear, they must be solved numerically after a suitable discretization. The discretization amounts to replacing the space and time derivatives in (1-6) by finite differences and / or functional transforms.

The time derivatives are always approximated by finite-differences. The time step is chosen as large as possible without inducing computational instability; as the resolution is increased, the time step must be decreased to maintain stability. Typical atmospheric GCM time steps are on the order of 10 to 20 minutes. The resulting time truncation errors are generally considered to be unimportant except for a few cases in which special methods have been developed to allow the use of unusually large time steps.

Vertical discretization is also accomplished with finite differences in almost all cases, although finite element methods are being experimented with. Vertical differencing schemes are designed, in many cases, to allow the vertically discrete equations to exactly reproduce some properties of the exact equations, such as energy conservation under adiabatic processes[26]. Most current atmospheric GCMs have about 20 layers from the surface to the lower stratosphere (20 or 30 km above the surface), giving an average vertical resolution through the troposphere on the order of 1 km. About 90% of the total mass of the atmosphere is contained within such a model; the rest lies "above the model top."

There are two widely used methods to discretize the equations in the horizontal.

The first consists of finite difference approximations, almost always applied on a fixed latitude-longitude grid. Typical resolutions used in climate models today are on the order of several degrees for both latitude and longitude. This translates into several hundred kilometers in each direction, except near the poles where the convergence of the meridians causes the east-west grid distances to become quite small. A typical grid is shown in Fig. 2. The small east-west grid distances near the pole would necessitate unacceptably small time steps to ensure computational stability. To avoid this problem, filtering procedures have been developed that effectively reduce the east-west signal velocity near the poles, for certain kinds of rapidly moving but meteorologically secondary waves[26].

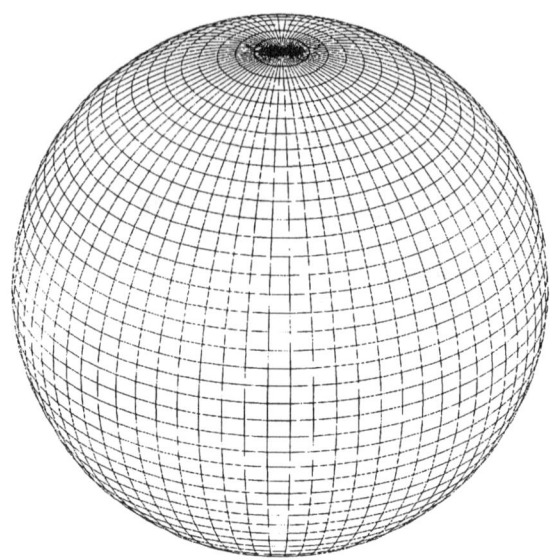

Figure 2: A typical grid used in climate models. The resolution in this case is five degrees of longitude by five degrees of latitude. Each small pseudo-rectangle represents a "grid cell" of these dimensions. The row of cells nearest each pole consists of pseudo--triangles, each six degrees "tall." The are 3168 cells altogether in this particular grid.

An advantage of finite difference methods is that there is an infinite variety of them, which means that the opportunity exists to choose or design the *best* one. This freedom can been used, for example, to guarantee exact conservation of mass and energy in the finite difference system[26].

The second approach to horizontal discretization is to expand the dependent variables in terms of a complete set of basis functions, usually taken to be spherical harmonics. The expansion is of course truncated after a finite number of modes. This approach completely avoids the "pole" problem referred to above, since the representation can be made uniform over the sphere. A second advantage of this so-called "spectral" method is that linear wave propagation is represented *exactly*, on a mode-by-mode basis, so that, for example, the linear phase speeds of individual modes are obtained without any error whatsoever.

Direct computation of nonlinear processes in the wave-number domain is computationally prohibitive, and would in any case be horrifically complicated. For this reason, the model variables are transformed onto a grid for the purpose of evaluating all nonlinear processes, including the advection terms and the physical parameterizations[27].

An important disadvantage of the spectral method is that conservation properties and non-negativity of such quantities as the water vapor mixing ratio cannot be guaranteed except in limiting cases. For example, spectral methods perform badly for advection of moisture, and would be even worse for such variables as cloud water. Many spectral GCMs are now being modified to do their advection processes using grid-point methods.

A second disadvantage of the spectral method is that it becomes computationally non-competitive with grid-point methods when the resolution becomes sufficiently high, say better than about 50 km. In fact, one of the strengths of the spectral method is that it converges rapidly at *low* resolution. This suggests that in the future spectral methods may be reserved for low-resolution models, and that grid-point methods will be used at moderate and high resolution.

At the present time, roughly half of the world's atmospheric GCMs are spectral, and the remainder are based on finite-difference methods.

Essentially all ocean GCMs are based on grid point methods, although in principle spectral methods could be used. An approximation used in most global ocean models is that the air-sea interface is a rigid lid rather than a free surface. This approximation has the effect of filtering rapidly propagating surface gravity waves that, if included, would require the use of a small time step. On the other hand, the rigid lid approximation complicates the treatment of "islands," so that ocean modelers often take liberties with the shapes of the ocean basins, connecting bits of land that are separated in reality, or replacing islands by undersea mountains that don't quite reach the surface. At the present time, there is a movement away from the rigid lid approximation, back to a more realistic and flexible (!) free-surface boundary condition.

As mentioned earlier, there are dynamically important eddies in the ocean with scales on the order of 100 kilometers. For this reason, ocean GCMs should, ideally, have resolutions on the order of 10 - 20 km or better (up to now, no ocean GCM has been run with a resolution finer than about 50 km, and most use resolutions of several hundred kilometers, with eddy mixing coefficients to represent the effects of the unresolved eddies). Contrast this with atmospheric GCMs, which have to deal with eddy sizes only down to about 1000 km, thus requiring resolutions on the order of a few hundred km.

For the sake of discussion, suppose that a world ocean GCM has a resolution 10 times finer than the atmospheric GCM that it is coupled to, and that the two models have the same number of layers in the vertical. Since the oceans cover about 70% of the Earth's surface, the ocean model will have about 10^2 x 0.7 = 70 times as many grid cells as the atmospheric GCM. If the time steps of the two models are comparable (a reasonable assumption) then clearly *the ocean model will strongly dominate the total computer time required for a coupled ocean-atmosphere simulation.*

Recall also that if only the atmosphere is being simulated, a few annual cycles are sufficient to define a model's climate; but if the deep ocean is included, thousands of years are required for the coupled system to equilibrate.

For both of these reasons, coupled ocean-atmosphere models are far more computationally demanding than atmospheric GCMs alone. Eddy-resolving coupled ocean-atmosphere models are just barely feasible with the computers of 1991.

A SURVEY OF PHYSICAL PARAMETERIZATIONS

Typically about half the computation time in an atmospheric GCM is given over to integrating (1-6), euphemistically referred to as the "dynamics," and the remainder to physical parameterizations, or "physics." In an ocean GCM, the split would be roughly 90% dynamics and 10% physics.

The methods used to parameterize solar and terrestrial radiation are discussed by T. Ackerman elsewhere in this volume, and will not be described here. Each of the parameterizations discussed below is described at greater length by Washington and Parkinson[28].

- *Turbulent transport of energy, moisture, and momentum in the planetary boundary layer.*

As mentioned earlier, turbulence is largely confined to the "planetary boundary layer," which occupies roughly the lowest kilometer of the atmosphere. The boundary layer may or may not contain clouds. The turbulence in the boundary layer arises both from instabilities associated with the shear of the mean wind, and from buoyancy-driven natural convection. The deepest, most turbulent boundary layers are maintained by convection.

When clouds occur in the upper portion of the boundary layer, they tend to invigorate the turbulence. One reason is that phase changes can help to generate buoyancy forces in the turbulent air. A second is that intense infrared radiative cooling occurs near the cloud tops, and this tends to cause convective overturning in the layer below.

Friction in the boundary layer destroys kinetic energy, and the motion tries to compensate for this energy loss by turning the winds down the pressure gradient, i.e. towards low pressure. Friction thus disrupts the geostrophic balance in which the air flows at right angles to the pressure gradient.

Boundary layer parameterizations typically take into account the various mechanisms that can generate turbulence, and parameterize the turbulent exchange rates in terms of the amount of turbulence kinetic energy present. Surface fluxes are constrained by semi-empirical similarity theories, while fluxes above the surface are often based on mixing length theories or simple empirical assumptions about the vertical structure of the boundary layer.

- *Sea ice*

Sea water freezes at about -2 °C. Sea ice forms when heat is removed from -2 °C water. The ice can, of course, become much colder than -2 °C. In effect, the ice serves to insulate the atmosphere from the -2 °C water below, thus limiting the rate at which the ocean loses energy. Second, the ice reflects much of the visible radiation that impinges on it, thus limiting the rate at which the ocean gains energy.

The ice moves under the influence of the wind and currents, and it typically contains at least a few percent of open water in the form of "leads." These are cracks and / or holes in the ice, filled with comparatively warm -2 °C water, through which the ocean can exchange heat and moisture with the atmosphere. The area-averaged heat flux from the ocean to the atmosphere is often dominated by the flux through the leads.

A good introduction to the large and interesting subject of sea ice and climate is given by Washington and Parkinson[28].

- *Cumulus convection.*

    Cumulus convection, as already mentioned, can accomplish very rapid mass, energy, and momentum exchanges between the lower and upper troposphere. Most of the precipitation that falls to Earth is produced during this convective overturning, and a large fraction of the cloudiness in the atmosphere is produced by cumulus convection, either directly in the cumulus clouds themselves, or indirectly in the cirrus and other debris that cumuli generate.

    Cumulus convection is a manifestation of a buoyancy-driven instability that occurs when the temperature decreases upward sufficiently rapidly (i.e., when the "lapse rate" of temperature is sufficiently strong) *and*, at the same time, sufficient moisture is available. Because of the latter condition, cumulus instability is often called "conditional instability."

    The degree to which buoyancy forces can drive cumulus convection thus depends on both the lapse rate and the humidity. Neutral states, i.e. those in which the system is marginally unstable with respect to cumulus convection, can have either steep lapse rates with low humidities or more modest lapse rates with higher humidities.

    Cumulus convection can release the instability by converting the potential energy of the large-scale thermodynamic structure of the atmosphere into cumulus kinetic energy. The time scale for this convective release is on the order of an hour-- very short, compared to the multi-day time scale of large-scale weather systems. This disparity of time scales implies that an ensemble of cumulus clouds always stays nearly in balance with the large-scale weather system. If a large-scale motion system or surface heating tries to promote cumulus instability, convection releases the instability almost as rapidly as it is generated, so that the system always stays close to a neutral state. Many cumulus parameterizations are based on this fact[18].

- *"Large-scale" precipitation.*

    Large-scale precipitation refers to a somewhat old-fashioned but still widely used parameterization that is supposed to represent the formation of stratiform clouds such as cirrus or stratus, when the mean state relative humidity reaches or tries to exceed a specified maximum value, such as 100%. Such large relative humidities can be produced, for example, by large-scale rising motion which leads to adiabatic cooling and a decrease of the saturation mixing ratio. The excess humidity is typically assumed to condense and fall out as precipitation. In many models, the falling precipitation is permitted to evaporate or partially evaporate on the way down.

An important mechanism for forming stratiform clouds is the outflow or "detrainment" of moisture from deep cumuli. Existing models do not handle this well. The coupling between convective and stratiform clouds is an important area of current research.

Stratiform clouds often form in the boundary layer over the oceans, in regions where the large-scale circulation is producing subsidence (sinking) of the air. In these cases, the large relative humidities result from the trapping of moisture near the surface, below strong capping inversions that are produced in part by the subsiding motion. Special parameterizations have been developed to deal with these cloud types[299].

Many modeling groups are abandoning conventional large-scale precipitation parameterizations in favor of prognostic cloud water variables which explicitly represent the amount of liquid water in each grid cell[30]. An advantage is that the predicted cloud water can be used to determine the optical properties of the clouds, as well as the precipitation rate. In addition, the "memory" provided by a prognostic cloud water variable makes it possible for clouds to persist long after the agencies that formed them have ceased. A prognostic cloud water variable thus represents a step towards more realistic physics, but not without difficulties. For example, there are serious numerical problems in advecting a cloud water field that is zero in many places and positive elsewhere.

- *Cloud formation, where "cloud" refers to a radiatively interactive mass of droplets or ice crystals.*

We have already discussed the parameterizations of cumulus convection and large-scale saturation, so it may seem a bit odd that the formation of radiatively active clouds must be covered separately. The reason is that in many GCMs the obvious physical links between convection, precipitation, and cloud-radiation interactions have not been properly taken into account in the past. The use of prognostic cloud water variables, discussed above, will allow us to overcome this difficulty.

- *Gravity wave drag.*

As already mentioned in the discussion following Eq. (1), gravity waves are associated with the "restoring force" due to buoyancy. A lifted parcel of air cools by adiabatic expansion and so, depending on the lapse rate, will often be cooler than the air it encounters aloft; it thus tends to sink back under the buoyancy force. Overshooting, it is warmed by adiabatic compression and finds itself warmer than the air below its level of origin, and so it rises again under the influence of buoyancy. This oscillatory motion can be modified by the effects of the Earth's rotation, particularly when the horizontal scale of the gravity wave is large.

Gravity waves are produced by many agencies, including flow over mountains. Depending on conditions, gravity waves forced near the Earth's surface can propagate upward through and beyond the depth of the troposphere. The waves can transport momentum vertically. For example, a flow past a mountain range exerts a force on the Earth that is accompanied by a downward momentum flux in the air near the surface; momentum is being exchanged between the atmosphere and the solid Earth. In the presence of gravity waves, the momentum flux in the atmosphere tends to be independent of height except at a level where the waves are destroyed[31]. As a result, the force that the solid Earth exerts on the atmosphere is felt, not near the Earth's surface, but aloft at the level of wave destruction. This is what is meant by gravity wave drag[32].

In recent years, the importance of this process for the large-scale circulation of the atmosphere has been recognized and parameterizations of gravity wave drag have been devised. These parameterizations try to evaluate the wave momentum flux and the level of wave destruction where the momentum is deposited. There is some evidence that gravity wave drag parameterizations must take into account the resolution of the GCM in which they are used[33].

- *Land-surface processes.*

Suppose that the land surface consisted of bare soil, with no vegetation. Immediately after a rain, the surface would be wet and evaporation would proceed rapidly. Once the uppermost centimeter or two of the soil became dry, however, further evaporation would be limited by the rate at which moisture can diffuse through soil, which is quite slow. As a result, evaporation from the land surface would be extremely weak except immediately after precipitation events.

Vegetation changes this picture by pumping water up from the deep soil, through roots and stems and out through "stomates," which are essentially pores in the leaves. The loss of water through the stomates is an unavoidable side-effect of the intake of carbon dioxide, which is needed for photosynthesis.

The rate at which moisture is transferred from the soil to the atmosphere is thus primarily controlled by biological processes. The plants can be viewed as solar-powered water pumps. For this reason, simple parameterizations of the land-surface biosphere[34, 35] are being included in an increasing number of GCMs. These parameterizations take into account vegetation type and cover, the response of stomates to available solar radiation and to the temperature and humidity of the air, and of course the availability of soil moisture. Further developments are under way and will be discussed briefly below.

BOUNDARY CONDITIONS

The lower boundary conditions on an atmospheric model include the arrangements of land and sea, the distribution of terrain heights, the albedo and other optical properties of the Earth's surface, the surface roughness, the surface temperature, and for land points the distribution and characteristics of the vegetation.

Information about the vegetation on the land surface is needed because, as explained earlier, the plants strongly influence the flow of energy and moisture across the land-atmosphere interface.

Terrain heights are typically determined as grid-box averages, although some models take into account that the winds can be blocked by the highest (not the average) terrain in a grid cell. In addition, models that include gravity wave drag require information about the sub-grid-scale variability of the terrain height.

For the oceans and lakes, the fractional cover and thickness of possible sea ice must be provided. For the land masses, the distributions of permanent ice, such as the ice caps of Greenland and Antarctica, are also needed.

In essentially all existing atmospheric models, the upper boundary condition is that no mass cross the model top. This can lead to spurious reflection of upward propagating waves. Efforts have been made to devise "open" upper boundary conditions that allow waves to pass upward out of the model domain. Progress has been slow, however, in part because such a wide variety of wave motions must be dealt with.

Ocean models provide lower boundary conditions for atmospheric GCMs; the commonly heard ocean modeler's lament is that he or she is "just a lower boundary condition."

Ocean models obviously require as input the shapes of the ocean basins, including bottom topography. The atmospheric model provides the solar and terrestrial radiation incident on the sea surface, as well as the atmospheric temperature, humidity, and wind speed that determine the surface fluxes of sensible and latent heat. In addition, the difference between evaporation and precipitation acts as an effective "salt source" for ocean models that include salinity.

WHAT CURRENT GLOBAL CLIMATE MODELS CAN AND CANNOT DO

Up to now, the vast majority of the GCM-based "climate simulations" have actually been simulations of the general circulation of the atmosphere with seasonally and geographically varying sea surface temperatures (SSTs) prescribed according to observations. So long as the SSTs are prescribed from observations, the simulated climate has to be fairly realistic. Of course, a poorly designed atmospheric GCM can produce a poor simulation of the atmospheric general circulation even with prescribed SSTs, but there are limits to how bad the result can be.

It was demonstrated during the 1960's that atmospheric GCMs with prescribed SSTs can be started from a resting, isothermal state, and within a simulated month or two develop a fairly realistic simulation of the atmospheric general circulation[36]. The realism of the models has improved markedly during the past several decades, as a result of improved physical parameterizations (based on improved physical understanding), better numerical methods, and more powerful computers that allow higher resolution.

A recent survey of the ability of GCMs to simulate the climate has been published by the Intergovernmental Panel on Climate Change[37]. Two figures from that document are reproduced here. Fig. 3 shows simulated and observed December-January-February distributions of temperature in the atmosphere and ocean, averaged with respect to longitude. The simulation was performed with the coupled ocean-atmosphere GCM of the National Center for Atmospheric Research. Again, the model does a reasonable job of reproducing the major features of the observations.

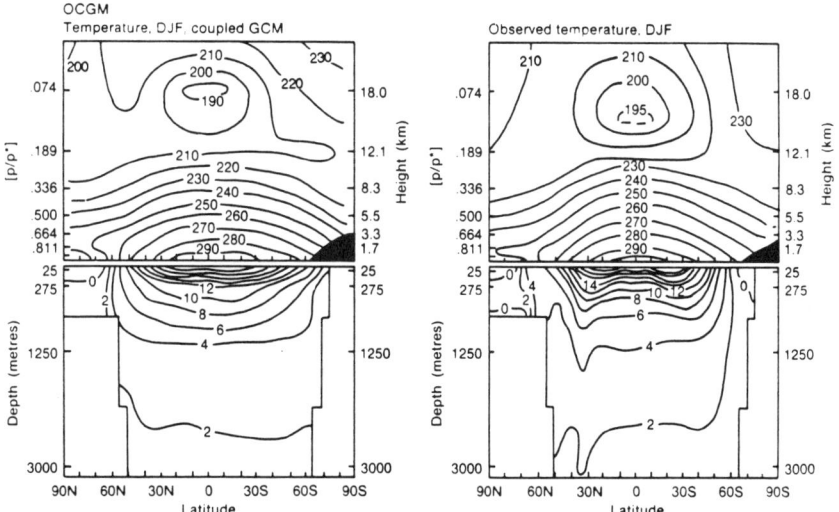

Figure 3: Simulated (left) and observed (right) distributions of longitudinally averaged temperature in the atmosphere and ocean, for December-January-February conditions. The model was developed at the National Center for Atmospheric Research. The horizontal axis is latitude. For the atmosphere, the vertical axis is labeled on the left with pressure normalized by the surface pressure, and on the right with height. . For the ocean, the vertical axis is depth.

Fig. 4 shows simulated and observed patterns of sea level pressure for January and July. The simulations were performed with a high-resolution version of the United Kingdom Meteorological Office atmospheric GCM, using prescribed SSTs. The model successfully reproduces the observed seasonally varying patterns.

In order to simulate the response of the climate to increasing CO_2 concentrations or other external forcing, it is absolutely necessary to allow the SSTs to vary in response to the forcing; unless the SSTs change, the climate really cannot change. The variability of sea ice must also be taken into account. Most simulations the global warming have, up to now, included only the upper 100 meters or so of the

40 Global Climate Models: What and How

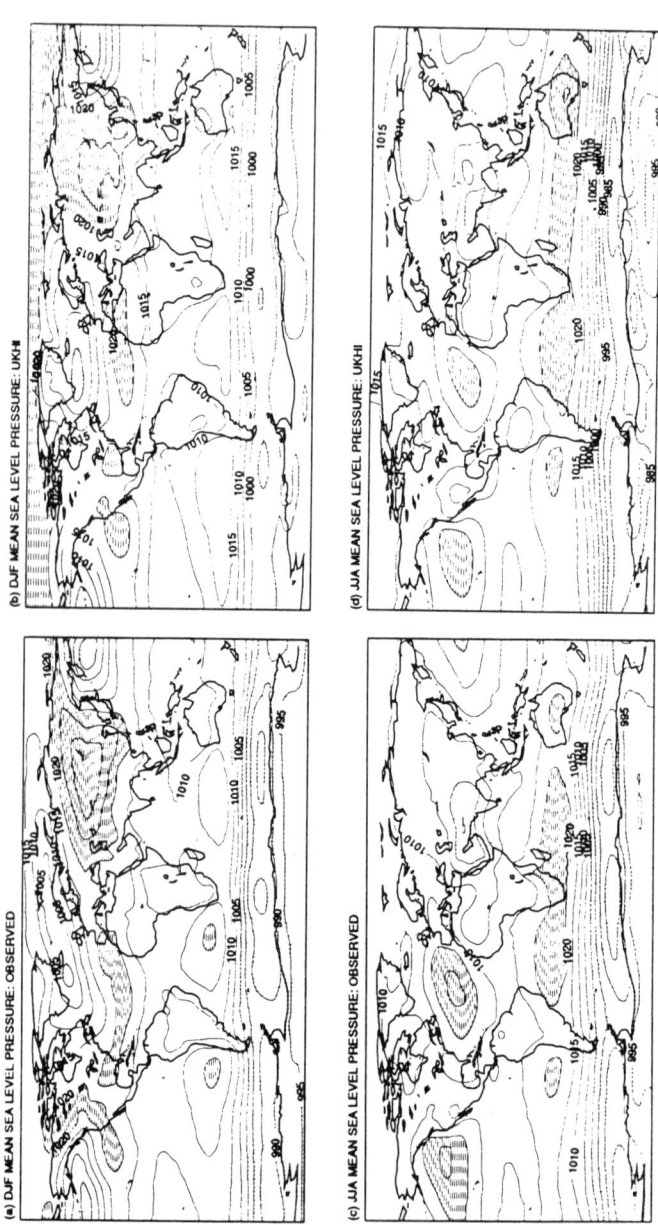

Figure 4: Maps of simulated (right) and observed (left) sea level pressure for December-January-February (top) and June-July-August (bottom). The simulations were performed with the atmospheric GCM of the United Kingdom Meterological Office, using prescribed SSTs. The units are millibars (mb). Values greater than 1020 mb are hatched, and values less than 1000 mb are shaded.

ocean, and this in simplified form[38]. Such models may be capable of simulating the transient climate response to forcing for a few decades, but on longer time scales the deep ocean becomes critically important.

Only a handful of simulations have been performed with "world" ocean models (a term that refers to true ocean GCMs that include the full depth of the ocean with realistic geometry) coupled to atmospheric GCMs. The primary reason is computational expense, but there are other obstacles. For example, the present (and past) state of the ocean is not very well known below the surface, so that it is difficult even to specify initial conditions with confidence. This is a serious problem. Suppose, for example, that we want to use a coupled ocean-atmosphere model to simulate the observed cooling of the globally averaged surface air temperature between 1940 and 1970. In order to do so, we have to specify the state of the ocean in 1940. The necessary data do not exist, so this experiment cannot be done.

Of course, before using such a coupled model for studies of climate change, it is important to verify that it can simulate the current climate. The fact is that existing coupled GCMs cannot do this as well as we would like. They may produce systematic SST errors on the order of several degrees, for example. Such errors are very serious and can arise from many problems, including insufficient resolution for the ocean model and unrealistic simulations of the effects of clouds on the radiation balance at the sea surface.

There are two schools of thought on how to respond to this problem. One view is that the errors should be tolerated until they can be removed by improved model design. The other is that artificial "flux corrections" should be applied to force the simulated SSTs to be close to observations in simulations of the present climate. A debate is raging on the relative merits of these two approaches. Of course, when the models become good enough the point will become moot.

In most cases, the results of climate change simulations cannot (yet) be compared with observations. Exceptions are simulations of *past* climates[39] for which at least some observational data is available. Simulations of past climates have so far been performed with atmospheric GCMs, using prescribed SSTs based on paleoclimate data. The results are encouraging.

CURRENT ISSUES AND A LOOK AHEAD

As discussed by R. Cess elsewhere in this volume, our inability to parameterize the effects of clouds is one of the most serious deficiencies of current climate models. The problem of developing an improved cloud parameterization can be broken down into two sub-problems: developing a successful theory of cloud formation and dissipation, and parameterizing the interactions of the predicted clouds with solar and terrestrial radiation.

My own somewhat biased opinion is that predicting the distribution of clouds is the more difficult of these two sub-problems. Part of the difficulty is that there are so many different types of clouds, each with its own distinctive formation and dissipation mechanisms. To give a few examples, clouds can be formed by flow over mountains, or by frontal lifting, or by the outflow from deep convection, or by trapping of moisture in a thin layer near the Earth's surface. They can be dissipated by large-scale sinking motion, or warming due to absorption of solar radiation, or precipitation, or turbulent mixing with dry air. Their liquid and/or ice concentrations must be predicted in order to physically determine their optical properties. Clouds rarely completely fill a GCM grid volume, i.e. they tend to be subgrid-scale both horizontally and vertically. When such fractional clouds occur in multiple layers, the degree to which they overlap is important for radiative transfer. These and various other problems of cloud parameterization are not well understood at present and will continue to challenge us for many years to come.

As discussed earlier, we are beginning to take into account the effects of vegetation on the exchange of moisture between the land surface and the atmosphere. The biosphere both on the land surface and in the oceans has many additional effects on the climate system, however. For example, the exchange of carbon between the atmosphere and the biosphere strongly influences atmospheric composition. Attempts are under way to incorporate simple representations of photosynthesis and respiration into the biospheric components of climate models. This will permit explicit simulation of carbon exchange processes. A logical next step would be the explicit prediction of biomass, so that the distribution of vegetation amount can influence and be influenced by the simulated climate. Eventually, vegetation type will also have to become interactive.

In order to simulate the climate with a coupled ocean-atmosphere GCM, it would be necessary to run the model for at least on the order of 10,000 years, to allow the deep ocean sufficient time to equilibrate. A few such runs have been made[40] with low-resolution models. The required computational resources are drastically reduced by using artificial coupling schemes (called "asynchronous coupling") in which the ocean evolves for thousands of years while the atmosphere evolves for only tens of years. For low-resolution models with more realistic ("synchronous") coupling, the longest runs that have been made up to now are on the order of a few hundred years[41] -- about a factor of 100 too short for true climate simulation. No high-resolution GCMs, capable of resolving the ocean's eddies, have ever been used to simulate the climate.

Of course, it is not clear that the present climate is in equilibrium, or to what extent the climate system is ever really in equilibrium, given the various quasi-random forcings such as the rise of industrial civilization, the explosions of major volcanoes, and the slow progressions of the Earth's orbital parameters, biological evolution and continental drift. It might turn out, then, that the *correct* equilibrium climate for "current conditions" is actually different from the climate that we observe.

Within the current decade, increasing computer power may permit the first true, synchronous, eddy-resolving GCM simulations of the fully equilibrated climate states of the atmosphere and ocean. This is, therefore, a unique and exciting time for those working in this area.

There are many other avenues for further development of climate models, including interactive atmospheric chemistry.

CONCLUDING REMARKS

The development of global climate models over the last thirty years or so can be viewed as our first attempt to simulate and predict the Earth as a system. Clearly, models of the Earth will continue to increase in complexity, realism, and practical importance for the indefinite future.

The major differences in the climate sensitivities of existing atmospheric general circulation models are, to a large extent, directly due to differences in their parameterizations of cloud-related processes. These differences in model design arise from disagreements, within the climate modeling community, concerning the most realistic way to formulate the relevant physics. Such disagreements are healthy because they serve to identify deficiencies in our collective physical understanding.

The wide range of model sensitivities cannot be narrowed simply by increasing model resolution. If at some future time all of the existing models could be run with drastically increased resolution, the differences in their climates would be quite comparable to those obtained today. Evidence for this was recently obtained by Tibaldi et al.[41], who found that although the forecast skill of the advanced GCM used at the European Centre for Medium Range Weather Forecasts progressively improves as the resolution is increased, the systematic error of the model, which represents the deficiencies of the simulated climate, does not improve much as the resolution increases beyond the moderate range.

These results indicate that dramatically increased computer power would not, by itself, be sufficient to greatly improve either our ability to simulate the present climate or our confidence in climate change simulations produced by existing models. Improvements in climate simulation and climate forecasting must come primarily from improved understanding of the physics of the climate system.

ACKNOWLEDGMENTS

Dr. T. Jensen provided valuable input. D. Randall's climate modeling research is supported by NASA's Climate Program under grant NAG 5-1058 to Colorado State University, by the U. S. Department of Energy under grant DE-FG02-89-ER69027 to Colorado State University, and by the National Science Foundation under Grant No. ATM-8907414 to Colorado State University.

REFERENCES

1. M. I. Budyko, *Tellus* **21**, 611 (1969).
2. W. D. Sellers, *J. Appl. Meteor.* **8**, 392 (1969).
3. G. R. North, *J. Atmos. Sci.* **32**, 2033 (1975).

4. R. S. Lindzen, *Dynamics in Atmospheric Physics* (Cambridge University Press, New York, 1990).

5. S. Manabe and R. F. Strickler, *Mon. Wea. Rev.* **93**, 769 (1965).

6. S. Manabe and R. T. Wetherald, *J. Atmos.Sci.* **24**, 241 (1967).

7. V. Ramanathan and J. A. Coakley, Jr., *Rev. Geophys. Space Phys.* **16**, 465 (1978)

8. L. Bengtsson, *Adv. Geophys.*, **28B**, 3 (1985).

9. E. Kalnay, M. Kanamitsu, and W. E. Baker, *Bull. Amer. Meteor. Soc.*, **71**, 1410 (1990).

10. I. M. Held and M. J. Suarez, *Tellus*, **26**, 613 (1974).

11. G. L. Stephens, G. L., G. G. Campbell, and T. H. Vonder Haar, *J. Geophys. Res.*, **86**, 9739 (1981).

12. A. H. Oort and T. H. VonderHaar, *J. Phys. Oceanogr.*, **6**, 781 (1976).

13. K. Masuda, K., *Tellus*, **40A**, 285 (1988).

14. R. S. Lindzen and A. Y. Hou, *J. Atmos. Sci.*, **45**, 2416 (1988).

15. M. L. Blackmon, J. M. Wallace, N.-C. Lau, and S. L. Mullen, *J. Atmos. Sci.*, **34**, 1040 (1977).

16. A. J. Semtner, Jr., and R. M. Chervin, *J. Geophys. Res.*, **93**, 15502 (1988).

17. H. Riehl and J. S. Malkus, *Geophysica*, **6**, 503 (1958).

18. A. Arakawa and W. H. Schubert, *J. Atmos. Sci.*, **31**, 674 (1974).

19. A. E. Gill, A*tmosphere-ocean dynamics. (*Academic Press, New York, 1982).

20. E. N. Lorenz, *J. Atmos. Sci.*, **20**, 130 (1963).

21. E. N. Lorenz, *Bull. Amer. Meteor. Soc.*, **50**,345 (1969).

22. K. Bryan and J. L. Sarmiento, *Adv. in Geophys.*, **28A**, 433 (1985).

23. N. A. Phillips, *J. Atmos. Sci.*, **23**, 626 (1966).

24. A. Kasahara, *Mon. Wea. Rev.*, **102**, 509 (1974).

25. K. V. Ooyama, *J. Atmos. Sci.*, **47**, 2580 (1990).

26. A.Arakawa and V. R. Lamb, *Meth. Comp. Phys.*, **17**, 173 (1977).

27. ·S. A. Orszag, S. A., *J. Atmos. Sci.*, **27**, 890 (1970).

28. W. M. Washington and C. L. Parkinson, A*n introduction to three-dimensional climate modeling. (U*niversity Science Books, Mill Valley, New York, 1986).

29. D. A. Randall, J. A. Abeles, and T. G. Corsetti, *J. Atmos. Sci.,* **42,** 641 (1985).

30. H. Sundqvist, *Quart. J. Roy. Meteor. Soc.*, **104**, 677 (1978).

31. A. Eliassen and E. Palm, *Geofys. Publ. Oslo*, **22**, 1 (1961).

32. T. N. Palmer, G. J. Shutts, and R. Swinbank, Quart. J. R. Met. Soc. **112**, 1001 (1986).

33. J. D. Mahlman, Y. Hayashi, and S. Miyahara, *J. Atmos. Sci.* **43**, 1844 (1986).

34. R. E. Dickinson, *Adv. in Geophys.*, **25**, 305 (1983).

35. P. J. Sellers, Y. Mintz, Y. C. Sud, and A. Dalcher, *J. Atmos. Sci.*, **43**, 505 (1986).

36. Y. Mintz, A*mer. Meteor. Soc. Meteorological Monographs* **8**, 20 (1965).

37. J. T. Houghton, G. J. Jenkins, and J. J. Ephraums, Eds. *Climate Change. The IPCC Scientific Assessment.* (World Meteorological Organization / United Nations Environment Programme. Cambridge University Press, 1990).

38. J. Hansen, I. Fung, A. Lacis, D. Rind, S. Lebedeff, R. Ruedy, and G. Russell, *J. Geophys. Res.*, **93**, 9341 (1988).

39. J. E. Kutzbach, *Science*, **214**, 59 (1981).

40. S. Manabe and R. J. Stouffer, *J. Climate*, **1**, 841 (1988).

41. S. Manabe, K. Bryan, and M. J. Spelman, *J. Phys. Ocean.*, **20**, 722 (1990).

42. S. Tibaldi, T. N. Palmer, C. Brankovic, and U. Cubasch, *Quart. J. R. Met. Soc.*, **116**, 835 (1990).

COMPARISON OF GENERAL CIRCULATION MODELS

R. D. Cess
State University of New York, Stony Brook, New York 11794-2300

INTRODUCTION

Many facets of the climate system are not well understood, and the significant uncertainty associated with modeling future climate change is largely due to interactive climate feedback mechanisms. Such feedback mechanisms can either amplify or damp the climate response resulting from a given radiative forcing. Emphasis here will be restricted to global-mean quantities, since the conventional concept of radiative feedback mechanisms applies only to global-mean quantities as well as to changes from one equilibrium climate to another.

RADIATIVE FORCING

Radiative forcing of the climate system is induced by an increase in greenhouse gases, and it is useful to place this in the context of the Earth's present greenhouse effect. Schematically illustrated in Fig. 1 is the radiative energy budget of the surface-atmosphere system. Although the surface absorbs 90 W m^{-2} more radiative energy than it emits, this is balanced by a 90 W m^{-2} (not shown) loss from the surface by sensible and latent heat

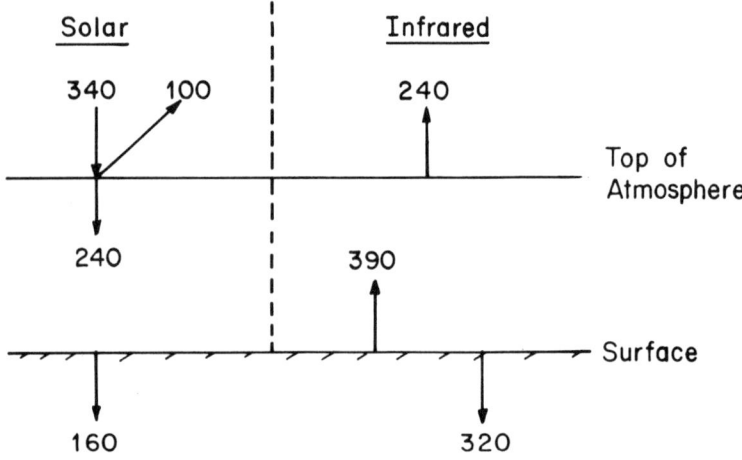

Fig. 1. Schematic illustration of the radiative energy budget of the surface-atmosphere system.

transfer. On a global and annual average, there is roughly 340 W m^{-2} of incident solar radiation at the top of the atmosphere (TOA). Of this, 100 W m^{-2} is reflected to space so that the surface-atmosphere system (i.e., the climate system) absorbs 240 W m^{-2}, of which 80 W m^{-2} is absorbed by the atmosphere and 160 W m^{-2} by the surface. Since the climate system must be in radiative equilibrium, it likewise emits 240 W m^{-2} of infrared radiation to space.

There are two important points to note from Fig. 1. The first is that the 240 W m^{-2} emission at the TOA is 150 W m^{-2} less than the 390 W m^{-2} emission from the surface. This radiative flux difference is the greenhouse effect of the Earth's present atmosphere, and it is caused by the absorption of infrared radiation by greenhouse gases and clouds. The second important point of Fig. 1 is that the atmospheric greenhouse gases and clouds in turn emit infrared radiation downward to the surface, and this direct radiative heating of the surface by the atmosphere (320 W m^{-2}) is twice the direct solar heating (160 W m^{-2}). By itself, the 320 W m^{-2} infrared surface heating produces about a 40°C surface warming, and it is this greenhouse effect that makes our planet habitable.

Increasing concentrations of greenhouse gases will provide a positive perturbation to the greenhouse effect, and a doubling of atmospheric CO_2 is used to demonstrate this. A doubling of atmospheric CO_2 produces a radiative forcing of roughly 4 W m^{-2}, although the definition of radiative forcing requires some clarification. Strictly speaking, it is defined as the change in net downward infrared radiation at the tropopause, so that the 4 W m^{-2} represents a radiative heating of the surface-troposphere system. If the stratosphere is then allowed to respond to the CO_2 forcing, while the climate parameters of the surface-troposphere system are held fixed, then the 4 W m^{-2} reduction in net downward infrared radiation applies both at the tropopause and at the TOA. It is in this context that radiative forcing is used in this presentation.

With reference to Fig. 1, the 4 W m^{-2} forcing, for a CO_2 doubling, increases the greenhouse effect by increasing the surface-to-TOA infrared flux difference from 150 to 154 W m^{-2}, which simultaneously reduces the TOA infrared emission from 240 to 236 W m^{-2}. Since solar absorption by the climate system remains fixed at 240 W m^{-2}, then the system is no longer in radiative equilibrium and must warm until a new equilibrium is achieved. It is this climate response that is governed by interactive climate feedback mechanisms.

RADIATIVE FEEDBACK MECHANISMS

In order to demonstrate radiative feedback mechanisms, it will be convenient to initially assume that climate change is manifested solely by temperature changes within the climate system, and that all other climate parameters remain fixed at their unperturbed values. In this framework, there will be no change in the climate system's 240 W m^{-2} solar absorption. Moreover, let G denote the 4 W m^{-2} radiative forcing, while ΔF is the change in the TOA infrared flux following the imposition of the forcing. The restoration of radiative equilibrium thus requires that $\Delta F = G$, or in terms of the increase in global-mean surface temperature ΔT_s,

$$\Delta T_s = \frac{G}{\Delta F/\Delta T_s} . \qquad (1)$$

For the present example, it may easily be shown[1] that $\Delta F/\Delta T_s$ = 3.3 W m^{-2} °C^{-1}, so that with G = 4 W m^{-2},

$$\Delta T_s = 1.2 \; °C . \qquad (2)$$

If it were not for the fact that this warming introduces numerous interactive feedback mechanisms, then $\Delta T_s = 1.2$ °C would be quite a robust global-mean quantity. Unfortunately, such feedbacks introduce considerable uncertainty in ΔT_s estimates, and three of the more commonly discussed radiative feedbacks are described in the following subsections.

Water-Vapor Feedback

The best understood feedback mechanism is water-vapor feedback, and this is intuitively easy to comprehend: a warmer atmosphere contains more water vapor, which itself is a greenhouse gas. Thus an increase in one greenhouse gas (CO_2) induces an increase in yet another greenhouse gas (water vapor), resulting in a positive (amplifying) feedback mechanism.

To be specific on this point, Raval and Ramanathan[2] have employed satellite data to quantify the temperature dependence of the water-vapor greenhouse effect. From their results it readily follows[3] that water-vapor feedback reduces $\Delta F/\Delta T_s$ from the prior value of 3.3 W m^{-2} °C^{-1} to 2.3 W m^{-2} °C^{-1}. From eq. (1) this in turn increases the global warming, for a CO_2 doubling, from 1.2°C to 1.7°C.

There is yet a further amplification. Because water vapor also absorbs solar radiation, water-vapor feedback leads to an additional heating of the climate system through enhanced absorption of solar radiation. With Q denoting solar absorption by the climate system (240 W m^{-2} for the present climate), this effect produces[4] $\Delta Q/\Delta T_s$ = 0.2 W m^{-2} °C^{-1}. To incorporate this into a ΔT_s estimate, extension of eq. (1) to include solar absorption yields

$\Delta T_s = \lambda G$, where λ is the climate sensitivity parameter defined by

$$\lambda = \frac{1}{\Delta F/\Delta T_s - \Delta Q/\Delta T_s}. \qquad (3)$$

It then follows that the inclusion of the solar component of water-vapor feedback results in $\Delta T_s = 1.9°C$, so that the net effect of water-vapor feedback is to amplify the initial $\Delta T_s = 1.2°C$ warming by the factor of 1.6.

The progressive forcing and feedback amplifications are summarized in Table I. Here H denotes the greenhouse effect (150 W m^{-2} for the present climate). The radiative forcing (process 1)

Table I. Forcing and response of the climate system caused by a doubling of atmospheric CO_2.

Process	$\Delta T_s(°C)$	H(W m^{-2})	F(W m^{-2})	Q(W m^{-2})
Present climate	0	150	240	240
1. Radiative forcing	0	154	236	240
2. Temperature response without water-vapor feedback	1.2	156.5	240	240
3. Including infra-red water-vapor feedback	1.7	159.2	240	240
4. Including solar water-vapor feedback	1.9	160.3	240.4	240.4

simultaneously increases H and reduces F, so that the planet emits 4 W m^{-2} less energy than it absorbs from the Sun. It is this imbalance that causes global warming and results in $\Delta T_s = 1.2°C$ (process 2). Although the climate system returns to its original radiation balance, with 240 W m^{-2} both absorbed and emitted, process 2 increases the greenhouse effect by 2.5 W m^{-2} because of enhanced surface emission.

Process 3 incorporates the infrared component of water-vapor feedback, with the 2.7 W m^{-2} increase in H being simultaneously due to the increase in atmospheric water vapor and to enhanced surface emission. The TOA radiation budget is only slightly modified by process 4. An important point is that the combined effects of water-vapor feedback and surface warming have amplified the 4 W m^{-2} greenhouse forcing to 10.3 W m^{-2}. As Raval and Ramanathan[2] have emphasized, this suggests that direct monitoring, from satellites, could reveal future changes in the greenhouse effect.

The most detailed climate models for the purpose of projecting climate change are three-dimensional general circulation models (GCMs), and these models seem to properly depict the infrared component of water-vapor feedback. In a recent intercomparison of atmospheric GCMs[4], it was found that 19 GCMs collectively produced

$$\Delta F/\Delta T_s = 2.3 \pm 0.2 \text{ W m}^{-2}, \quad (4)$$

such that the models are both self consistent and in agreement with the observational result of Raval and Ramanathan[2].

Snow-Ice Feedback

An additional feedback mechanism is snow-ice feedback, by which a warmer Earth has less snow and ice cover, resulting in a darker planet that in turn absorbs more solar radiation. While this conventional albedo feedback description is quite obvious, and by itself constitutes a positive feedback, it appears that the retreat of snow and ice cover might activate other interactive processes. To demonstrate this possibility, diagnostic results for a single GCM (M.-X. Zhang, private communication) are illustrated. This GCM is the OSU/IAP model described by Cess et al.[4], and it constitutes a variant of the Oregon State University GCM. The climate change simulation was similar to the perpetual July simulation used in the GCM intercomparison[4] in which sea-surface temperature was perturbed by 4°C, except that a perpetual April was adopted here so as to activate snow feedback. In order to isolate one feedback at a time, sea-ice cover was held fixed.

Analogous to Wetherald and Manabe's[5] (1988) study of cloud feedback, two climate-change simulations were performed, one in which snow cover was fixed and one in which it retreated with climate warming. Comparison of these two simulations showed that variable snow cover produced a 1.6 factor amplification of the climate sensitivity parameter λ, but this positive feedback is only partially the result of the change in surface albedo. The percentage contributions to this 1.6 factor amplification are summarized as follows:

Snow-albedo feedback: 41%
Lapse-rate feedback: 60%
Water-vapor feedback: -19%
Cloud feedback: 17%

The point of the above is that snow-albedo feedback provides less than half of the net feedback due to the change in snow cover. The dominant effect is lapse-rate feedback by which there is a change in the atmosphere's vertical temperature gradient. This occurs, within the model, as the result of surface warming, caused by the reduction of snow cover, producing a decrease in net precipitation (convective precipitation increases but is more than offset by a decrease in large-scale precipitation). There is thus a reduction of latent heat release within the atmosphere, with a concurrent steepening of the lapse rate that, by itself, increases the greenhouse effect[2] and so acts as a positive feedback mechanism.

It is cautioned that this snow feedback diagnosis is probably very model dependent, and the above results are presented solely to illustrate the complexity of the interactive nature of this feedback. It is probable that sea-ice feedback includes even more interactive effects, particularly since sea ice insulates the ocean surface.

Cloud Feedback

Feedback mechanisms related to clouds are extremely complex phenomena. To demonstrate this, it will be useful to first consider the impact of clouds on the present climate. Summarized in Table II are the radiative impacts of clouds on the global climate system for annual-mean conditions. These radiative impacts refer to the effect of clouds relative to to a "clear-sky" Earth. The presence of clouds heats the climate system by 31 W m^{-2} through increasing the greenhouse effect. Note the similarity to trace-gas radiative forcing, which is why this impact is referred to as cloud-radiative forcing. Through reflection of solar radiation, clouds also result in cooling of the system. As demonstrated in Table II, the latter dominates over the former, and the net effect of clouds on the annual climate system is a 13 W m^{-2} radiative cooling.

Table II. Infrared, solar and net cloud-radiative forcing (CRF). These are annual-mean values estimated from data for January, April, July and October[6].

Component	CRF(W m^{-2})
Infrared	31
Solar	−44
Net	−13

Although clouds produce net cooling of the climate system, this must not be construed as a possible mechanism for offsetting global warming due to increasing greenhouse gases. As discussed in detail by Cess et al.[4,7], cloud feedback constitutes the change in net CRF associated with a change in climate. To emphasize the complexity of this feedback mechanism, three contributory processes are summarized as follows:

●Cloud amount: If cloud amount decreases because of global warming, as occurs in typical GCM simulations[7], then this decrease reduces the greenhouse effect attributed to clouds and so acts as a negative feedback mechanism. But there is a related positive feedback; the solar radiation absorbed by the climate system increases because the diminished cloud cover causes a reduction of reflected solar radiation by the atmosphere. There is no simple way of appraising the sign of this feedback component.

●Cloud altitude: A vertical redistribution of clouds will also induce feedbacks. For example, if global warming displaces a given cloud layer to a higher and colder region of the atmosphere, this will produce a positive feedback because the colder cloud will emit less radiation and will thus enhance the greenhouse effect.

●Cloud water content: There has been considerable recent speculation that global warming could increase cloud water content, thereby resulting in brighter clouds and hence a negative component of cloud feedback. Cess et al.[7] have suggested that this might be an oversimplification. In one case, they demonstrated that this negative solar feedback induces a compensating positive infrared feedback, and in a more recent study[4] they indicate that in some models the net effect might be that of positive feedback (see also Schlesinger[8]).

The above discussion clearly illustrates the multitude of complexities associated with cloud feedback; indeed, differences in models' depictions of this feedback largely account for the significant differences in climate sensitivity among 19 GCMs[4]. As previously discussed, this intercomparison employed a perpetual July simulation in which the climate was changed by imposing a $4°C$ perturbation to the global sea-surface temperature while holding sea ice fixed. Since a perpetual July simulation with a GCM produces very little snow cover over land, this effectively eliminates snow feedback. The details of this simulation are given elsewhere[4,7]; the main point is that it was chosen to minimize computer time and thus allow a large number of modeling groups to participate.

Cess et al.[4] have summarized climate sensitivity parameters (λ as defined by eq. 3) for the 19 GCMs, and these results are reproduced in Fig. 2. The important point is that cloud effects were isolated by separately averaging the models' clear-sky TOA fluxes, so that in addition to evaluating the climate sensitivity parameter for the globe as a whole (filled circles), it was also possible to evaluate it for an equivalent "clear-sky" Earth (open circles). Note the remarkable agreement of the clear-sky sensitivity parameters, and this is due to the agreement of water vapor feedback as previously discussed. There is, however, a nearly threefold variation of the global (clear plus overcast) sensitivity parameter, and considering the clear-sky agreement, then clearly most of the variation in the global sensitivity parameter can be attributed to cloud feedback. Certainly improvements in the treatment of cloud feedback are needed if GCMs are ultimately to be used as reliable climate predictors.

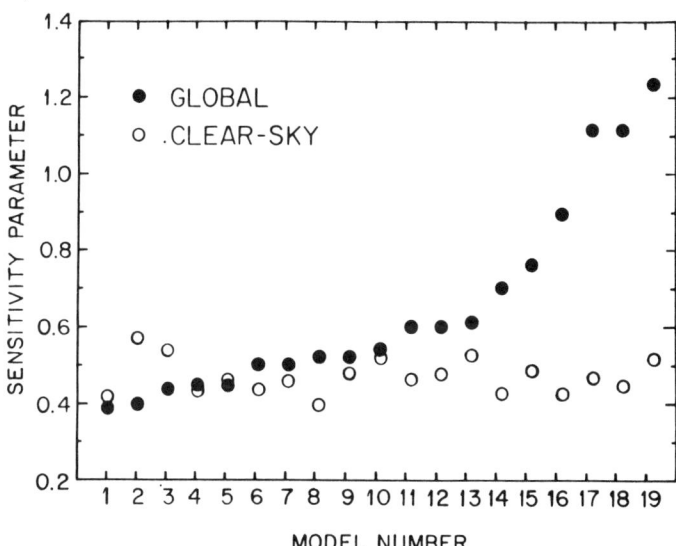

Fig. 2. Clear-sky and global sensitivity parameters ($^\circ C\ m^2\ W^{-1}$) for 19 GCMs.

REFERENCES

1. R. D. Cess, J. Atmos. Sci. <u>33</u>, 1831, (1976).

2. A. Raval, and V. Ramanathan, Nature <u>342</u>, 758, (1989).

3. R. D. Cess, Nature <u>342</u>, 736, (1989).

4. R. D. Cess, G. L. Potter, J. P. Blanchet, G. J. Boer, A. D. Del Genio, M. Déqué, W. L. Gates, S. J. Ghan, J. T. Kiehl, A. A. Lacis, H. Le Treut, Z.-X. Li, X.-Z. Liang, B. J. McAvaney, V. P. Meleshko, J. F. B. Mitchell, J.-J. Morcrette, D. A. Randall, L. Rikus, E. Roeckner, J. F. Royer, U. Schlese, D. A. Sheinin, A. Slingo, A. P. Sokolov, K. E. Taylor, W. M. Washington, R. T. Wetherald, and I. Yagai, J. Geophys. Res. (in press).

5. R. T. Wetherald, and S. Manabe, J. Atmos. Sci. <u>45</u>, 1397, (1988).

6. V. Ramanathan, R. D. Cess, E. F. Harrison, P. Minnis, B. R. Barkstrom, E. Ehmad, and D. Hartmann, Science 243, 57-63, 1989.

7. R. D. Cess, G. L. Potter, J. P. Blanchet, G. J. Boer, S. J. Ghan, J. T. Kiehl, X.-Z. Liang, J. F. B. Mitchell, J.-J. Morcrette, D. A. Randall, M. R. Riches, E. Roeckner, U. Schlese, A. Slingo, K. E. Taylor, W. M. Washington, R. T. Wetherald and I. Yagai, Science 245, 513, (1989).

8. M. E. Schlesinger, Nature 333, 303, (1988).

Climate and the Earth's Radiation Budget

V. Ramanathan,[a] Bruce R. Barkstrom,[b] and Edwin F. Harrison[c]

A NASA multisatellite experiment has determined that clouds cool the planet more than they heat it and identified them as a major source of uncertainty in three-dimensional models used for studying the greenhouse effect and global warming.

Among the first payloads aboard satellites in the early 1960s were instruments for measuring the Earth's radiation budget.[1] The radiation budget consists of the incident and reflected sunlight and the longwave (infrared and far infrared) radiation emitted to space. The source for the recent spurt in scientific and public interest in the greenhouse effect and global warming is the alteration of the radiation budget by the anthropogenic emission of trace gases into the atmosphere.

After two decades of progress in satellite instrumentation, the Earth Radiation Budget Experiment[2] began in the 1980s. ERBE instruments are carried on three satellites: the Earth Radiation Budget Satellite, NOAA-9 and NOAA-10. The instruments on ERBS

---

[a] V. **Ramanathan** is a professor in the department of geophysical sciences at the University of Chicago.
[b] **Bruce R. Barkstrom,** science team leader of the Earth Radiation Budget Experiment, is a senior scientist in the atmospheric sciences division of NASA's Langley Research Center, Hampton, Virginia.
[c] **Edwin F. Harrison** is a senior scientist and head of the radiation sciences branch in the same division.

This chapter originally appeared in the May 1989 issue of *Physics Today*. It is reprinted here with the permission of the American Institute of Physics.

(launched by the space shuttle Challenger) began observing the Earth in November 1984; those on NOAA-9, in February 1985; and those on NOAA-10, in December 1986. These instruments are still collecting data critical to understanding the greenhouse effect. The data are also fundamental to defining the role of human activities in climate change.

Global radiation energy balance

To understand more fully the role of these satellites and to appreciate the implications of their observations of the radiation budget for theories of climate, it is useful to examine a few simple, conceptual models of climate. Such models illustrate the strong links among the radiation budget, the climate and the circulation of the atmosphere and the oceans. As we shall see, the effects of cloud cover may be the greatest sources of uncertainty in our understanding of climate and how it responds to human activities.[3]

The simplest zero-dimensional model of climate considers the long-term average (on a time scale greater than a year) of the global and annual mean temperature. A balance between the absorbed solar energy and the emitted energy governs this temperature. We can write this symbolically as

$$H = \frac{S_0}{4}(1-\alpha) - \sigma T_e^4 = 0.$$

Here H is the net energy input to the climate system. The solar irradiance S_0 is the solar power per unit area intercepted at the mean Earth–Sun distance; recent satellite measurements show that S_0 is about 1365–1372 W/m². In the equation, the geometric factor 4 is the ratio of the Earth's surface area to the area of the Earth's disk. The planetary albedo α is the fraction of the solar irradiance reflected by the planet's surface and atmosphere; past satellite measurements show that α is 0.30±0.03. The surface–atmosphere system is assumed to emit like a blackbody at a temperature T_e. (The coefficient σ is the Stefan–Boltzmann constant.) Although the surface emission is close to that of a blackbody, the atmosphere emits in specific wavelength bands, and so the emission departs significantly from that of a blackbody. T_e is still a useful parameter, however, provided we think of it as a planet's *effective* radiating temperature. Thus, we expect T_e to be 255 K if H is zero.

What does this effective radiating temperature, T_e, mean? If the atmosphere did not impede the radiative energy flow, the surface tem-

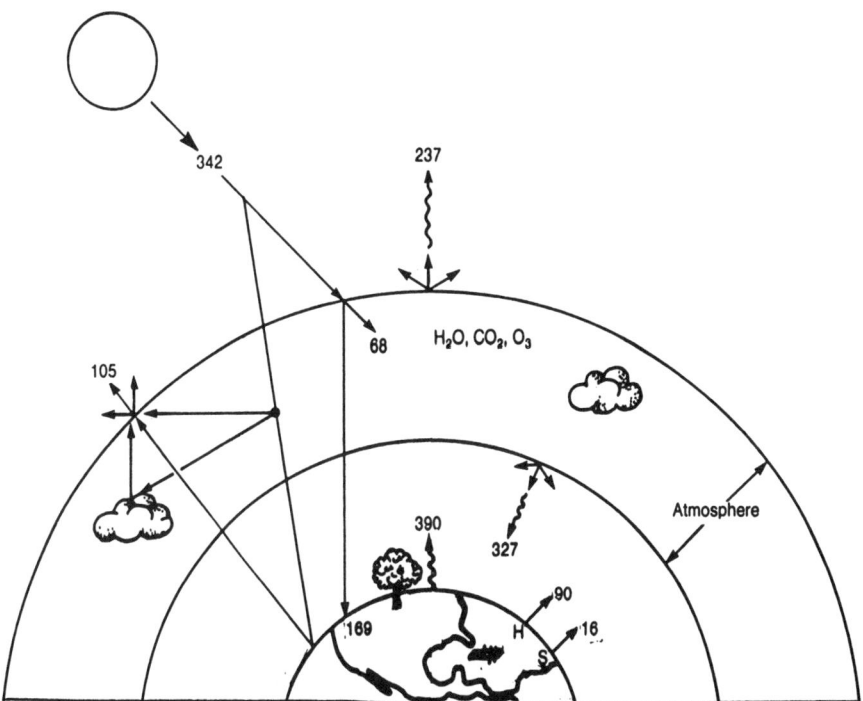

Figure 1. Global energy balance for annual mean conditions. For the top of the atmosphere, the estimates of solar insolation (342 W/m^2), reflected solar radiation (105 W/m^2) and outgoing long-wave radiation (237 W/m^2) are obtained from satellite data. The other quantities are obtained from various published model and empirical estimates. These quantities include atmospheric absorption of solar radiation (68 W/m^2); surface absorption of solar radiation (169 W/m^2); downward long-wave emission by the atmosphere (327 W/m^2); upward long-wave emission by the surface (390 W/m^2); H, the latent heat flux from the surface (90 W/m^2); and S, the turbulent heat flux from the surface (16 W/m^2). H and S are averaged over both ocean and land.

perature T_s would be nearly the same as T_e. T_s, however, is observed to be 288 K. The 33-K difference is attributed to the greenhouse effect.

Satellite measurements of the radiation budget show this difference more directly. Let us make an illustrative calculation based on the numbers in figure 1. At a temperature of 288 K, the surface emits 390 W/m^2. Only 237 W/m^2 escapes to space. The energy trapped in the atmosphere is the 153 W/m² difference between the surface emission and the total energy loss.

Because the atmosphere is generally colder than the ground, we know from the blackbody radiation law that a molecule in the atmosphere will absorb more energy than it emits. The net result of these absorption and emission processes is that part of the infrared radiation emitted by the ground is trapped. The infrared trapping by the atmosphere—familiarly known as the greenhouse effect, is due primarily to water vapor, clouds and CO_2, with a smaller, 5% contribution from the gases O_3, N_2O and CH_4. But several anthropogenic gases, such as the chlorofluorocarbons like $CFCl_3$ and CF_2Cl_2, are now beginning to make an appreciable contribution.

Radiative–convective equilibrium

Now let us turn to a one-dimensional model rather than a zero-dimensional model. The atmosphere constantly loses energy (see Figure 1) because it emits 327 W/m² to the surface. The atmosphere traps 153 W/m² of long-wave radiation (as we showed before) and absorbs only 68 W/m² from the Sun. Hence the atmosphere loses 106 (that is, $-327 + 153 + 68$) W/m² of radiation energy. In other words, there is radiative cooling of the atmosphere and a corresponding radiative heating of the Earth's surface. (The surface must gain 106 W/m² to balance the 106 W/m² atmospheric loss.)

Heating the lower boundary of a fluid while cooling its interior is the classical mechanism for inducing convective instability and turbulence. In the Earth's atmosphere, evaporation of water from the surface and condensation elsewhere complicates the heat exchange. Turbulent transfer of heat and condensation of water make up for the atmosphere's radiative energy deficit (see Figure 2). The combination of these nonradiative processes is loosely called convective heat transport.

The convective stirring of the atmosphere is so efficient that it drives the atmosphere toward a neutral thermal lapse rate ($-dT/dZ$, or the negative change in temperature with height). The troposphere is defined as the region in which nonradiative processes govern the lapse rate; the stratosphere is the region in which the radiative-equilibrium lapse rate agrees with the observed lapse rate. The boundary between these two regions is the tropopause. Once the lapse rate is prescribed, the surface temperature is the only degree of freedom for the tropo-

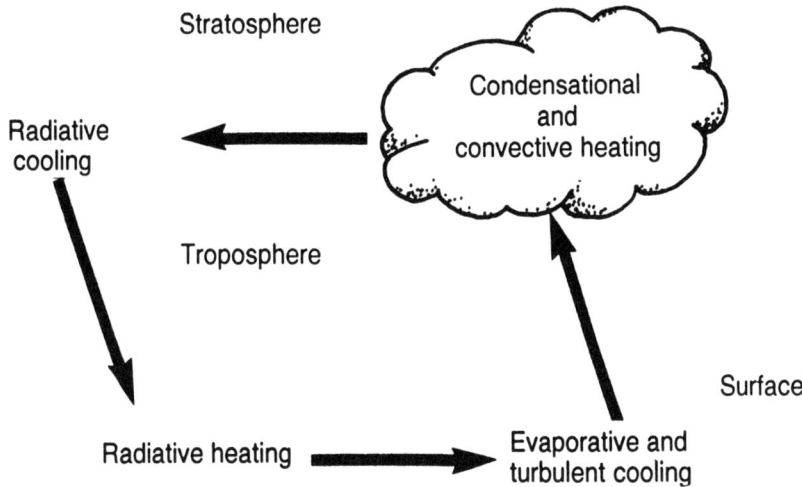

Figure 2. Radiative–convective interactions between the surface and the atmosphere.

sphere; it is determined by the net (down minus up) flux of the solar and infrared radiation at the tropopause.

The radiation fluxes at the upper boundary are influenced strongly by internal parameters such as the distribution of water vapor, clouds and other gases; by the lapse rate; and by surface properties such as ice and snow cover, vegetation types and soil moisture. The dependence of these parameters on the surface temperature T_s gives rise to several feedback loops, of which the interaction between water vapor and T_s is the best understood and that between clouds and T_s the least understood.

The concept of radiative–convective equilibrium in a one-dimensional model enables us to formulate the climate problem in terms of forcing and feedback. The fundamental climate-forcing term is the radiative flux at the tropopause. We will describe later the feedbacks that govern the response of the climate to radiative forcing.

Radiative–convective–dynamic interactions

Including interactions between radiation, convection and planetary-scale dynamics yields two-dimensional climate models. The radiation energy is not balanced at each latitude. Regions at low latitudes receive more solar energy than do the polar regions. Sea ice and snow cover at high latitudes steepen this gradient. Such surfaces reflect much more sunlight than does the darker, open water in low- and midlatitude oceans. The tropical energy surplus and the polar deficit give rise to a strong pole-to-equator temperature gradient. The long-wave emission

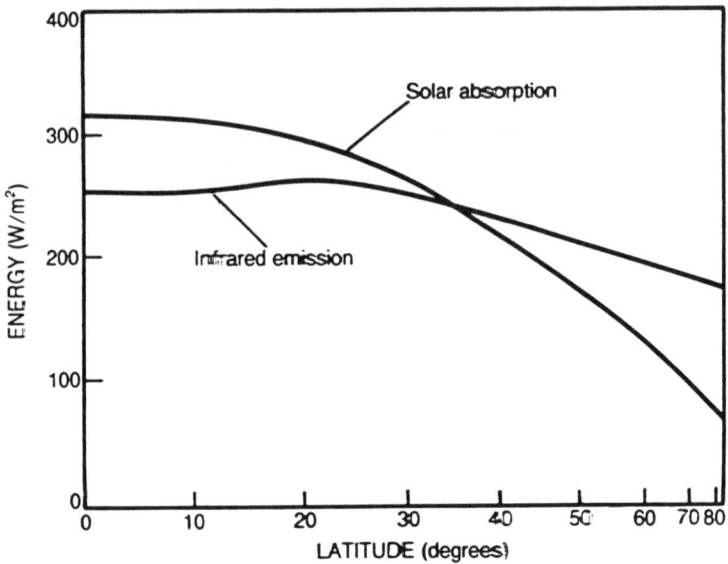

Figure 3. Annual zonal mean estimates of absorbed solar radiation and outgoing long-wave flux (infrared emission) obtained by satellites. Net heating takes place at low latitudes; net cooling, at higher latitudes. Equal-area projection is shown. (Adapted from ref. 10.)

is much more uniform, but it does not compensate for the excess solar heating in the tropics. Thus there is a net radiative heating at low latitudes, accompanied by net cooling at high latitudes (see Figure 3).

The imbalance in heating and cooling acts as the fundamental energy source for the atmospheric and ocean circulations. They must transport the excess heat from the equator toward the poles. The process is extremely complex. The net radiative heating in the tropics manifests itself in the atmosphere as latent heat released within deep cumulonimbus clouds. The heat release drives the tropical Hadley cell (see Figure 4). Near the surface the flow toward the equator carries the moisture evaporated from the subtropical oceans and deposits it as tropical rain. At the surface the net solar and long-wave fluxes provide the energy required to compensate for the evaporative cooling of the oceans. The Coriolis force on the poleward motion of the Hadley cell leads to strong west-to-east winds having speeds of 20–50 m/sec in the lower 10–15 km of the atmosphere. Strong westerlies and mountain barriers give rise to dynamical instabilities that break the mean west-to-east motion into eddies, particularly at midlatitudes. Furthermore, the atmosphere does not transport all the heat: Oceans carry some through wind-driven and thermohaline circulation. (Thermohaline

V. Ramanathan et al. 61

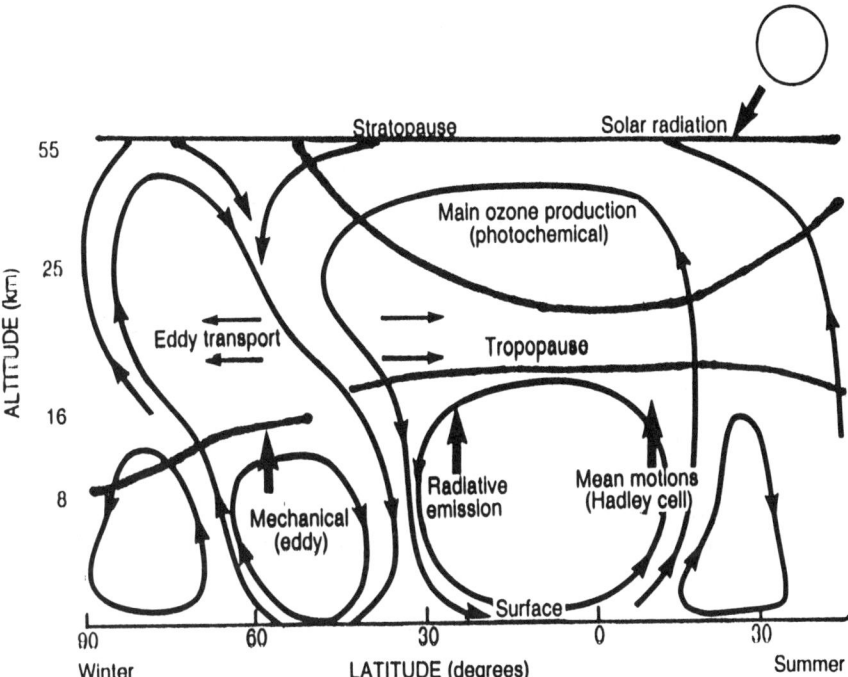

Figure 4. Radiative–dynamical–chemical coupling between the troposphere and stratosphere. The tropopause, which is the boundary between the troposphere and stratosphere, extends from about 8 km near the poles to as high as 17 km at low latitudes. It shifts abruptly around the latitude of the jet stream. The mean meridional circulation has been adapted from a global circulation model simulation carried out by Syukuro Manabe and Jerry Mahlman[11] of Princeton University's geophysical fluid dynamics laboratory.

circulation results from the density variations in the oceanic fluid due to the combined effects of temperature and salinity.)

More quantitatively, the divergence of heat flow in the atmosphere and oceans balances the net radiative heating H. Direct estimates of the oceanic heat transport have large uncertainties because of severe spatial sampling problems. Radiation budget measurements of H and atmospheric circulation measurements determine two of the three terms in the balance equation. Thus the oceanic heat transport is the difference between these two observations. The principal finding[4] is that in the Northern Hemisphere the oceans transport 40% of the required total heat transport. The numerical values are not yet definitive, however, because of the possibility of sampling errors in both H and atmospheric transport.

So far we have restricted our attention to the troposphere. The stratosphere is an important and integral part of the climate system as well. Radiative, dynamical and photochemical processes (see Figure

4) couple it to the lower atmosphere. Absorption of solar ultraviolet radiation is a major energy source for the middle and upper stratosphere. For the lower stratosphere ozone absorption of infrared radiation at wavelengths of 9–11 μm is a significant heat source. The O_3 radiative heating rate increases with altitude and is sufficiently large to explain the observed thermal inversion in the stratosphere. The major energy sink is emission of radiation by CO_2 in the 15 μm region. Globally the solar and infrared heating by O_3 is largely balanced by the CO_2 emission, which maintains the global stratosphere in radiative equilibrium.

The O_3 radiative heating also plays an important role in governing latitudinal gradients in temperature and the circulation of the stratosphere. In winter, radiative heating is maximal in the tropics and zero near the pole tilted away from the Sun. This creates a strong temperature gradient and a correspondingly strong west-to-east jet. Countering this effect is one arising from a mechanical energy source set up by tropospheric eddies. Their energy propagates vertically and poleward into the stratosphere to heat the stratosphere near the poles and thereby reduce the equator-to-pole temperature gradient. Thus the eddies act to oppose the radiatively driven temperature gradient and circulation. For reasons not yet understood, the Antarctic stratospheric circulation is driven more radiatively than is the Arctic. As a result, the lower stratosphere of the Antarctic is significantly colder than the Arctic stratosphere and the west-to-east jet is significantly stronger. This cold and the strong jet explain the severity of the ozone hole in the Antarctic. (See the news story in PHYSICS TODAY, July 1988, page 17; also August, page 21.)

Photochemical processes in the middle and upper stratosphere (see Figure 4) produce ozone that is transported poleward and downward by winds. The ozone absorbs Sunlight, and the result is a strong effect on the temperature gradients in the stratosphere.

Inadequate understanding of the strong coupling of radiation, dynamics and chemistry in the stratosphere is a major source of uncertainty in recent model projections of ozone change.

Radiation budget and climate change

Human activities alter the radiation budget. This is the central fact behind the current debate over the greenhouse effect. As is customary, we discuss the climate change problem for the change produced by doubling the atmospheric Co_2 concentration.[5]

If we could immobilize the atmosphere and suddenly double its CO_2 concentration, the long-wave flux F would decrease by about 4 W/m$_2$ at the tropopause. This decrease is easy to explain. Recall from our earlier discussion that an absorbing gas decreases F because the atmosphere is colder than the surface. Thus increasing the CO_2 concentration decreases the energy loss from the troposphere to space so that the heating increases by 4 W/m^2. According to our zero-dimensional climate model, the global situation will restore the radiation energy balance. In other words, the climate system will force H to zero. The planet's surface and troposphere could warm up until it radiates to space the excess 4 W/m^2. The increase in F effected by a higher temperature balances the decrease in F caused by the increase in CO_2 concentration.

The discussions here ignore the stratosphere, which in fact will cool because of the increased CO_2 infrared emission. If the infrared emission were only a function of T and of CO_2 concentration, the total change in H would be a sum of an initial forcing term and a direct response term:

$$\Delta H = \frac{\partial F}{\partial [CO_2]} \Delta [CO_2] + \frac{\partial F}{\partial T} \Delta T.$$

Initially ΔT is zero. So ΔH equals the initial forcing of F (by a change in CO_2 concentration). The climate system's tendency to equilibrium again drives ΔH to zero. Then the resulting temperature change (in the troposphere or at the surface) is

$$\Delta T = \frac{\partial F / \partial [CO_2]}{\partial F / \partial T} \Delta [CO_2].$$

The rate at which the emission increases with increasing T governs the temperature change.

Now let us include a well-known feedback in the above thought experiment. The emission depends upon water vapor as well as carbon dioxide. As the Earth's surface warms, water evaporates more rapidly from the surface. To keep the process near equilibrium, more water must condense. However, the net result is increased water vapor in the atmosphere. This vapor will further decrease F—and increase heating. For these simple climate models, we usually assume that humidity is only a function of temperature. The energy balance equation changes accordingly to become a sum of three terms—initial forcing, direct response and indirect response:

Table 1. Climate feedback from models and satellite observations.[a]

	Climate models	Satellite studies
	(W/m²/K)	(W/m²/K)
Infrared feedback (dF/dT)	1.3–2.3	1.6–2.2
Snow or ice albedo feedback $(S_0/4)(d\alpha/dT)$	– (0.3–0.6)	– 0.4

[a] Adapted from ref. 6.

$$\Delta H = \frac{\partial F}{\partial [CO_2]} \Delta [CO_2] + \frac{\partial F}{\partial T} \Delta T + \frac{\partial F}{\partial [H_2O]} \frac{\partial [H_2O]}{\partial T} \Delta T.$$

When the system reaches equilibrium, the radiation balance perturbation vanishes (that is, $\Delta H = 0$). The final temperature change for water vapor feedback is

$$\Delta T = - \frac{(\partial F/\partial [CO_2])\Delta [CO_2]}{\partial F/\partial T + (\partial F/\partial [H_2O])(\partial [H_2O]/\partial T)} = \frac{\Delta Q}{\lambda}.$$

The numerator ΔQ is the forcing term. The denominator λ is the feedback parameter.

With a warmer atmosphere, the albedo of the planet might change. For example, the ice caps could melt or clouds could change. In other words, the climate processes would produce other feedbacks that would modify the response of the system to the initial forcing. For a change in albedo, we would call this an albedo feedback. A general expression that takes this feedback into account is

$$\lambda = \left(\frac{dF}{dT}\right) + \frac{S_0}{4}\left(\frac{d\alpha}{dT}\right).$$

Current three-dimensional climate models yield global warming by amounts in the range of 3–4.5 K. For a doubling of CO_2, the model estimate of ΔQ is about 4 W/m². Thus, the theoretical value of λ lies in the range 0.9–1.5 W/m²/K. There have been many empirical estimates[6] of λ from satellite measurements of Earth's radiation budget. In general, these studies use the latitudinal and seasonal changes in the observed F, α and T to estimate λ. As shown in Table 1, the model feedback parameters are consistent with those derived from observations. This consistency would be a satisfactory proof of the model, however, only if the seasonal climate variation mimics a climate change caused by CO_2.

What is the significance of the numbers in the table? Let us begin with a climate system devoid of all feedbacks except for the increase in temperature. An increase in emission with an increase in surface temperature is a negative feedback. In this case, models yield[6]

$$\frac{dF}{dT} = \frac{\partial F}{\partial T} \approx 3.3 \text{ W/m}^2/\text{K}.$$

The difference between 3.3 and the lower values shown in the table is due to the water-vapor feedback. Thus the water-vapor-temperature coupling is a positive feedback. The albedo feedback also is positive because an increase in T causes a melting of sea ice and snow cover. The decrease in the area of ice and snow lowers the albedo and thereby increases the absorbed sunlight. The whole process amplifies the surface warming.

Thus radiative interactions govern both climate forcing and the response of the climate to the forcing. It is obvious that observations of the radiation budget are important for testing theories of climate change.

Outstanding problems

The fundamental limitation of the models described above is that they lack explicit formulations for the interaction between the distribution of sources and sinks of diabatic energy (radiation and latent heat) and the dynamics of the atmosphere and oceans. Because of the limitations of simple models, a class of three-dimensional models known as general circulation models is becoming a basic tool for studying climate and climate change.

The GCMs were developed initially for forecasting weather, but are now being modified to address climate problems. The most significant modification involves the treatment of physical processes (for example, radiation, sea-ice formation and processes involving soil hydrology and vegetation) that were either ignored altogether or treated very crudely. Such models lead to significant insights into the causes of important climatological phenomena, such as the role of mountains and land–ocean asymmetries in the amplitudes and phases of planetary-scale stationary and transient waves (weather disturbances);

the location and seasonal variation of the jet streams; and monsoon circulation, to name a few.

Attention is now shifting toward climate changes on time scales of decades or longer, and challenging problems such as cloud–radiation feedback (described later) and ocean–atmosphere interactions have emerged as major issues. We need an observational base of sufficient accuracy to allow us to develop and constrain our three-dimensional models. For example, we need to know how the absorption of solar radiation by the oceans, the sea ice, the tropical forests and the deserts (among other geographical features) vary with the zenith angle of the Sun and with the seasons in order to model their effects. The required observation is the clear-sky albedo of the planet. Next, we need to know how clouds modulate the solar absorption of the various regions of the world. The required observation here is the cloud radiative forcing (to be defined later). It is the desire for such detailed insights that motivated the ERBE studies.

ERBE has several unique features. First, the three satellites sample the Earth at different local times to minimize time sampling errors and systematic biases, which were present in earlier measurements. Second, ERBE's preflight and on-board calibrations significantly improve the accuracy and precision of the measured radiation. Third, ERBE has treated data processing much more rigorously than previous missions have.

Measurement challenges

The fundamental inference from the model studies is that climate should be extremely sensitive to small variations in radiative forcing. For example, a 1% increase in the solar irradiance will increase the absorbed solar radiation by 2.4 W/m^2. According to the λ values from GCMs, this increase would lead to a 1.6–2.6 K warming of the globe.

These sensitivities pose very stringent accuracy requirements on observations. Thus the measurement of the radiation budget has been a story of increasingly sophisticated instruments and increasingly rigorous data processing. The ERBE detectors include active-cavity radiometers and thermistor bolometers. Both types of detectors use heat to measure radiant energy. If the detectors are to be accurate, the data reduction must quantitatively relate absorbed radiation to detected heat.

As an example, consider an ideal cavity designed to accept radiation from the Earth or the Sun. This electrically heated cavity is attached to a massive heat sink, which operates at a constant temperature. The energy balance of the cavity requires that the conduction heat loss from the cavity to the heat sink be the sum of two terms, electrical heating and absorbed radiation:

$$K(T_{\text{cavity}} - T_{\text{heat sink}}) = \frac{V^2}{R} + A(1-\alpha)S_0.$$

The term on the left-hand side expresses the conductive heat loss from the cavity to the heat sink. K is the heat conductivity between these two masses. Because the thermostat maintains $T_{\text{cavity}} - T_{\text{heat sink}}$ at a constant value, this heat loss is constant. Measurement of the heater voltage V determines the electrical heat released in the cavity. The heater resistance R comes from preflight measurements. The radiative power into the cavity enters through an aperture of area A, with a fraction α being reflected.

Instruments to detect radiant energy become complicated because they need to maintain heat flows. On ERBE the cavity detectors have surrounding field-of-view limiters to prevent confusion between stray radiation from the satellite or from the Sun and radiation from the Earth. The calibration and data reduction processes must remove the effect of possible heat exchanges with the surroundings.

Quantitative understanding is required in processing data from radiation budget measurements. The first step is "inversion," which involves relating the satellite measurement of radiance to the energy loss from the top of an atmospheric column. The satellite can view a particular location from only one direction at a time, so the angular space is undersampled. The second step is interpolation between the satellite measurements to average data over specific time scales. The satellites do not see the entire globe simultaneously; thus, both time and space are sparsely sampled.

Instruments, inversion techniques and the averaging process were all advanced for the purposes of ERBE. Mathematical models of the detector heat flow guided the development and calibration of the instruments. The precursor measurements of Earth's radiation budget taken by the Nimbus-7 satellite provided data for empirical models of angular dependence. As a further aid to accurate data reduction, the angular models classify areas of the Earth into various categories of cloudiness. The ERBE time averaging took into account the detailed dependence of the albedo on solar position. For the first time, the averaging also included the diurnal variation of emission.

Where are we now with respect to producing an accurate measure of the Earth's radiation budget? First, for the instantaneous irradiance, we expect uncertainties of about 1% for long-wave and 2–3% for shortwave-radiation from the consistency of the three spectrally overlapping scanner channels. Second, for instantaneous fluxes from $2.5° \times 2.5°$ (roughly 300 km $\times$ 300 km) geographic regions, ERBS and NOAA-9 intercomparisons offer observational uncertainty estimates of ± 5 W/m^2 in the long-wave and ± 15 W/m^2 in the shortwave. Third, on a monthly average, regional basis, the uncertainties in the scanner data are about ± 5 W/m^2 for shortwave and ± 5 W/m^2 for long-wave. Simulations with geostationary operational environmental satellite data provide this estimate. Fourth, the uncertainty in global, annual average net radiation is probably about ± 5 W/m^2 as estimated from the imbalance obtained using scanner data over four months. A definitive error analysis is under way.

The battle for accuracy is far from won. However, the improvements are impressive enough that understanding some of the climate problems is now within our grasp.

Solar irradiance and its variability

Changes in the solar "constant" constitute the primary external forcing of the climate system. The most apparent phenomenon that could cause the solar output to vary is sunspots. These are small portions of the solar disk with lower emission that appear black against the brilliant background of the solar disk. During the sunspot minimum, there may be no sunspots at all. During the sunspot maximum, spots may cover 1–2% of the disk. Thus if the spots emitted no energy, the solar irradiance might vary by 10–20 W/m^2. However, the spot temperatures are only 10–20% lower than the usual temperatures of the disk. This means that the actual modulation[7] of the solar irradiance by sunspots is probably 0.1–0.2%. Until spaceborne solar irradiance monitors began observing in 1979, the actual modulation was not measurable.

Detailed investigations have shown that the solar irradiance is also perturbed by bright areas known as faculae, which surround the spots. They cover a larger area than do the spots and contrast less with the disk. Evidently, after sunspots develop, faculae increase and raise the solar irradiance. The solar irradiance is indeed lowest at the sunspot minimum, when the radiant contribution by faculae is also a minimum. Figure 5 shows this behavior over three years of ERBE solar

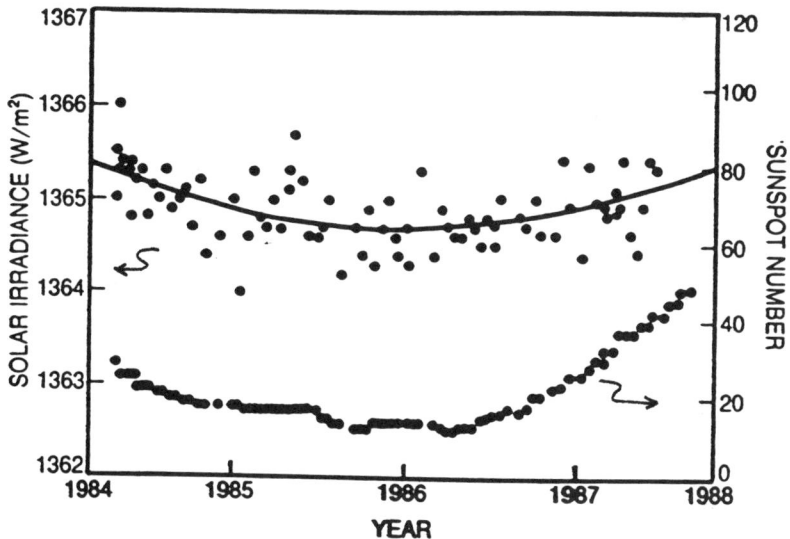

Figure 5. Solar constant as a function of time, derived from the solar monitor on the Earth Radiation Budget Satellite. The curve represents a second-order polynomial fit to the solar constant values; the bottom rows of dots represents smoothed (12-month running mean) sunspot numbers. (Data from ref. 7.)

observations. Before the sunspot minimum in 1986, the observed solar irradiance decreased at a rate of 0.02%/year. After the minimum, it increased at the same rate. All three of the solar irradiance experiments have measured similar decreases and increases.

The observational record for sunspots extends far enough into the past that it is a prime candidate for empirical modeling. The time series of facular observations is not as long. However, researchers are developing empirical models of the correlations between sunspots and faculae. These would allow us to investigate the correlation between climate and the Sun.

Diurnal variations in Earth's radiation fields

Poor diurnal sampling was one of the principal sources of uncertainty in earlier radiation budget measurements. Most of the Earth radiation budget instruments before ERBE flew on single, Sun-synchronous satellites. Satellites in these orbits sampled the Earth at only one local time during the day and once at night. ERBE is the first multiple-satellite system to provide diurnal sampling capability for global radiation budget studies.[8] Two NOAA satellites are in Sun-synchronous orbits, and NASA's Earth Radiation Budget Satellite is in a midinclined orbit. ERBS's orbit precesses through all local hours at the

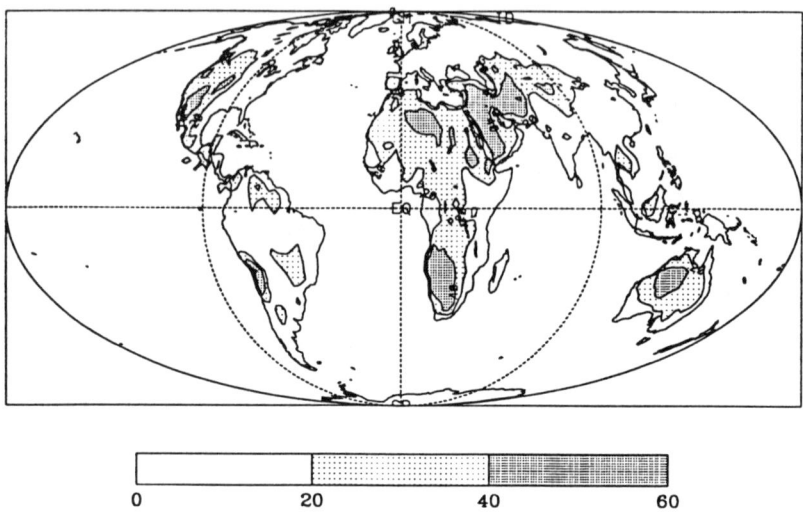

Figure 6. Diurnal range of long-wave flux (in W/m²) as seen by ERBE for July 1985. (See ref. 8 for additional details.)

equator in 36 days (when one counts both ascending and descending nodes). This 56° inclination orbit provides diurnal coverage over the tropics. It also provides coverage of the midlatitudes, where the maximum diurnal range of long-wave radiation occurs. The two NOAA satellites supplement this diurnal coverage at low latitudes. They also provide the necessary coverage over the polar regions.

In addition to minimizing sampling errors, diurnal cycle observations provide critical insights into climate feedback processes involving the land surface, meteorology and solar heating, because several types of variability are synchronized with the Sun. The global distribution of the diurnal range of long-wave radiation for July 1985 (see Figure 6) reveals patterns that can be related not only to the heating and cooling of the surface but also to diurnal changes in cloud cover and in the vertical structure of clouds.

The largest diurnal variation of long-wave radiation (about 60 W/m²) occurs over deserts (the Sahara, Saudi Arabia, the Gobi, the Great Sandy, the Atacama, the southwestern US, and Kalahari). In fact, with a few exceptions, the diurnal range is closely correlated with the aridity and vegetation cover of the soil. The drier the soil and the sparser the vegetation, the more the Sun heats the ground (instead of evaporating moisture) and so enhances both the amplitude of the diurnal temperature variation and the long-wave emission. The major exceptions are the vegetated regions of Central America, and the northern Amazon and Congo basins, which have diurnal variations as large as 40 W/m^2. These regions experience intense convective cloud activity, and the cloud shield undergoes a strong diurnal cycle. Because ocean surface temperatures are relatively constant, diurnal variations of long-wave flux over the oceans are generally small (less than 10 W/m^2) and are due primarily to changes in clouds. An intriguing possibility raised by the global pattern shown in Figure 6 is that regional scale shifts in soil hydrology (due, say, to deforestation) can be detected by monitoring the diurnal range of long-wave flux.

Clouds and climate

Do clouds heat or cool the planet? This question has perplexed many working on the theory and modeling of climate. An unambiguous theoretical estimate is still lacking.

When atmospheric water vapor condenses to a liquid or a solid, it scatters ultraviolet and visible radiation significantly compared with cloudless skies. The scattering is both forward to the ground and backward to space. An individual droplet scatters 85% of the incident energy in the forward direction. A cloud of drops, however, can scatter 75% or more of the energy backward. The resulting enhancement in the surface–atmosphere albedo reduces the solar radiation absorbed by the atmospheric column.

Clouds also significantly enhance the long-wave opacity of the atmosphere. Like the gaseous absorption, this reduces the radiation emitted to space. Thus, while the greenhouse effect of clouds warms the planet, the albedo effect cools it. The problem is further complicated by the significant dependence of cloud radiation on cloud microphysics. These properties include the density of liquid water and droplet size distribution, both of which vary significantly from one cloud to another. As a result the albedo and the greenhouse effects are subject to significant variability.

Cloud-radiative forcing: a simple approach

It is a challenge to measure the two competing effects of clouds on the radiation budget. The major problem is that cloud structures vary in scale from meters to thousands of kilometers. This means that many of the "satellite pixels" image mixtures of clear and overcast regions (or "scenes"). As a result it is nearly impossible to unscramble the radiances to produce an overcast radiance.

ERBE has found a rather simple approach to this problem,[9] one that is quite successful in obtaining the net radiative effect of clouds on climate. It starts with the observation that the spatial variability is considerably smaller in clear-sky fluxes than in mixed-scene fluxes. The clear-sky flux is also an extremal value: The "hottest" long-wave radiances and the "darkest" shortwave radiances come from clear skies. Therefore clear skies are easier to identify than mixed skies are. Spatial homogeneity also makes it easier to estimate what the clear-sky fluxes would have been if the clouds were not present.

Let us make this discussion more quantitative by considering a region partially covered by clouds. The region can be the entire planet, an entire latitude belt or a specific region of the globe. Let f be the fraction that is covered by clouds. We define F as the average flux emitted to space by this region from the cloud condition *in toto*. F_c is the flux from the clear-sky portion. F_0 is the flux from the overcast sky. We can then write

$$F = (1-f)F_c + fF_0.$$

How can we observe the clouds' influence on F and, ultimately, on the net heating H? We can rewrite F as

$$F = F_c - f(F_c - F_0) = F_c - C_{\text{LW}}.$$

For a region with multilayer clouds, C_{LW} is redefined as

$$C_{\text{LW}} = \sum_i f_i(F_i - F_{0,i}),$$

where the summation extends over i different cloud layers.

To find C_{LW} from observations of cloudy regions, we have to identify the individual overcast areas and their radiative properties. As we discussed earlier, this has many difficulties. However, C_{LW} can also be obtained from the equation for F:

$$C_{\text{LW}} = F_c - F.$$

Thus obtaining the cloud effects on climate reduces to obtaining the clear-sky flux, a considerably simpler problem. For reasons that will become obvious later, we refer to C_{LW} as the long-wave forcing.

Likewise, we can define C_{SW} as the difference in the reflected solar flux between clear and cloudy skies. The net cloud-radiative forcing, C, is then the sum of the long-wave and the shortwave cloud forcing.

In the long-wave, cloudy-sky fluxes F_0 are usually lower than the clear-sky fluxes F_c, giving $C_{LW} > 0$. In other words, clouds produce an additional greenhouse effect that forces the surface temperature to be higher than it would be otherwise. In the shortwave, cloudy fluxes are usually higher, so $C_{SW} < 0$. In this case, the clouds force the climate system to be cooler. Clearly, "forcing" is an appropriate name. Note that if clouds were absent, f would be zero and so would C. (The total flux in the absence of clouds is the clear-sky flux $F = F_c$.)

An exciting outcome of this approach is that it produces a global estimate of the clear-sky albedo of the planet. The albedo distribution reveals[9] the oceans to be the darkest region of the globe. Albedo values range from 6–10% in low latitudes to 15–20% in the polar oceans. The brightest regions of the globe are, of course, the snow-covered Arctic and Antarctic. Next in brightness are the deserts, with the Sahara reflecting as much as 40% of the incident solar radiation. The other major deserts, such as those of Saudi Arabia, the Gobi and the Great Sandy, reflect about 25–35%. The darkest land surfaces are the tropical rain forests of South America and central Africa. Thus clearing of forests in the tropics should have a significant impact on the regional heat balance.

As of this writing ERBE has processed four months of data: April 1985, July 1985, October 1985 and January 1986. Figure 7 shows the long-wave and shortwave cloud forcing for July 1985. These measurements provide a global perspective that was previously lacking.

The cloud-forcing patterns mirror the major climate regimes and the organized cloud systems of the planet. For example, the peak values of the long-wave cloud forcing (Figure 7, top) reveal organized convective- and cirrus-cloud systems in the tropics. These are present over the monsoon regions of the tropical Pacific and Indian Oceans, over central Africa and over the Amazon basin. At midlatitudes the cloud forcing coincides with storm tracks and jet-stream cirrus systems.

The organized pattern of time-averaged cloud forcing stands in sharp contrast to the more chaotic spatial structure within individual cloud systems. This pattern of cloud forcing is encouraging, since GCMs are quite successful in simulating the time-averaged patterns of the large-scale circulation. With a physically realistic model of clouds, it should be possible to use our climate models to estimate their radiative effects.

In contrast to the long-wave forcing, shortwave forcing (Figure 7, bottom) peaks in the midlatitudes. The shortwave forcing is as large as

74 Climate and the Earth's Radiation Budget

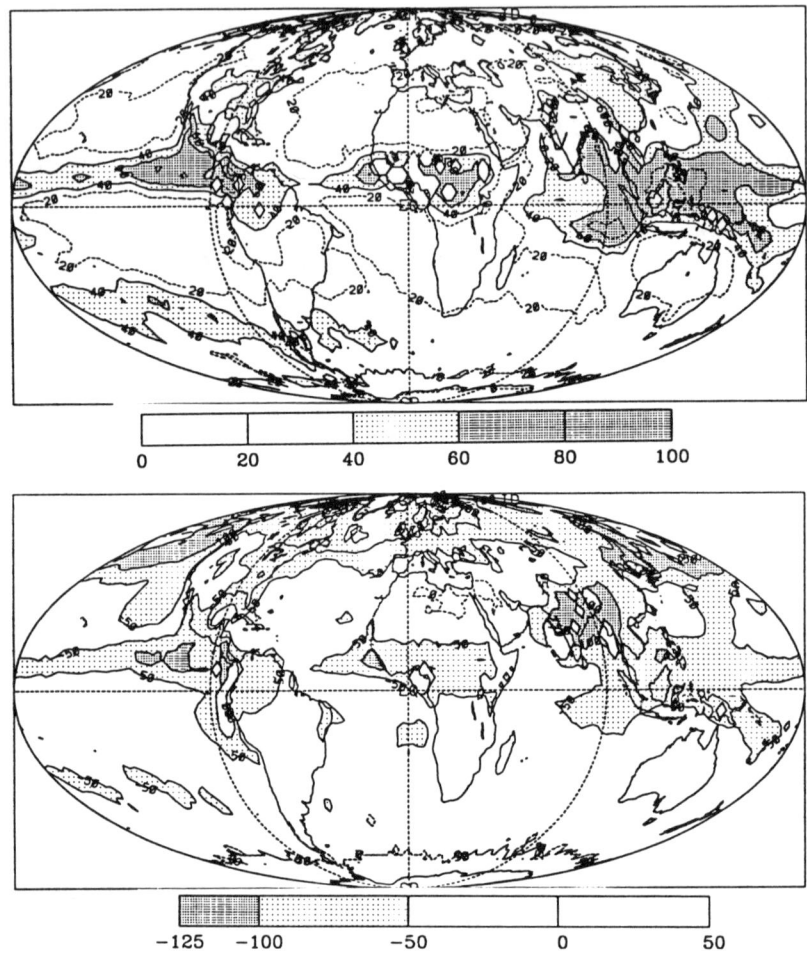

Figure 7. Cloud-radiative forcing (in W/m^2). Monthly averages derived from ERBE for July 1985 are depicted for long-wave (top) and shortwave forcing (bottom). Long-wave cloud forcing is the reduction by clouds in the long-wave radiation that is emitted to space; hence it is the greenhouse effect of clouds. Clouds reduce emission to space because at their bases they absorb radiation emitted by the warmer surface and at their tops they emit to space at colder temperatures. Deep cirrus clouds such as the monsoon cloud systems over the Indian Ocean and Indonesia and the jet-stream cirrus clouds at midlatitudes give a large greenhouse effect. Because clouds reflect more shortwave solar radiation than the adjacent clear skies, the shortwave forcing is negative—a cooling effect. Surprisingly, the magnitude of the cooling is as large as the long-wave forcing over the tropical cirrus systems, and is even larger over the mid- and high-latitude oceans. The net cloud-radiative forcing is the sum of long-wave and shortwave cloud forcing. The averages range from $-(100–140)$ W/m^2 to 10–40 W/m^2. The net effect is largely negative; hence clouds have a cooling effect on the planet. The strongest cooling is caused by persistent stratus and storm-track clouds over mid- and high-latitude Atlantic and Pacific oceans.

the long-wave forcing in the monsoons and quasistationary convective regions. In the more poleward oceans, the shortwave forcing exceeds the long-wave forcing by more than 100 W/m^2.

The net cloud-radiative forcing is the sum of the long-wave and shortwave cloud forcing. The global distribution of net cloud-radiative forcing reveal two intriguing features:

There are regions in the tropics where net cloud forcing nearly vanishes. However, clouds produce major changes in both the long-wave and shortwave fluxes. The two forcing terms nearly cancel each other within ±10 W/m^2 (the uncertainty in the estimate). We do not understand the physical and dynamical constraints that require these systems to have such a delicate balance.

There are regions of large negative cloud forcing over the midlatitude and polar oceans. Clouds reduce the radiative heating in these areas by as much as 100 W/m^2. In these regions, the dominant cloud systems are storms associated with cyclones and extensive layers of stratus clouds. These systems are very sensitive to surface temperatures and temperature gradients in the troposphere. Hence these oceanic clouds can have a significant feedback effect on climate change.

Global effects

The global averaged long-wave forcing for July 1985 is 30.1 W/m^2; it is the greenhouse effect of clouds. It is larger than that resulting from a doubling of CO_2 by a factor of about 7. The CO_2 concentration in the atmosphere has to be increased by more than two orders of magnitude to produce a greenhouse effect comparable to that of the clouds. The shortwave cloud forcing for July 1985 is -46.7 W/m^2. The net cloud forcing, which is the sum of long-wave and shortwave cloud forcing, is -16.6 W/m^2. A negative cloud forcing of similar magnitude was also obtained for three other months that have been analyzed so far: April 1985, October 1985 and January 1986. Thus the ERBE results reveal that clouds have a global *cooling* effect.

The global mean cloud-radiative cooling, when averaged over all seasons, should be balanced by a corresponding global mean radiative heating under clear skies to maintain global energy balance. Without the -16 W/m^2 cloud forcing, the planet would be significantly warmer. The magnitude of the warming would depend on the model we use to convert the forcing into a temperature change, but it could be as large as 10–15 K. The longwave cloud forcing heats the surface–

atmosphere column, while the shortwave forcing cools it. In the current climate, the shortwave effect dominates. Why this is so is not at all obvious.

The atmospheric and ocean circulations govern the generation of clouds. These circulations respond to the sources and sinks of radiative and thermodynamic energy that are governed by the distribution of clouds. Hence a climate change could change the net cloud forcing, and that change could in turn feed back into the climate. The exploration of how these changes might occur is one of the central questions in climate theory today.

We can understand specific components of the feedback by studying the evolution of cloud forcing during climate changes on shorter time scales. One spectacular example is the El-Niño phenomenon. The warm sea-surface temperature anomaly in the Pacific results in a perturbation of the convective cloud system, which is accompanied by large changes in long-wave and shortwave cloud forcing. This natural experiment provides a unique opportunity to examine the link between sea-surface temperature changes and cloud-radiative forcing. Likewise, a comparison of the shortwave cloud forcing over North America during the drought year of 1988 with the forcing during the previous years would help establish the link between soil moisture, cloudiness and regional radiative heating.

We now have the necessary (but not sufficient) observational base to develop the theory of climate change. The effects of human activities, such as the increase in trace gases or the alteration of the surface albedo by deforestation, alter the clear-sky radiative heating. Hence observations of changes in clear-sky radiative forcing on time scales of a few to several decades would document the influence of human activities. Observations of long-term changes in cloud–radiative forcing would help establish the importance of cloud-climate feedback.

The principal investigators of the ERBE science team are Barkstrom, R. Cess, J. Coakley, Y. Fouquart, A. Gruber, Harrison, D. Hartmann, B. Hoskins, F. House, F. Huck, R. Kandel, M. King, A. Mecherikunnel, A. Miller, Ramanathan, E. Raschke, L. Smith, W. Smith and T. Vonder Haar. Ramanathan was supported by a NASA ERBE grant and NSF grant ATM87002B6.

References

1. F. B. House, A. Gruber, G. E. Hunt, A. T. Mecherikunnel, Rev. Geophys. **24**, 357 (1986).

2. B. R. Barkstrom, Bull. Am. Meteorol. Soc. **65**, 1170 (1984).

3. R. D. Cess, G. L. Potter, Tellus **39A**, 460 (1987).

4. T. H. Vonder Haar, A. H. Oort, J. Phys. Oceanogr. **3**, 169 (1973).

5. M. E. Schlesinger, J. F. B. Mitchell, Rev. Geophys. **25**, 760 (1987).

6. V. Ramanathan, J. Geophys. Res. **92**, 4075 (1987).

7. R. Lee, B. R. Barkstrom, R. D. Cess, Appl. Opt. **26**, 3090 (1987).

8. E. F. Harrison, D. R. Brooks, P. Minnis, B. A. Wielicki, W. F. Staylor, G. G. Gibson, D. F. Young, F. M. Denn, ERBE Science Team, Bull. Am. Meteorol. Soc. **69**, 1144 (1988).

9. V. Ramanathan, R. D. Cess, E. F. Harrison, P. Minnis, B. R. Barkstrom, E. Ahmad, D. Hartmann, Science **243**, 57 (1989).

10. J. S. Ellis, T. H. Vonder Haar, Atmospheric Sciences Report 240, Dept. Atmos. Sci., Colorado State U., Fort Collins (1976).

11. S. Manabe, J. Mahlman, J. Atmos. Sci. **33**, 2185 (1976).

12. F. S. Rowland, I. S. A. Isaksen, eds., *The Changing Atmosphere*, Wiley, New York (1988). R. E. Dickinson, ed., *The Geophysiology of Amazonia: Vegetation and Climate Interactions*, Wiley, New York (1987); see the articles by J. E. Lovelock, p. 11; P. J. Crutzen, p. 107; R. C. Harriss, p. 163.

13. V. Ramanathan, Science **240**, 293 (1988).

14. J. Hansen, I. Fung, A. Louis, D. Rind, S. Lebedeft, R. Rudey, G. Russell, J. Geophys. Res. **93**, 9341 (1988).

15. R. T. Wetherald, S. Manabe, J. Atmos. Sci. **45**, 1397 (1988).

16. M. G. Tomasko *et al.*, J. Geophys. Res. **85**, 8187 (1980).

17. J. B. Pollack, O. B. Toon, R. Boese, J. Geophys. Res. **85**, 8223 (1980).

18. R. J. Charlson, J. E. Lovelock, M. O. Andrae, S. G. Warren, Nature **326**, 665 (1987).

19. P. J. Crutzen, L. Donner, V. Ramanathan, R. C. Srivastava, "Clouds, Chemistry and Climate," research proposal submitted to NSF (1988).

TEMPERATURE AND SEA LEVEL CHANGE

George A. Maul
National Oceanic and Atmospheric Administration
Atlantic Oceanographic and Meteorological Laboratory
Miami, Florida 33149-1097

ABSTRACT

Instrumental measurements of air and sea temperature and of sea level are rarely more than a century in length, and are characterized by numerous observational inconsistencies. Truly "global" data sets do not exist, except from satellite measurements during the last decade, primarily because most of the 71% of Earth's surface covered by the oceans is not sampled. Surface air land temperature records are plagued by changes in measurement techniques, location of thermometers, and microclimate changes, notably urbanization. Surface marine air temperatures too are affected by changing ship design, height of thermometers, and particularly daytime biases. Sea surface temperature data have similar difficulties plus the added problems of changed ship routes, and shifting from bucket to engine intake observations. Sea levels from coastal tide gauges have notably shorter record-lengths than temperature records, and often are dominated by diastrophism and subsidence at the lowest frequencies. Determining statistically significant climatic trends in any of these geophysical time-series leads to uncertain results due to natural variability such as El Niño-Southern Oscillation events.

INTRODUCTION

Debate in recent years over the question of anthropogenic activities affecting the natural operation of the atmospheric "greenhouse effect" has been lively and heated.[1,2,3] In an attempt to synthesize the scientific opinion on the issue, the World Meteorological Organization and the United Nations Environment Programme sponsored an Intergovernmental Panel on Climate Change (IPCC). The IPCC report[4] too is being debated, and was the topic of a special session of the Fall 1990 Meeting of the American Geophysical Union. Table I lists the several organizations and other acronyms that are common in the literature of global warming and climate change, but by no means is it a complete survey of all the participants in the debate. The issues of climatology are much broader than the scope of temperature and sea level change, but that is the focus of this chapter.

Most of what is known about Earth's climate history comes from geological evidence and geochemical analysis of deep ocean and ice cores. While there are some temperature measurements that started during the 17th century (*e.g.*, the "weather" of London has been recorded daily since 1668), most land air temperature records can be considered to have been systematically collected starting in the 19th century. Records of temperatures at sea, both of air and water, were also systematized in that century, mostly through the efforts of an American naval officer, Matthew Fontaine Maury in 1853. Similarly, relative sea level from coastal tide gauges was not actually recorded in the instrumental sense until the 19th century, although readings of "tide poles" are known from ancient Egypt and Greece. It is the instrumental records with which we are concerned.

Although there are individual records of temperature or sea level that give some indication of long term change, constructing a "global" record is a late 20th century endeavor. Each of these record sets, land air temperature (LAT), marine air temperature (MAT), sea surface temperature (SST), and relative sea level (RSL), have unique characteristics, coverage,

Table I. Definitions.

AGU:	AIP American Geophysical Union
COADS:	Comprehensive Ocean-Atmosphere Data Set
GLOSS:	IOC Global Sea Level Observing System
ICSU:	International Council of Scientific Unions
IOC:	UNESCO Intergovernmental Oceanographic Commission
IPCC:	Intergovernmental Panel on Climate Change
JOI:	Joint Oceanographic Institutions
NCAR:	National Center for Atmospheric Research
NCDC:	NOAA National Climatic Data Center
NOAA:	National Oceanic and Atmospheric Administration
PSMSL:	Permanent Service for Mean Sea Level
UKMO:	United Kingdom Meteorological Office
UNEP:	United Nations Environment Programme
UNESCO:	United Nations Educational, Scientific and Cultural Organization
WCRP:	World Climate Research Programme
WMO:	World Meteorological Organization
WOCE:	WCRP World Ocean Circulation Experiment

advantages and disadvantages, and error sources both random and systematic. Truly "global" LAT, MAT, and SST records are only available for the last decade by remote sensing from meteorological satellites, and these too are limited in their applicability to the problem of determining anthropogenic effects on Earth's climate. Sea level from oceanographic and geodetic satellites is an emerging science, but satellite altimeter measurements of long term changes for climate purposes probably is several generations of instrumentation in the future.

The outline of the chapter is as follows. First, a note on the record of climate during the Holocene is given to place the period of instrumental record into context with the last 18,000 years. Next a global LAT record is examined and compared with the regional record of the contiguous U.S.A. historical climate network. Both the MAT and SST records are considered and compared, with emphasis on changing ship routes and vessel design. Satellite measurements of LAT/MAT and SST are discussed next, followed by a review of the error sources and of attempts to construct combined LAT/MAT/SST global data sets. Finally the issue of RSL is discussed with emphasis on the difference between relative sea level and absolute sea level, including as an example, an analysis of one century-long RSL record for climate signals.

In order to focus on the observational record of the last century, the scope of this chapter has not included predictions of future RSL, SST, MAT or LAT. The IPCC report is replete with predictions, but the interested reader should also study other opinions, in particular, those of the U.S. National Academy of Sciences.[5] Ultimately each of us must make a judgment about the future, and the purpose of this chapter is to lay a foundation for that judgment based on physics and facts.

HOLOCENE TEMPERATURE AND SEA LEVELS

Earth's climate is in a constant state of flux and has been changing throughout the geological record. Only in the last few centuries, however, have quantified estimates been available from the instrumental record. Figure 1 summarizes estimates of surface air temperatures in three time blocks:[6] 0-900,000 years before the present (ybp), 0-11,000 ybp (the Holocene), and 900-1990 AD. In all but the last 150-200 years one can argue that the effect of human

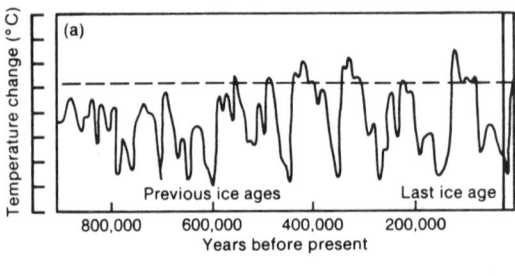
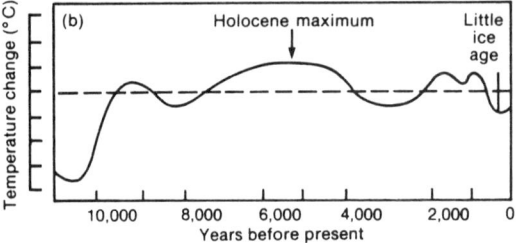
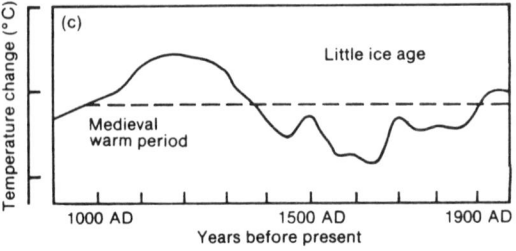

Figure 1. History of Earth's air temperatures during the last 10^6 years (from Folland et al.[6]). Dashed lines $\simeq$ LAT in year 1900.

activity on climate has been negligible, and yet the records show a rich signal on all time scales. Much of the variance on 100,000 year time scales can be accounted for by the orbital parameters of the sun–earth system. During the Holocene, some of the details (such as the Younger Dryas cool period 10,500 ybp) are being unraveled and are hypothesized as being linked to the thermohaline circulation of the North Atlantic Ocean. The Little Ice Age (1350–1850 AD) marks a period of decidedly colder weather in Europe that led to the collapse of the Norse colonies on Greenland. It has been associated with atmospheric circulation "blocking" patterns over the Northern Hemisphere, which are prolonged changes in spatial wavenumber in the Polar Front.

Temperatures during the Little Ice Age and other earlier times are coming from proxy techniques such as isotopic oxygen analysis. Coral core samples, for example, are being used to quantify the temporal variability of Earth's sea water temperature patterns. During warmer times, ^{18}O is more abundant in surficial sea water because ^{16}O more readily evaporates into the atmosphere. Corals then have an enriched ^{18}O in their $CaCO_3$ skeletons during milder epochs in Earth's climate cycles. Using $\delta^{18}O$ geochemical methods, where

$$\delta^{18}O = [\{(^{18}O/^{16}O)_{sample}/(^{18}O/^{16}O)_{standard}\}-1]\cdot 10^3,$$

stratigraphy, and ^{14}C dating, geologists have pieced together a remarkable record of our climate. Surface temperature patterns, of course, do not vary in isolation from other geophysical evidence of climate change.

Associated with the general warming from the last glaciation through the Younger Dryas to today has been a retreat of continental glaciers and a concomitant rise in global sea level. Figure 2 is an example of the change in Holocene sea levels from 18,000 ybp to today.[7] This example is from the island of Barbados (13°N, 60°W) at the eastern margin of the Caribbean Sea. The $\Delta\delta^{18}O$ data (right-hand ordinate) shows the change in $\delta^{18}O$ of mean ocean water as a solid line through the radiocarbon-dated coral depths from Barbados (dots) and other nearby islands (circles). As Earth warmed from the last major ice age, sea level rose over 120 meters at rates in excess of 2.5 mm/yr. For the last 3,000 years the rate of rise has been remarkably slower, typically 0.4 mm/yr in many places, and has given civilization the false notion that sea level is fairly steady. Many oceanographers estimate that global (eustatic) sea level today is rising at a

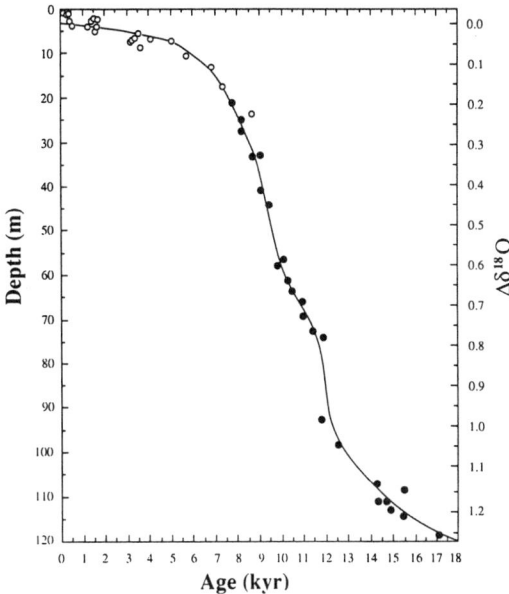

Figure 2. Holocene sea level curve from Barbados for last 2 × 10⁴ years (from Fairbanks[7]).

rate of 1-2 mm/yr, although at some tide gauge locations the RSL rate is 2-5 times these values due to local subsidence.

A great deal of corroborative evidence exists to support the warming seen during the last centuries of the Holocene. For example, observations of mountain glaciers in Europe, North and South America, New Zealand, Africa, and Asia since *ca.* 1650 show steady or advancing termini locations during the Little Ice Age, followed by significant retreat since the middle of the 19th century. Similarly, archaeological studies have shown markedly different patterns of human and wildlife habitations in prehistoric times, and ample evidence of the effects of the Little Ice Age are recorded in the historical documents of Europe and China.[8,9] But it is the thermometric record of the last hundred years that 20th century scientists turn to for quantitative evidence of global warming.

LAND AIR TEMPERATURE CHANGE 1880-1990

A global record of temperature change is sought because climatic variations on the regional scale, such as the contiguous U.S.A., may show too much spatial and temporal variability. One of the most prominent studies of global temperature is that of Hansen and Lebedeff.[10] Although there are many others,[6,11,12] the study by Hansen and Lebedeff amply illustrates the nature of the problem. Figure 3 from their study shows the location of the land stations used, and the changing coverage from 1870 to 1960 (land data sets from the U.K.[12] and U.S.S.R.[13] also are available). At this point it must be emphasized that the Hansen and Lebedeff study only uses land air temperatures to make a global estimate, thus leaving the 71% of Earth's surface covered by water unsampled.

Hansen and Lebedeff combined the individual LAT records from each of the stations by computing the mean temperature for the common periods and then adjusting the records to a common scale. In this way the increasing number of stations could be added to the total. They did this by regions, and then combined all the data into hemispheric and global averages, year by year, from 1880 to 1985. Their study used the data from the (U.S.) National Center for Atmospheric Research, supplemented by data from the (U.S.) National Climatic Data Center of NOAA. Essentially they calculated the averages without further quality checks. The result is shown in Figure 4.

The upper curve in Figure 4 is the global LAT record from Hansen and Lebedeff;[10] the middle and lower curves are their Northern and Southern Hemisphere curves, respectively. All three curves are characterized by a quasi-steady rise from 1880 to about 1940, decreasing temperatures from 1940 to about 1965, and rapidly rising values for the last two decades. The

82   Temperature and Sea Level Change

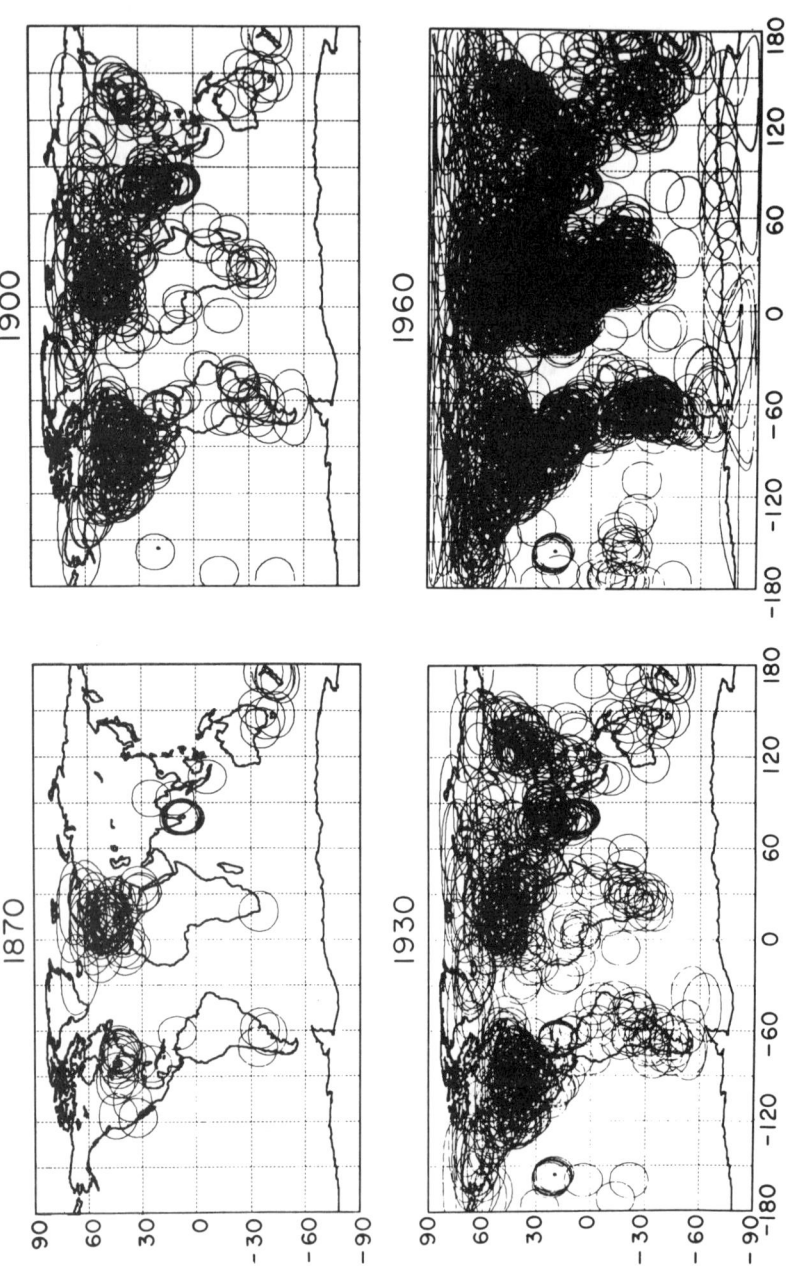

Figure 3. Distribution of meteorological stations with air temperature records used by Hansen and Lebedeff[10] for four time frames.

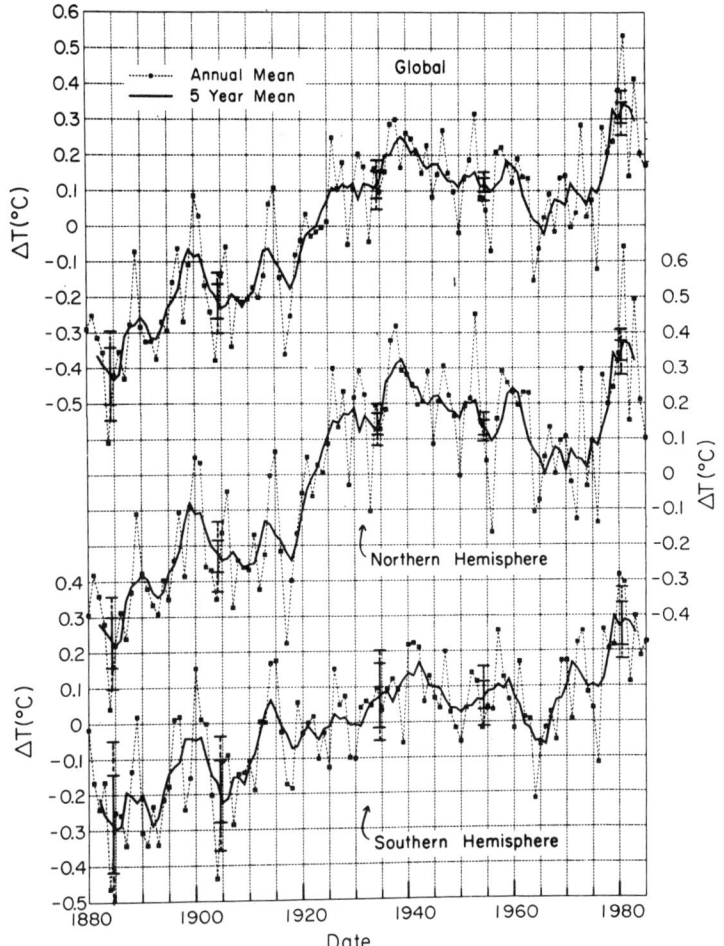

Figure 4. Global and hemispheric land air temperatures estimated from meteorological stations shown in Figure 3 with ±1σ and ±2σ confidence limits shown as vertical bars (from Hansen and Lebedeff[10]).

Southern Hemisphere LATs seem to have the least amount of variance about the linear trend, which is physically satisfying since the Southern Hemisphere is 90% ocean. Trends in these LAT data are about 0.005°C/yr–0.007°C/yr, and the signal is about the same in either hemisphere. Hansen and Lebedeff recognized the potential errors in these results due to incomplete coverage and due to (in particular) large city heat island effects, but concluded that the trends were significant and represented a global warming.

For comparison on a regional scale, the study of Hanson et al.[14] for the contiguous U.S.A. is shown in Figure 5. These data differ from those of Hansen and Lebedeff[10] in that they are the NCDC Climate Division data and are specifically designed to record climate change. Over 6,000 stations were used in these data in 1985 (Figure 5b), and they form a subset of the NCAR

84   Temperature and Sea Level Change

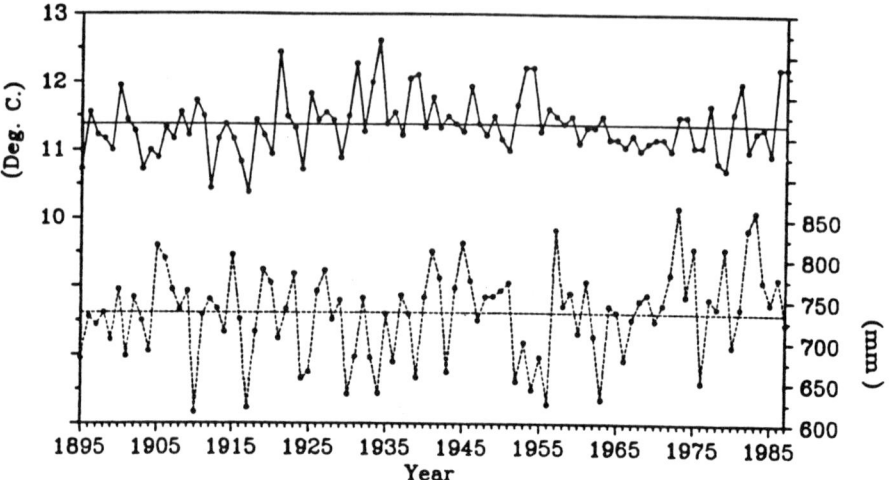

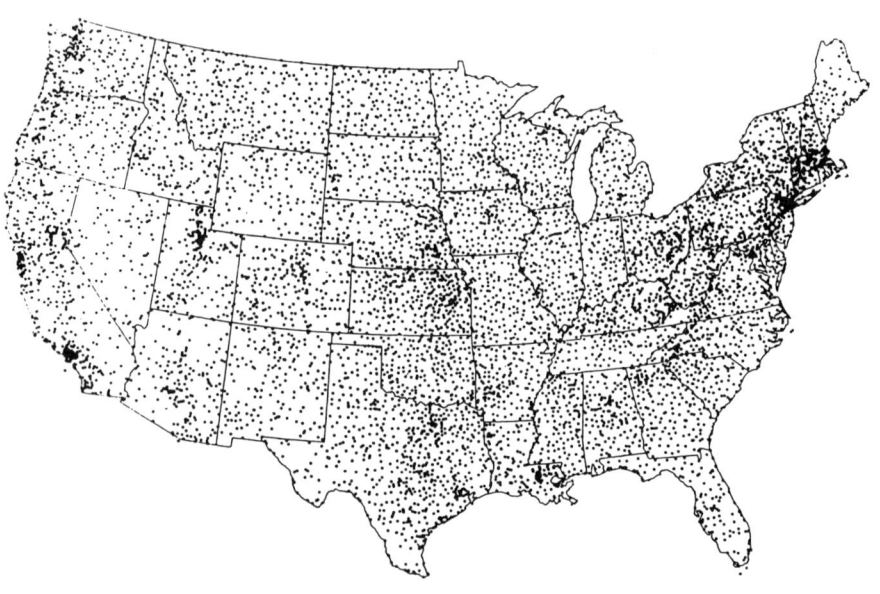

Figure 5. a: (upper panel) Land air temperature (solid) and land precipitation (dashed) for the contiguous U.S.A. using the NOAA Climate Division data (from Hanson et al.[14]). b: (lower panel) Cooperative NOAA National Weather Service station network used to construct the U.S.A. LATs above.

data discussed for the globe in Figure 4. Hanson et al. chose the time frame 1895-1987 for their study because these data were considered more reliable than data prior to 1895. These NCDC data (Figure 5a) are combined in a manner to preserve the actual temperature values, the mean and standard deviation of which for the U.S.A. are 11.38°C ± 0.18°C respectively. The most obvious difference between Figures 4 and 5a is that the LATs for the contiguous U.S.A. appear to be without a trend.

Hanson et al.[14] confirmed that the Climate Division data were without any statistically significant trend using two tests: the Spearman rank test and the two-phase regression.[15,16] The two-phase regression showed a marginally (5% level) significant change in trend (+0.024°C/yr prior to 1940 and -0.010°C afterwards). In applying the two-phase regression analysis technique one must ask the question, is there any physical reason to expect that a change in trend occurs at a particular time? If not, then the null hypothesis of no trend may still be valid even though the trend change is statistically significant at say 95% confidence. Hanson et al. concluded that the null hypothesis of no trend could not be rejected based on these two tests.

Are the results of Hanson et al.[14] contradictory to those of Hansen and Lebedeff?[10] There is ample evidence that "global" climate change cannot be tested using regional data, because LAT patterns are associated with atmospheric general circulation changes, and these can be very different from region to region.[6] The better question to ask is whether the NCAR LAT data per se are adequate to demonstrate global warming. Jones et al.[12] note that the amount of warming in Hansen and Lebedeff is about 0.2°C greater over 1880-1987 than in the U.K. Meteorological Office (UKMO q.v. Table I) data set. Error sources in these LAT data sets will be discussed below, but it should be noted (cf. Figure 3) that a lack of coverage over the oceans is potentially of greatest concern in the Hansen and Lebedeff LAT analysis; studies by Jones et al. and the IPCC[6] have greatly emphasized the necessity of blending LAT with marine air and sea surface temperatures.

MARINE RECORDS 1870-1980

Creation of the Comprehensive Ocean-Atmosphere Data Set (COADS[17]) is a milestone in attempting to collate all marine observations from over a century of ship observations. The COADS includes information other than MAT and SST (winds, etc.) but emphasis herein will be on the temperatures. Other data sets such as the COADS exist, notably the UKMO data set, but for illustrating the limitations of air temperature and SST for detecting global change, the COADS is quite adequate. As with the LAT discussed above, creation of the COADS was an NCAR effort, the limitations of which are discussed next.

Figure 6, from Oort et al.,[18] shows the marine coverage for three time periods: 1880-1889, 1920-1929, and 1960-1969. The most noticeable thing is the increased coverage with time, a situation similar to that encountered in Figure 3. Perhaps of equal significance is the change in ship routes, particularly after the opening of the Panama Canal in 1914. One consequence of post-1914 routing is that the southeastern South Pacific Ocean has many fewer samples; this is distinctly different from the LAT issue where coverage is increasing almost everywhere. Also, before the COADS, there was a great deal of duplicate data in the archives much of which has been culled out only in the last decade.

As with measurements on land, MAT and SST are affected by changing observational techniques, instrumentation, and microenvironment. Most notable for MAT is the change from small wooden ships to large steel vessels, and the concomitant increase in speed. Air thermometers in some older ships were kept in screens similar to those found at some land meteorological stations, and there are questions about the usefulness of daytime marine observations due to observational practices. Larger ships have also affected the MATs because of

86 Temperature and Sea Level Change

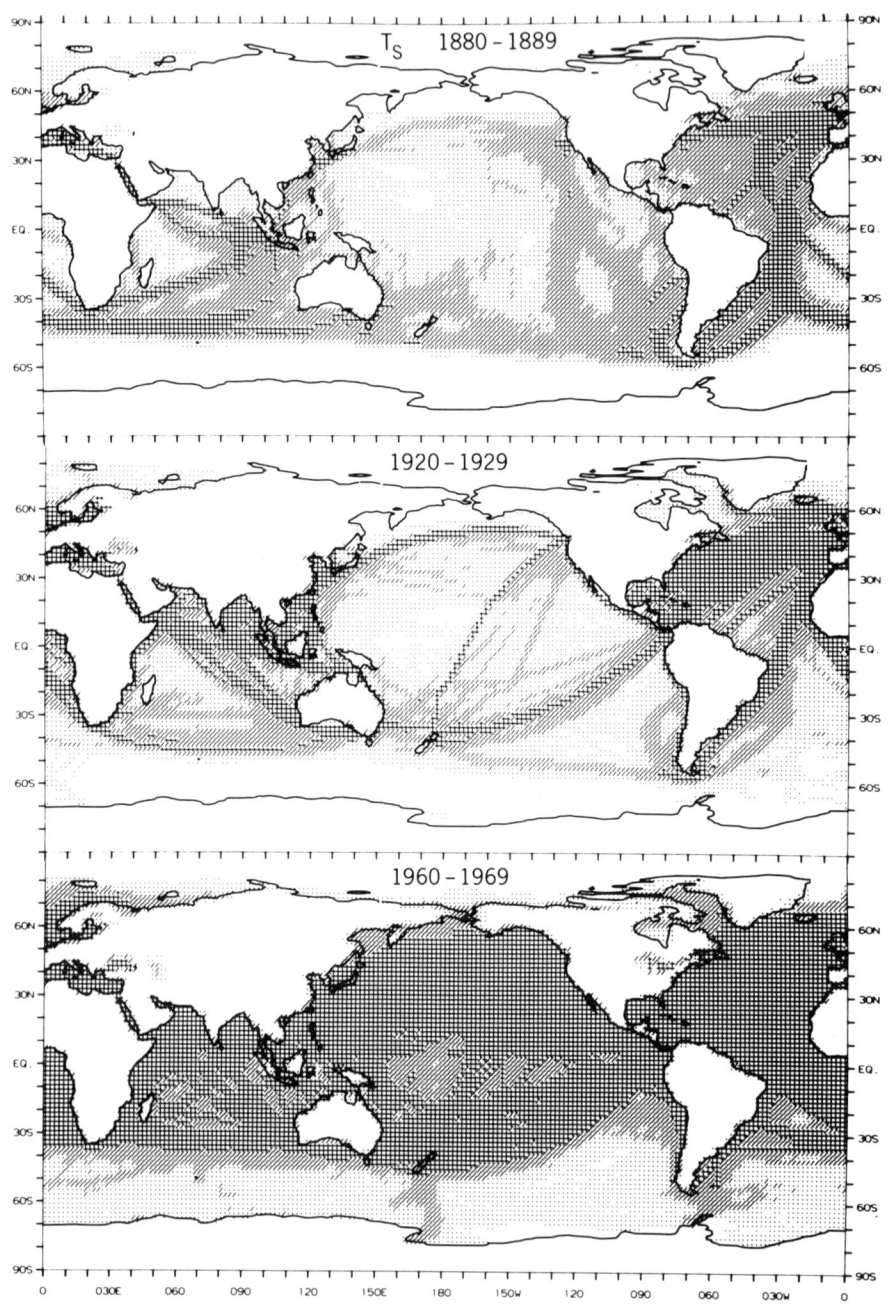

Figure 6. Distribution of 2° × 2° latitude/longitude squares over the ocean from the COADS (from Oort et al.[18]).

navigation bridges being farther above the water than on smaller ships (lowers temperature), and because of the thermal inertia from more massive steel hulls and superstructure (raises temperatures). It must also be borne in mind that marine observers are working under different conditions and with different motivation than land observers, and in high stress circumstances may be prone to making other errors.

Perhaps the best method of illustrating the problematic nature of MAT data is to show (Figure 7a) the COADS marine air temperatures, and to call particular attention to the period 1940-1945. The warm temperatures during World War II are almost certainly due to a change in observational behavior at sea, and is compounded by far fewer data points than at other times. Jones et al.[12] plotted the difference between coastal and island LAT and nighttime MAT from the UKMO archive, and show differences that range from +0.2°C to -0.9°C, and note that the difference signal is not perfectly symmetrical about the equator. Although one might expect that the COADS MATs and the UKMO MATs would be similar, Jones et al.[12] conclude that the data sets are not the same, and that there are important differences particularly prior to 1940 between them.

Sea surface temperatures are the other major source of climatological data from over the oceans (Figure 7b).[18] Here too, shipboard observations are affected by different issues, in particular, the changes from sail to steam power, and the shift from using buckets to obtain SST samples to using thermometers in the seawater intake line of engine cooling systems. Jones et al.[12] summarize many discussions in the literature concerning the effects of various buckets (wooden, canvas, metal, rubber) on the determination of SST, and some interesting models to correct for the differing effects have been developed; the dashed line in Figure 7b is the uncorrected SST and the solid line is the result of applying one such correction model. Unfortunately, the corrections depend on knowing the mix of bucket types used, and detailed records are not available. Personal experience with the SST issue suggests that the engine intake readings are probably much less reliable than most researchers suspect because of the perceived purpose the engineroom Oiler or Wiper attaches to making the readings; in the best of circumstances, the data are perceived to have ancillary engineering purposes. Nevertheless, estimates of the difference between engineroom readings and bucket readings range from +0.1°C to +0.6°C increase in SST.

Figure 8 from Folland et al.[6] summarizes both the MAT and SST global records and provides a convenient comparison of the IPCC opinion on LAT during the last 130 years. There are two SST curves, and the differences arise primarily from interpreting how the mix of bucket types evolved over the period before ca. 1947. Uncertainties of ∼0.2°C can be seen during the 19th century, and there is essentially no SST difference in the last few decades. From a small 2° x 2° latitude/longitude box in the heavily trafficked shipping lanes of the Straits of Florida off Miami (not shown), the COADS SST data show a temperature rise of 1.5°C per century, but if the record is divided into pre- and post-World War II parts, the earlier half is offset from the later half by about 0.7°C and the trend since 1947 is slightly negative. Oort et al.[18] broke the COADS SST data into 10° latitude bands and also showed, for many zonally averaged SSTs, that there is little or no trend in the last four decades. Thus, with all the meticulous work by the COADS, UKMO, and other groups, there are significant differences in the data analyses and uncertainty in the meaning of the published results.

Also shown in Figure 8 is the global LAT, computed as a simple average of the results from Hansen and Lebedeff,[10] Vinnikov et al.[13] and Jones et al.;[12] the UKMO nighttime MAT global average is also shown. In many respects the air temperatures and the SSTs are quite similar (particularly at the higher frequencies), but the apparent agreement in the 1950-1980 30 year "normal" period is due to using that epoch to compute anomalies. Also, it is very important to note that the marine data has been made homogeneous with the land data, and therefore offers no independent information (i.e., the degrees of freedom are not increased). It seems certain,

88 Temperature and Sea Level Change

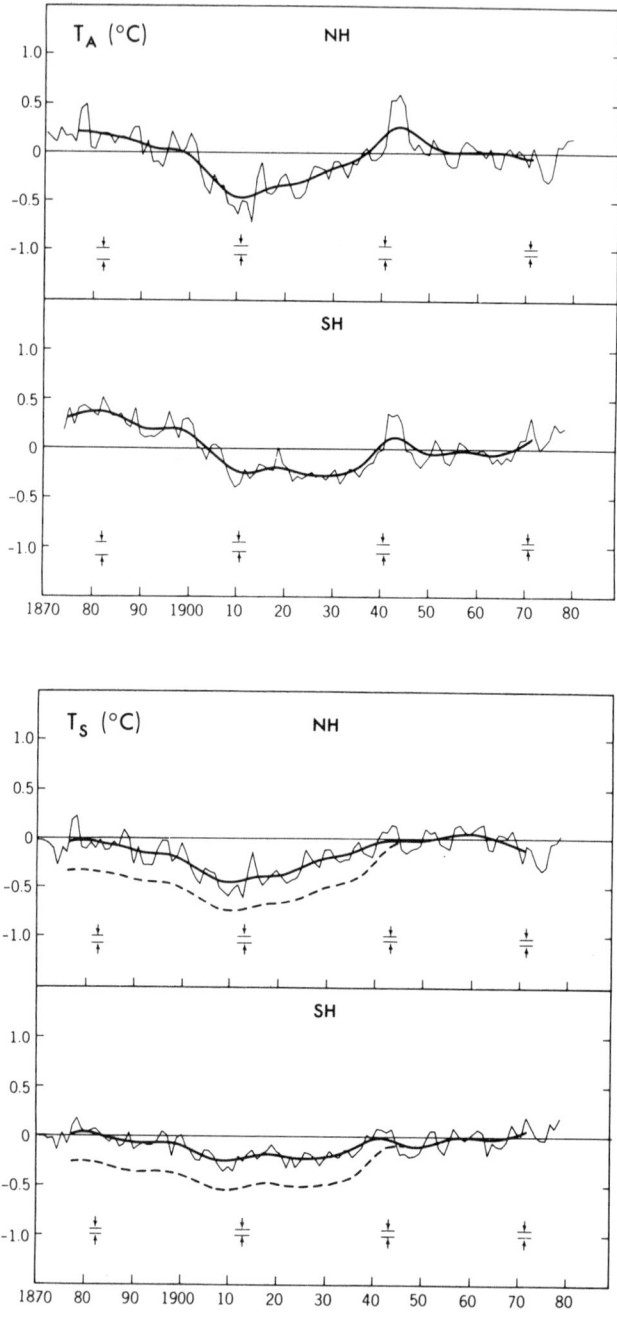

Figure 7. a: (upper panel) MAT from COADS, and b: (lower panel) COADS SST; curves show both northern (NH) and southern (SH) hemispheres (from Oort et al.[18]). Light line shows annual means; heavy line is five-year running mean.

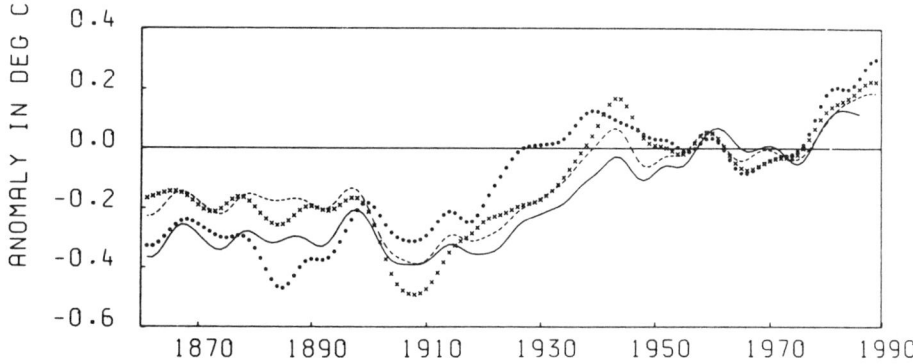

Figure 8. Two estimates of global SST (solid, dashed), nighttime MAT (crosses), and LAT (dot) anomalies relative to the 30-year norm 1951-1980 (from Folland et al.[6]). Homogenization of the marine data with land data has been affected, and the marine and land data are not independent data sets in this figure.

however, that these *measurements* show that the 1980's are ~0.1°C warmer than the 30-year norm, and the post 1940's are ~0.2°C warmer than the previous 80 years. Whether or not the *measurements* are without significant residual error is the subject of the next section.

ERROR SOURCES

As an introduction to the topic of error sources in LAT, the table of maximum-minimum thermometer location information for the meteorological station at Key West, Florida, is presented in Table II. From 1830 to 1974 the MAX-MIN thermometer had 17 different locations; it was replaced in 1974 with an electronic thermometer and the data are now hourly readings as well as daily minimum-maximum values. For many years the thermometers were in Cotton-Region shelters (used since first employed in 1872), but prior to that time "window units" were used, usually on the north side of a building. The second issue is that these were MAX-MIN thermometers, and daily means are calculated from the arithmetic average of the two values. Among other things this may introduce systematic errors in the daily means, and whether or not these cancel when global averages are calculated is still a matter of debate. Comparison of Key West LAT with COADS MAT from the Straits of Florida (not shown) uncovers other data dissimilarities: the significant cooling in Key West LAT from 1885-1920, for example, is not repeated in the COADS MATs except for a brief period in 1905-1910. This dissimilarity introduces the topic of homogenization.

Jones et al.[19] argue that "data reliability and long-term homogeneity can be far more convincingly demonstrated for the gridded land data than for the marine data because land station inhomogeneities can be more easily identified, explained and corrected." In view of Table II, this assumption should be considered with caution. However, they chose 15 regions where LAT and MAT were compared (Figure 9a) and the differences between coastal LAT minus COADS MATs were computed for the Northern and Southern Hemispheres (Figure 9b). For the period of World War II, the analysis of Oort et al.[18] and Jones et al.[19] are clearly in agreement, and there is good physical reason to accept Figure 9b as a correction for MATs between 1940-1945. After 1950 or so the differences are negligible, but one should not construe that to mean these latter data are necessarily correct. An area such as the Gulf of Mexico,

Table II. History of locations and heights of maximum-minimum thermometers at Key West, Florida, for the period 1830-1974.

Location	Begin	End	Elevation (m)
U.S. Army Surgeons	1830	1870	Unknown
Russell House	1870	1870	Unknown
Tift & Co. Building	1870	1871	4.9
Duval Street	1871	1872	5.5
Louvre Hotel	1872	1882	13.1
Wall Building	1882	1886	6.1
U.S. Naval Depot	1886	1886	14.6
Waite Building	1887	1903	12.8
Weather Bureau Building	1903	1911	3.4
Island City Bank Building	1911	1913	12.5
Weather Bureau Building	1913	1931	3.0
Airways Station Building	1931	1942	1.2
Boca Chica Airport	1942	1944	1.5
Boca Chica Airport	1944	1953	5.8
U.S. Post Office Building	1953	1957	1.5
Key West Airport	1957	1958	7.3
Key West Airport	1958	1974	5.5

for example (*q.v.* Figure 9a), has wintertime SST differences of 10°C or more depending upon whether the ship was in the Gulf Stream or a detached anticyclonic eddy, or in the cooler upwelling water of Campeche Bank, or in the plume of the Mississippi River. Added to these oceanic complexities are the issues of changing ship routes (*q.v.* Figure 6) and observational practices. Clearly, homogenization reduces the differences between LATs and MATs; the cost is that independent assessment of the global temperature is lost (*q.v.* the caveat surrounding Figure 8).

Another major concern in understanding long-term temperature trends is the effect on the measurements of growing population centers. This is known as urbanization and is typically a non-linear function of population (P) given by an equation of the form

$$\delta LAT\ (°C) = 0.00182 \cdot P^{+0.45},$$

where δLAT is the average annual land air temperature difference between a station in an urban location with a known P and a rural location with $P \simeq 750$ persons (sign of the exponent corrected from Karl and Jones[20]). Urbanization is a problem in homogeneity and is typically done for LATs by comparing records kept at primary observing sites, such as airports, with records in the surrounding countryside or in smaller population centers.

An example of the urbanization problem is discussed by Karl and Jones[20] wherein they compared the difference in the data over the contiguous U.S.A. between Hansen and Lebedeff[10] and Jones *et al.*[12]. Figure 10 shows the result of Karl and Jones from comparing Hansen and Lebedeff minus the Historical Climate Network (HCN) (*q.v.* Figures 4 and 5), and Jones *et al.*[12] minus the HCN. Urbanization effects in the Hansen and Lebedeff data were found to be between 0.3°C and 0.4°C when compared to the HCN; the Jones *et al.*[12] urbanization effect was approximately 0.1°C for the same 1900-1985 time frame. The solid lines in Figure 10 are the result of applying a two-phase regression[16] to show the statistically significant (5% level) trends. In both cases no significant change point was found, and the overall trend magnitudes were 0.38°C and 0.15°C for Hansen and Lebedeff[10] minus the HCN and for Jones *et al.*[12] minus the HCN, respectively.

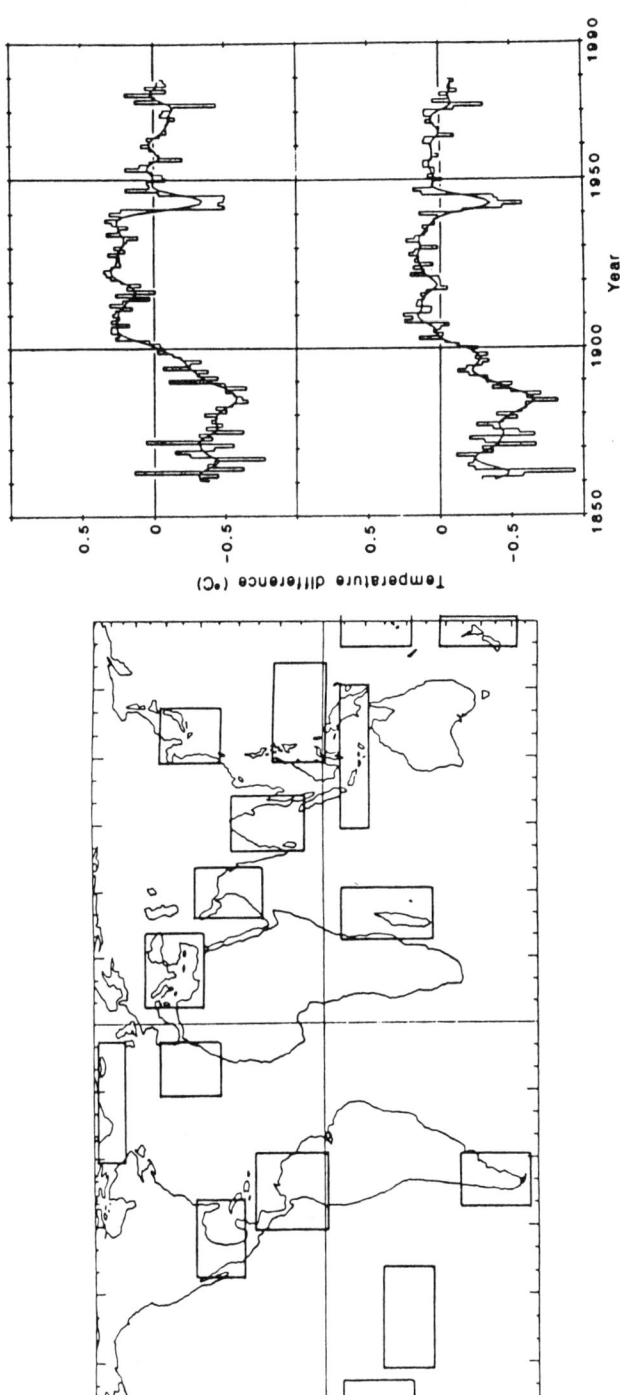

Figure 9. a: (left panel) Regions where MAT and LAT were compared. b: (right panel) Temperature differences coastal LAT minus uncorrected COADS MAT for the northern (upper) and southern (lower) hemispheres (from Jones et al.[19]).

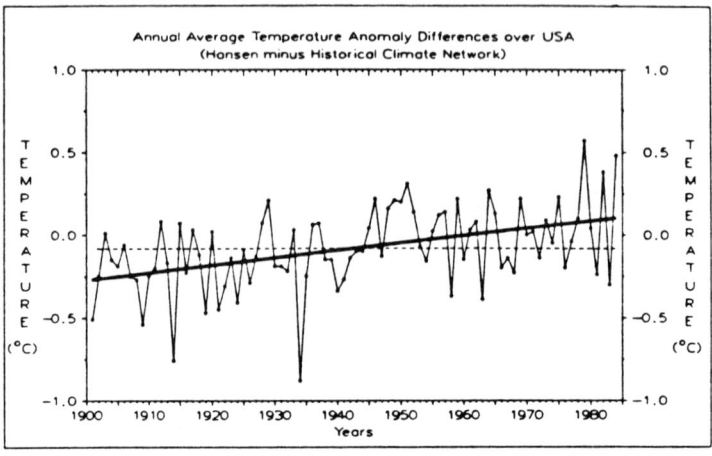
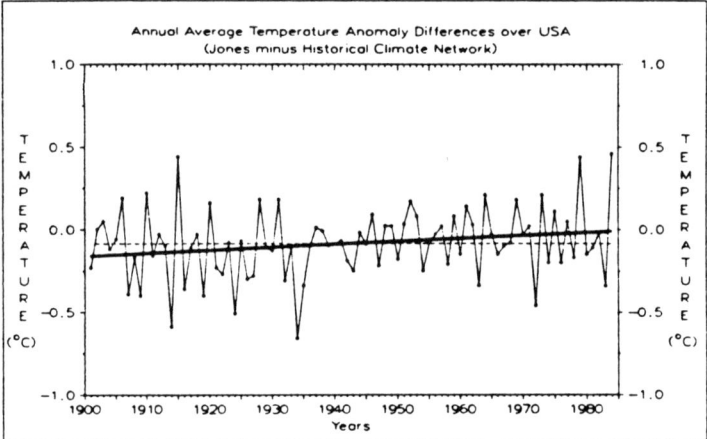

Figure 10. Urbanization effects over the contiguous U.S.A. shown by Karl and Jones[20] in the Hansen and Lebedeff[10] data (upper panel) and the Jones et al.[12] data by respectively subtracting from their data the Historical Climate Network LATs and applying the two-phase regression of Solow.[16]

Changing coverage is the last type of error to be considered, and this has been already introduced for LAT (q.v. Figure 3) and for MAT/SST (q.v. Figure 6). Folland et al.[6] report on the changing coverage effect using blended (LAT/MAT/SST) global data using a "frozen grid" analysis wherein the coverages in 1861-70, 1901-10, and 1921-30 are independently used to compute three curves of global temperatures. Although the coverage sites change for each of these three decades, the final time-series of global temperatures (not shown) are remarkably similar. Nevertheless, Folland et al.[6] estimate that the bias due to this effect is at least ±0.05 °C, and may be particularly acute in the Southern Hemisphere. Of concern is that the frozen grid analysis uses blended LAT/MAT/SST data, and that the homogenization process is forcing all land and marine data into agreement with some unexpected biases. One would hope that the global coverage offered by measurements from meteorological satellites would overcome many of these errors, but as will be shown in the next section, no such simple panacea exists in that technology as yet.

SATELLITE MEASUREMENTS 1979-1988

Although meteorological satellites have been operational since the early 1960's, quantitative estimates of SST and LAT/MAT are limited to the last decade or so with the advent of multispectral remote sensing. The basic problem is that of radiative transfer through an absorbing, emitting, and scattering atmosphere. In the case of SST, the radiation from a reflecting and emitting surface complicates the problem. Two cases will be considered in this section: SST by visible/infrared detecting optical sensors[21,22] and LAT/MAT from passive microwave radiometry.[23] The theoretical presentation given below is necessarily brief, but the reader interested in satellite oceanography is referred to Maul[24] and for remote sensing of the atmosphere, Deepak[25] is a good choice.

The basic physics is given by the radiative transfer equation (RTE) for a non-scattering atmosphere, in which the spectral radiance (L_O) (Wm^{-2} μm^{-1} sr^{-1}) is the sum of the surface radiance (L_S) term and the atmospheric emittance (L_E) term respectively,

$$L_O = L_S \cdot \tau - \int L_E \cdot (\partial \tau / \partial p) \cdot dp,$$

where τ is the atmospheric transmissivity at a particular wavelength (λ), L_E is given by Planck's Law, and the integration is from (pressure) $p = 0$ to $p = p_S$, *i.e.*, over the total atmosphere. In this formulation, the surface is considered a blackbody with emissivity $(1 - \text{reflectivity}) = 1$ but in reality, none of Earth's surfaces are non-reflecting. Conceptually, the entire RTE is then integrated over the response function (ϕ) of the optical or microwave filter, and inversion of the integral $\int \phi \cdot L_O \, d\lambda$ (Wm^{-2} sr^{-1}) gives an equivalent blackbody temperature. In the infrared (IR), ϕ is typically 1 μm wide, and individual absorption lines are averaged over; in the microwave, ϕ is often chosen to be 0.1 GHz or so, and absorption line shape is important.

Multispectral determination of SST is accomplished by viewing the same oceanic pixel at two wavelengths λ_1 and λ_2 through a region of the IR "window" (spectral region where $\tau \geq \sim 0.7$), typically 10.5 μm $\leq \lambda \leq$ 12.5 μm, and where τ is differentially affected by the same gas. Rewriting the RTE in τ coordinates and using the mean value theorem to define a mean emitted radiance, $\overline{L_E}$,

$$L_O(\lambda_1) = L_S \cdot \tau_1 + (1-\tau_1)\overline{L_E} \quad \text{and} \quad L_O(\lambda_2) = L_S \cdot \tau_2 + (1-\tau_2)\overline{L_E}$$

can be combined to solve for SST from the equivalent blackbody temperature (T_b) in each multispectral channel

$$\text{SST} - T_{b_1} = (T_{b_1} - T_{b_2}) \cdot \Gamma - (\overline{T_{E_1}} - \overline{T_{E_2}}) \cdot (1-\tau_2) \cdot \Gamma$$

where $\Gamma = (1-\tau_1)/(\tau_1 - \tau_2)$. In this linear form of the RTE, Γ is a slope, and the right-hand term is the intercept. For two juxtaposed channels, such as on the TIROS-n series of NOAA satellites, Γ and the intercept term are assumed to be "constant" for Earth's atmosphere. In practice, a linear equation of this form is least squares fitted to an ensemble of *in situ* SST data, and is given the name MCSST for Multichannel SST analysis.

Strong[21] used MCSST data from the NOAA operational satellites to study global SST change between 1982 and 1988, and came to the conclusion that global SST has been

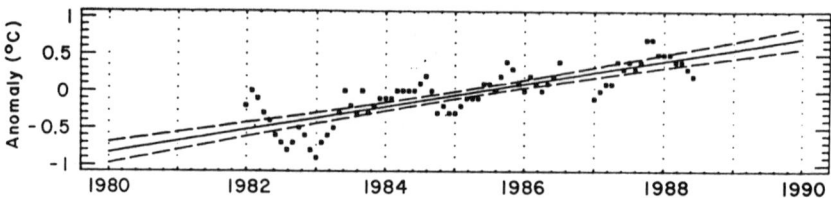
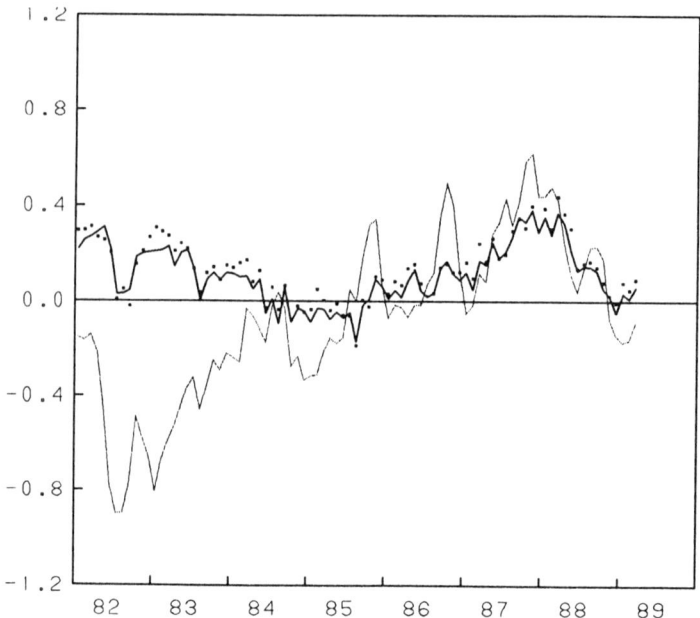

Figure 11. a: (upper panel) Multichannel satellite SST estimates for 1980's from Strong[21] for the northern hemisphere, and b: (lower panel) using the blended MCSST (heavy line) from Reynolds et al.[22] for the same area; light line in (b) is Strong's estimate redrawn to another scale; dots show *in situ* temperature anomalies.

increasing at a rate of ~0.1°C per year. The upper panel in Figure 11 shows the trend for the Northern Hemisphere (0°-60°N), according to Strong, which had a one-time calibration applied. Reynolds et al.[22] use simultaneous *in situ* data to create a "blended" global SST, and their result for the same area and timeframe (1982-1988) is shown in the lower panel of Figure 11. Strong's conclusions were refuted by Reynolds et al. because the shipboard observations did not reflect the warming seen in the unblended MCSST data. Strong argues that the blended data do not give global coverage and could be misleading. The issue of blending is essentially that of homogenization, already considered for creating LAT/MAT/SST ensembles, but what are the physical issues surrounding the controversy?

First is the RTE formulation itself. Although modeling of the RTE suggests that for a given L_s, $L_o(\lambda_1)$ and $L_o(\lambda_2)$ are linearly related for a wide range of atmospheric conditions, in fact the term Γ which determines the slope and intercept is not constant either for varying atmospheric conditions or for varying nadir angles. The MCSST formulation also depends on a well behaved relation between the absorbing gas (primarily H$_2$O vapor in the 10.5 µm $\leq \lambda \leq$ 12.5 µm window) and L$_E$. Atmospheric windows are, in fact, "dirty" in that other gases are actively absorbing and emitting IR radiation, but more so, the windows are affected by clouds, aerosols and other particles that scatter. The eruption of the volcano El Chichón in 1982 added significant stratospheric dust particles and probably made MCSSTs appear too low in the early part of the Strong[21] and Reynolds et al.[22] study timeframe. In fact, the MCSST formulation of the RTE has seasonal and geographic biases, and although a vast improvement over the earlier satellite SST analyses, the MCSST analysis is not without important limitations, particularly when applied to problems of climate and global change.

Remote sensing of global air temperature is also based on application of the RTE. A spectral region is chosen where the term $L_S \cdot \tau = 0$, i.e., an opaque region of the atmospheric spectrum is used, for example, the 60 GHz O$_2$ absorption line. A passive microwave receiver senses M1 radiation from atmospheric O$_2$, and the mean temperature of a tropospheric slab several kilometers thick is inferred. For this problem, the RTE term $\int L_E \cdot (\partial \tau / \partial p) \cdot dp$ can be expressed as an exponential, $\tau = \exp(-p/p_m)$, where p_m is the pressure at the peak of the microwave channel weighting function, and the upwelling radiance $L_O(p_m)$ becomes the Laplace ($\mathcal{L}$) transform of the Planck function L_E:

$$L_O = -\int L_E \cdot \partial \tau / \partial p \cdot dp = \int L_E \cdot e^{-p/p_m} \cdot dp/p_m = (p_m)^{-1} \mathcal{L}_{1/p_m} [L_E(p)].$$

The Planck function then is just the inverse Laplace transform

$$L_E(p) = \mathcal{L}^{-1} [L_O(1/k)/k] = (2\pi i)^{-1} \int [L_O(1/k)/k] \cdot e^{kp} dk$$

where $k = 1/p_m$. Measurements are made by the microwave sounding unit (MSU), also aboard the TIROS-n series of NOAA meteorological satellites. It has four channels, one of which (53.74 GHz) is used in the work of Spencer and Christy[23] to estimate middle tropospheric LAT and MAT globally.

The center of the MSU 53.74 GHz weighting function is at $p_m \simeq$ 60 mb, or at a height of about 4,500 meters. The troposphere is typically considered 10 km thick, so these MSU measurements are called middle troposphere values. The assumption with MSU observations is that Earth's surface is not sensed, which is just the opposite of IR measurements of SST. In practice, some MSU measurements are contaminated by sensing the land or sea surface, as some IR measurements in very humid regions encounter $\tau \simeq 0$ conditions. These limitations seem to affect the MCSST T_b's more than MSU T_b's. Each remote sensing approach (IR vs. microwave) has advantages and disadvantages, but for the problem of sensing global temperatures, the microwave approach seems to be the most suitable method.

LAT/MAT from the MSU on two NOAA TIROS-n type satellites were compared by Spencer and Christy for a two-year overlap period in the early 1980's, and the precision ($\pm 1\sigma$) in the monthly mean temperatures was $\sigma = \pm 0.01$°C for the globe. Furthermore, they report that there was no discernible drift between MSUs on any of the six NOAA satellites studied to date. In a comparison with contiguous U.S.A. LATs for the period 1979-1987 (Figure 12a), a linear correlation coefficient $r = 0.89$ was found with monthly mean MSU brightness temperatures (T_b in the parlance of microwave remote sensing). For comparisons with the global data collected using the methods of Hansen and Lebedeff[10] and Jones et al.[12], the percent variance explained ($100 \cdot r^2$) using annual mean temperatures was 53% and 74%, respectively. In com-

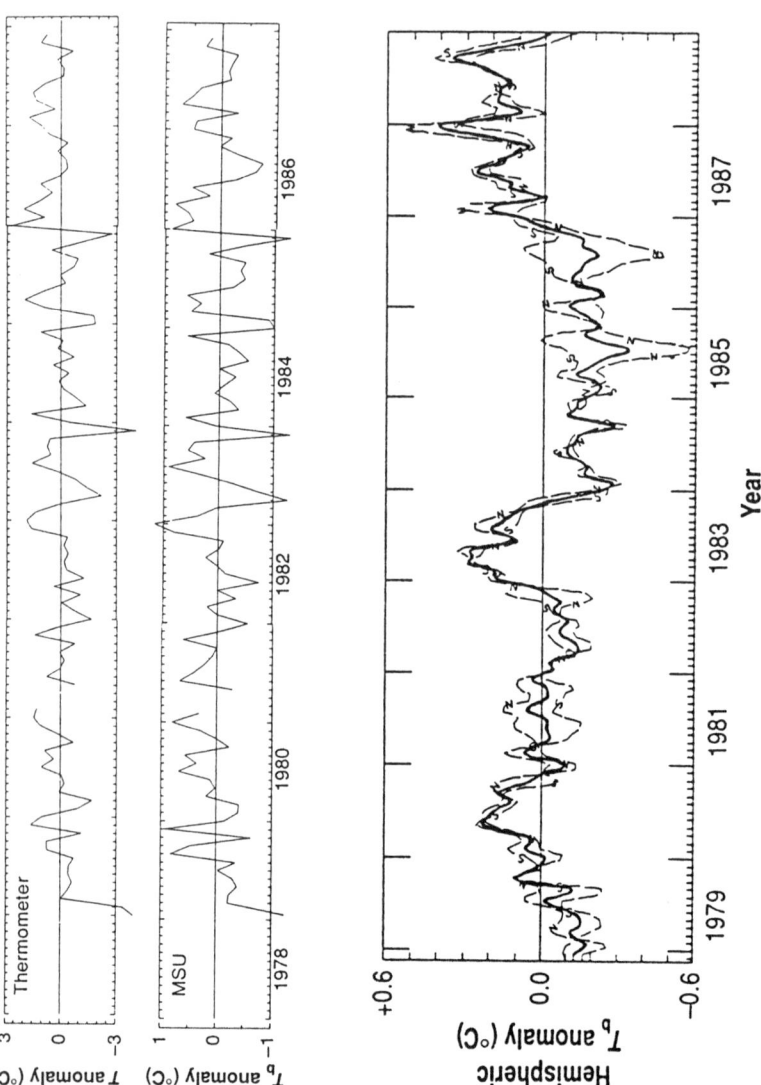

Figure 12. a: (upper panel) Comparison between monthly mean contiguous U.S.A. land air temperatures and microwave sounding unit (MSU) middle troposphere temperatures over the same region. b: (lower panel) MSU hemispheric anomalies for northern (N) and southern (S) hemispheres and the global mean (heavy line). Both figures from Spencer and Christy.[23]

paring these statistics, recall that if there are trends in data sets, the correlation will be higher than if detrended sets are used.

Figure 12b shows the MSU results for each hemisphere during 1979-1988 as well as the global average. These temperature anomalies from the Spencer and Christy[23] analysis show no trend over the decade of data availability, but as with the SST studies of Strong[21] and Reynolds et al.[22], 10 years is much too short a timeframe in which to discuss climate trends. Many of the features in the Reynolds et al. analysis of SST (q.v. Figure 11b) are also apparent in Spencer and Christy's results. The MSU does seem to have great potential for monitoring Earth's middle tropospheric temperatures, but a simple interpretation of MSU data should be viewed with caution because the measurement is not a direct analog for classical LAT/MAT observations.

SEA LEVEL MEASUREMENTS 1880-1990

In many respects the climate history of Earth is reflected in the waxing and waning of sea level because water bound up on continental and mountain ice sheets comes from and returns to the sea. During the Holocene (q.v. Figure 2), sea level rose ~120 meters or more in 18,000 years. The English Channel came into being only 8,000 ybp, and Florida was more than twice its current surface area when New York and Moscow were under tons of ice. Sea level change then is as much a part of the natural cycle of climate change as are differing patterns of temperature and precipitation and the rich history of human migration that seems to accompany these changes. In this section the notion of sea level change is explored with respect to the debate of whether or not global sea level is capable of sensing climate change.[26] To judge the scale factor of relative sea level changes of centimeters to be discussed one should keep the Barbados sea level curve shown in Figure 2 in mind.

Relative sea level is defined as the change in water level measured at the coast by an instrument that is geodetically tied to survey benchmarks on land; RSL is distinguished from "absolute sea level" which is measured with respect to the geodetic surface known as the ellipsoid. The water level measuring instrument is known as a tide gauge, and a sketch of one is given in Figure 13.[27] Readings from the tide staff are recorded on the marigram either by a human observer or through some instrumentation; it is the tide staff that is connected to the vertical geodetic datum. Although some of these electromechanical gauges are being replaced by acoustic gauges, the historical record of sea level relative to the fixed survey benchmarks forms the data base from which changes are being deduced. Records of monthly and annual mean sea level for hundreds of stations around the globe are in the archives of the Permanent Service for Mean Sea Level (PSMSL, Table I). The longest PSMSL record is from Brest, France, and dates back to 1807.

Relative sea level can be expressed as the sum of eight height (H) terms including (respectively) tides, air pressure, winds, eustatic change, steric change, circulation, and geologic motion:

$$RSL = H_T + H_P + H_W + H_E + H_S + H_C + H_G + R,$$

where R is an unexplained residual height. For studies of climate change the high and medium frequency tides (semidiurnal, diurnal, fortnightly, monthly, and annual) are filtered out, and H_T is mostly due to the 18.61-year lunar nodal tide. H_P is due to the static inverted barometer effect where RSL rises ~1 cm for a decrease in atmospheric pressure of 1 mb. Wind effects, H_W, are often seen during storms, but certain semi-enclosed oceanic basins such as the Caribbean Sea have a considerable quasi-permanent effect due to the prevailing winds. Eustatic change is that caused by global changes in oceanic water volume due to glaciation or basin volume changes; the curve in Figure 2 is an example of H_E affecting RSL. H_S is height change

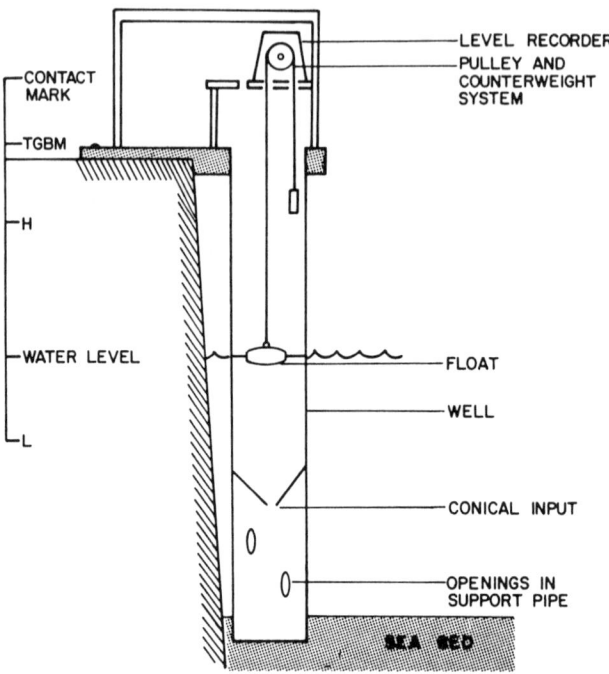

Figure 13. Elements of a stilling well tide gauge showing the water level recorder where heights are recorded on the marigram. A tide staff is affixed to the gauge support structure against which all levels and geodetic control are referenced (after Pugh[27]).

due to changes in temperature and/or salinity of the water column itself, and is of particular interest in global warming discussions. RSL also changes in response to the dynamic effects of oceanic circulation, and if the currents change due to changed wind-forced or thermohaline motions, H_C will reflect that too. Finally the issue of H_G or geologically-caused changes in RSL, either due to coastal subsidence or tectonic effects, adds to the final accounting.

Figure 14 shows the variety of RSL annual mean values along the west coast of North America.[28] Most of the variability in RSL due to H_T, H_P, and probably H_W and H_C, is removed by computing annual mean values, but the contribution from H_G probably dominates the signal. RSL is seen to be falling at several sites, rising at others, and fairly steady at at least two others. This is typical of the global situation and has led some researchers to organize the data by broad regions or to select tide gauge sites that are least likely to have H_G dominate the RSL signal.[29] When broad regions of the coast have vertical motion, the geodetic ties between the tide staff and the benchmarks will not reveal local motion because all the benchmarks are moving together. Attempts to account for this motion by direct measurement[30] are being implemented,[31] and will be discussed later.

In attempts to quantify the value of H_G, geophysicists are constructing models of Earth's crustal motion due to removal of the ice sheets associated with Holocene glaciation. Figure 15 (C.G.A. Harrison, U. Miami, personal communication) shows two modeled rates of vertical land motion due to glacial rebound along the east coast of the U.S.A.[32,33] The sign convention on Figure 15 is that positive values signify land that is sinking (*i.e.*, RSL rising if all other terms are zero). Peltier's model[32] in general shows positive rates of rebound south of Delaware, whereas Nakiboglu and Lambeck (N&L)[33] show negative values. Maul and Hanson[34] found that Miami RSL and oceanic steric height (H_S) east of Abaco Island, The Bahamas (see map insert on Figure 15) had linear trend differences for the 1950–1987 time frame that agreed well with the H_G value from Peltier.[32] Thus, for the Miami/Abaco study, RSL $\simeq H_S + H_G$, using two lunar nodal 18.61-year cycles to minimize H_T in the balance equation. These glacial rebound models do not account for the massive rates of RSL rise (> 10 mm/yr) seen in deltaic coasts such as the Mississippi or the Nile, nor on other coasts where subsidence is prevalent.

Much of the material on global sea level rise and climate change has been reviewed by Warrick and Oerleman[35] in the IPCC report, and Gornitz *et al*.[36] made an assessment of 1.2 mm/yr rise that seems to be close to the median range given by various authors. As with the problems of defining "global" LAT or MAT or SST, sea level stations are not uniformly distributed, and naturally cluster in the Northern Hemisphere. Figure 16 shows the distribution of sea level gauge sites with records of 20 years or longer in the upper panel, and two estimates of "global" RSL change quoted by Warrick and Oerlemans. Even though the methods of analysis and station acceptance or rejection criteria differ, there is almost universal agreement that global RSL has been rising since the turn of the century, but (*q.v.* Figure 14) RSL change at individual stations may not reflect the "global" trend. From a human socioeconomic perspective[37] it is the relative change that is important but, for the scientific problem of global change, the RSL data may still be dominated by H_G, *i.e.*, by diastrophism.

Trend analysis in these data is also a non-trivial issue. Many reports give the linear regression coefficient but neglect to discuss the error bounds or the percent variance explained. Maul and Hanson[34] used several statistical techniques to address these issues, and applied the "bootstrap" Monte Carlo approach of Efron[38] to the problem. In the bootstrap, a random number generator is used in a variance preserving sample and replace scheme from which the frequency distribution of linear trends is calculated. The bootstrap suggests that standard statistical analysis underestimates the regression coefficient errors, but this is probably

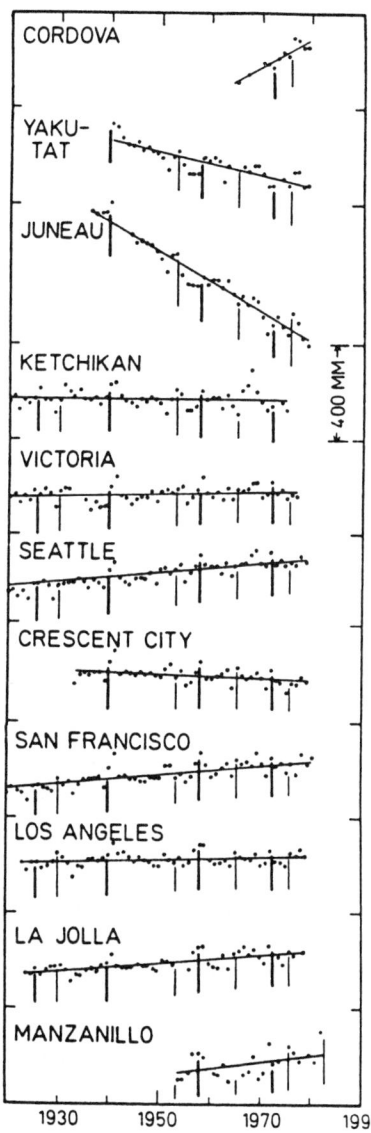

Figure 14. Annual mean relative sea level along the west coast of North America (dots) with linear regression lines fitted to each tide gauge station's data (from Emery and Aubrey[28]).

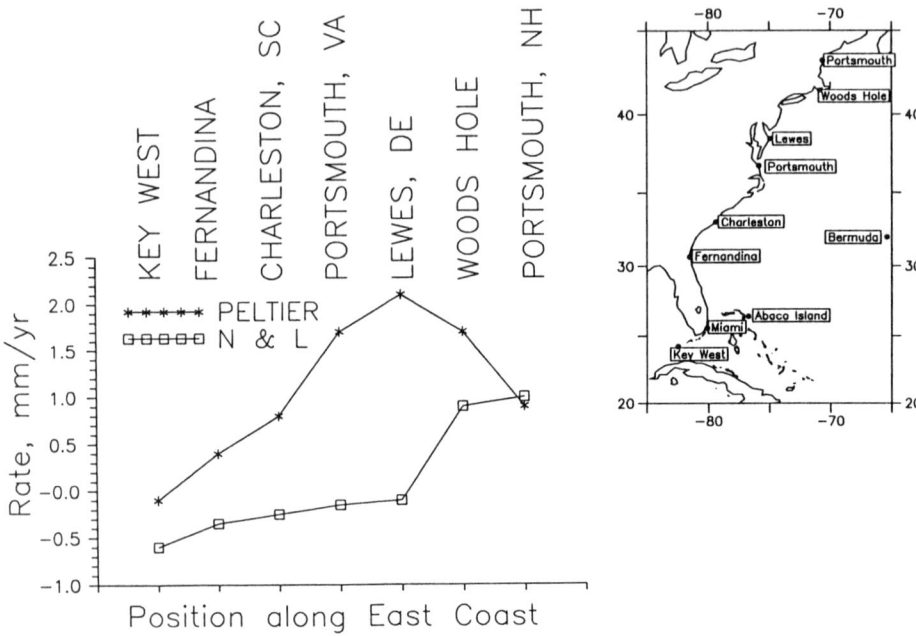

Figure 15. Modeled post-glacial rebound for the U.S.A. east coast from Peltier[32] and Nakiboglu and Lambeck;[33] re-drawn from C.G.A. Harrison (personal communication).

due to serial correlation in these geophysical data sets.[15] The other issue Maul and Hanson addressed was that both the record length and the central date of the record were important in estimating trend. Choosing multiples of $H_T = 18.61$ years eliminates much of the trend due to tidal-frequency cyclical variations, but even with a 37-year record window (two nodal cycles), linear RSL trends for the southeastern U.S.A. varied from −1 mm/yr to over +4 mm/yr depending on the central date of the data window. Note that few of the PSMSL stations have 20 year record lengths (q.v. Figure 16), and short records are notorious in giving trends that reflect interannual variability rather than climate change (the same southeastern U.S.A. data give trends ranging from −3 mm/yr to +8 mm/yr using 19-year record segments).

Regardless of the difficulties in determining rates of RSL, the total sea level rise over the last 100 years seems to be accepted at 10-20 cm. Table III[35] summarizes the contributions to sea level rise from thermal expansion (H_S in our parlance), glaciers and small ice caps, the Greenland ice sheet, and the Antarctic ice sheet; the latter three items are contained in the H_E term of the RSL equation. Many glaciologists think that Antarctica has not contributed to RSL rise, and in fact may (with global warming) be a net sink of water due to increased precipitation; Hanson[39] observes no temperature trend at South Pole Station, from 1957-1989. There is little disagreement that alpine glaciers are receding, and have contributed as much as 40% to sea level rise, and that Greenland is also an important source to H_E. Oceanographically, the H_S term is the most interesting because steric change can reflect changes in the thermohaline circulation as well as simple expansion of the water column due to heating or freshening.

There are very few places in the ocean where data are available to study decadal scale change in the temperature and salinity from the ocean's surface to the bottom. Off Bermuda (q.v. Figure 15), a deep oceanographic location has been resampled fortnightly by ship since

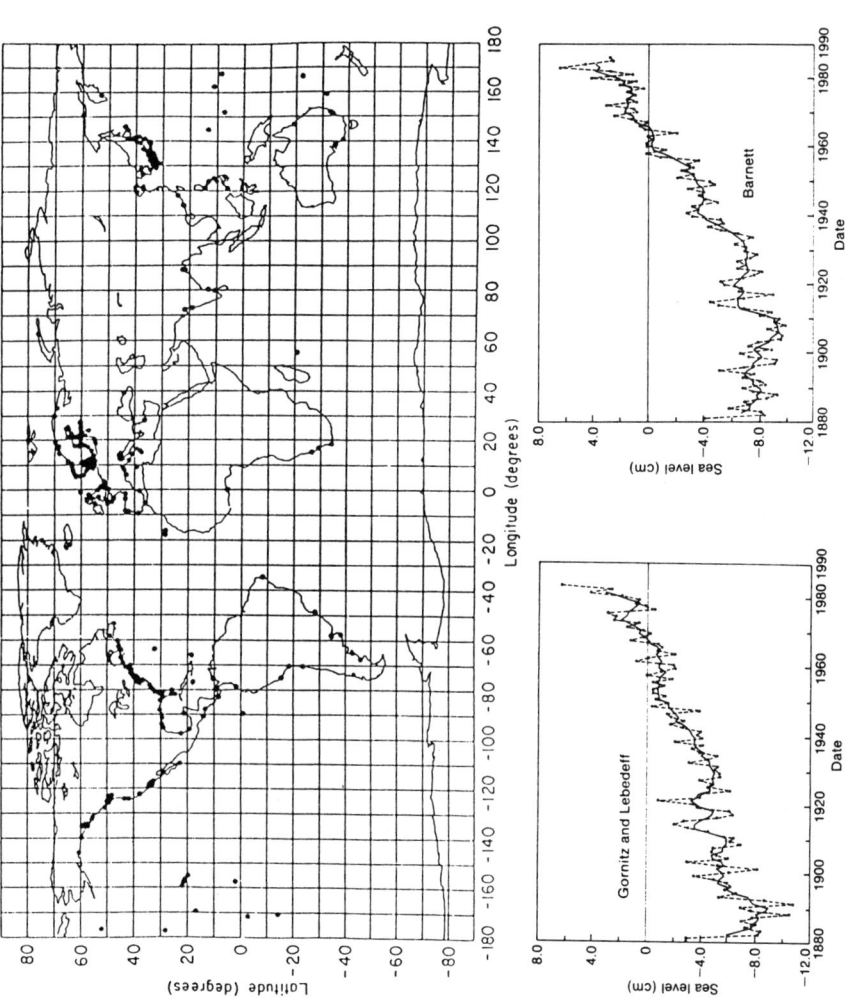

Figure 16. a: (upper panel) Distribution of sea level stations with 20 years or longer records. b: (lower panels) Two independent estimates of global RSL, both quoted by Warrick and Oerlemans.[35]

Table III. Estimated contributions to sea level rise (cm) over the last 100 years (from Warrick and Oerlemans[35]).

Source	Low (cm)	Best (cm)	High (cm)
Thermal Expansion	2.0	4.0	6.0
Alpine Glaciers	1.5	4.0	7.0
Greenland Ice Sheet	1.0	2.5	4.0
Antarctic Ice Sheet	−5.0	0.0	5.0
TOTALS	−0.5	10.5	22.0

1954. From the measurements of temperature and salinity at numerous pressure (p) levels (surface to bottom), a time history of H_S can be constructed from integration of the hydrostatic equation:

$$H_S = \int g^{-1} (a - a_0) \cdot dp$$

where a is the specific volume (reciprocal of density) of sea water for a given temperature, salinity, and pressure, a_0 is the specific volume for 0°C, 35 PSU (practical salinity units; 35 PSU $\simeq$ 35 grams salt per kilogram water) as a function of p, and g is the acceleration of gravity. The units of H_S are dynamic centimeters (dyn-cm), and are approximately equal to geometric centimeters; the range of H_S globally is ±100 dyn-cm.

A study in 1990 by Roemmich[40] (*cf.* Schroeder and Stommel[41]) illustrates the value of comparing H_S calculated as in the above equation with RSL from the Bermuda tide gauge. The left-hand panel of Figure 17 shows the range in H_S at the Bermuda oceanographic station from the surface to p = 2000 db (1 db $\simeq$ 1 m water depth). These values are the ensemble of annual means from 1955 to 1981. Roemmich fitted empirical orthogonal functions (EOFs) to these yearly steric height curves and found that most of the year to year variability comes from two depths: the upper ($\sim$100 m) mixed layer, and the main thermocline at $\sim$800 m. The annual EOF describes about 62% of the variance in H_S, and shows that most of the long-term trend in these data also comes from 100 m and 800 m, respectively. The right-hand panel of Figure 17 shows time series plots of five year running means of RSL at Charleston and Bermuda, of H_S at Bermuda, and [RSL (Bermuda)−H_S] with a linear trend line. The trend line has a slope of 1 dyn-mm/yr, which is approximately the value Peltier[32] models for glacial rebound at Bermuda. While there does seem to be a phase difference between RSL (Bermuda) and H_S, the spectrum between them (not shown) is highly coherent at all frequencies from the Nyquist to the lowest. One possible cause of the phase difference is that the Bermuda H_S data only go to 2,000 meters depth, and much of the ocean thermohaline change is contained in North Atlantic Deep Water, which has a bottom depth of $\sim$5,000 meters.

Folland *et al.*[6] compared the changes in the areally averaged open water of the North Atlantic and the North Pacific Oceans from 1957 to 1981 (Figure 18). The upper layers of both oceans (< 500 m) are cooler in 1981 than they were 25 years earlier. However, the deeper waters of the North Atlantic show marked warming between 1957 and 1981, whereas the North Pacific is essentially unchanged below the main thermocline. In another study, Levitus[42] showed that the steric height of the western North Atlantic had markedly decreased ($\sim$15 dyn-cm) between pentadal averages of 1955-1959 and 1970-1974. This is clear in Bermuda RSL (*q.v.* Figure 17), and probably is best interpreted as interannual variability in the natural functioning of the oceanic circulation rather than being indicative of long-term change.

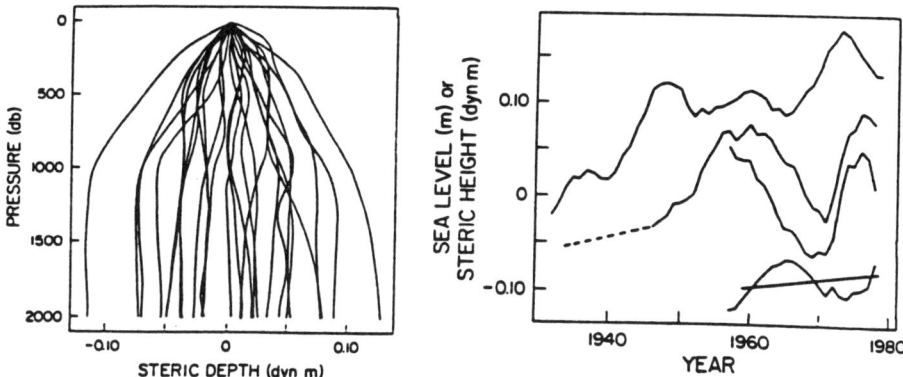

Figure 17. a: (left panel) Annual mean steric heights (H_S) from the Bermuda serial oceanographic station (known as the "Panulirus" station or station "S"[41]). b: (right panel) Top to bottom: RSL at Charleston; RSL at Bermuda; H_S at Bermuda; Bermuda RSL minus Bermuda H_S.[40]

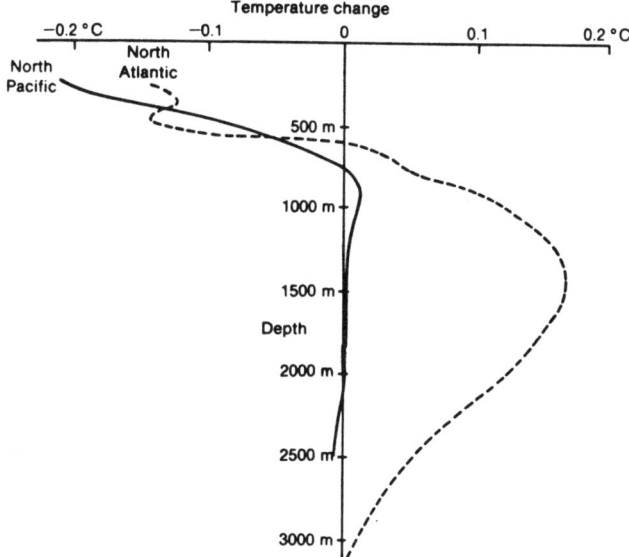

Figure 18. Difference in subsurface temperature in the North Atlantic (dashed) and the North Pacific (solid) for 1981 minus 1957 (from Folland et al.[6]).

The point to be made here is that deep ocean data are scarce and studies of climate scale change are probably better served by analysis of RSL for the near future.

Is there then evidence of acceleration in sea level rise since the start of industrialization? Woodworth[43] attempted to answer the question by least squares fitting an equation of the form

$$RSL = C_1 + C_2(y-1900) + C_3(y-1900)^2,$$

where y-1900 is year minus 1900, to numerous RSL records from western Europe. Woodworth evaluated the regression coefficient C_3 and its standard error (ϵ), and applied standard statistical tests to the null hypothesis that $C_3 \pm \epsilon$ was not significantly different from $C_3 = 0$. Results were mixed, as one might understand considering the tectonic complications of European glacial rebound, and Woodworth concluded that there was no clear evidence for acceleration in RSL.

Studying the geophysical record for climate change is so important that the procedure recommended by the World Meteorological Organization is summarized in Table IV, and illustrated in Figure 19. The upper panel of Figure 19a is the RSL curve from Fernandina Beach, Florida (see Figure 15 for location), which is remarkably similar to RSL at Charleston (Figure 17). Three least squares curves are fitted to Fernandina RSL: linear, second order (as in the above equation), and third order. If the Woodworth[43] question is asked, then $C_3 = 0.0017 \pm 0.006$, and Student's t-test would indicate that C_3 was significantly different from zero. On the other hand, if the 95% confidence interval in the linear correlation coefficient ($0.88 \geq r \geq 0.73$) is compared with the second order 95% interval ($0.89 \geq r \geq 0.76$), the conclusion is that there is no significant difference. A similar test with the third-order regression ($0.90 \geq r \geq 0.79$) leads to the same conclusion. If the linear trend is first removed, then standard statistical testing suggests that $C_3 \pm \epsilon$ for Fernandina RSL is within the confines of the null hypothesis.

Inset into Figure 19a is an example of the "bootstrap" of Efron[38] applied to the linear trend at Fernandina. Fernandina RSL has N = 91 years of annual mean values. In this calculation, 91 random sample and replace selections were made, the linear regression coefficient

Table IV. Summary of optimum procedure for analysis of climatological time-series (abstracted from WMO[15]).

---

1. Verify the homogeneity or uniform representativeness of the time-series.
2. Establish the probable form of the time-series' frequency distribution; specifically is it a Gaussian distribution?
3. Compute the power spectrum.
4. Compute the autocorrelation function and examine the serial correlation coefficient (r_1) for Markov-type persistence.
5. If r_1 is found to be consistent with persistence, derive the theoretical Markov spectrum appropriate to the value of r_1.
6. Test the time-series' spectrum for statistically significant departure from the null-hypothesis Markov spectrum.
7. If the lowest frequencies exceed the local value of the Markov spectrum by statistically significant amounts, accept the alternative hypothesis that a secular trend is present in the series.
8. If the spectrum at wavelengths considerably shorter than the record length are found to exceed the local value of the Markov spectrum by statistically significant amounts, accept the alternative hypothesis that quasi-periodicity is present in the record.
9. Search for the physical significance of any non-randomness in the time-series.

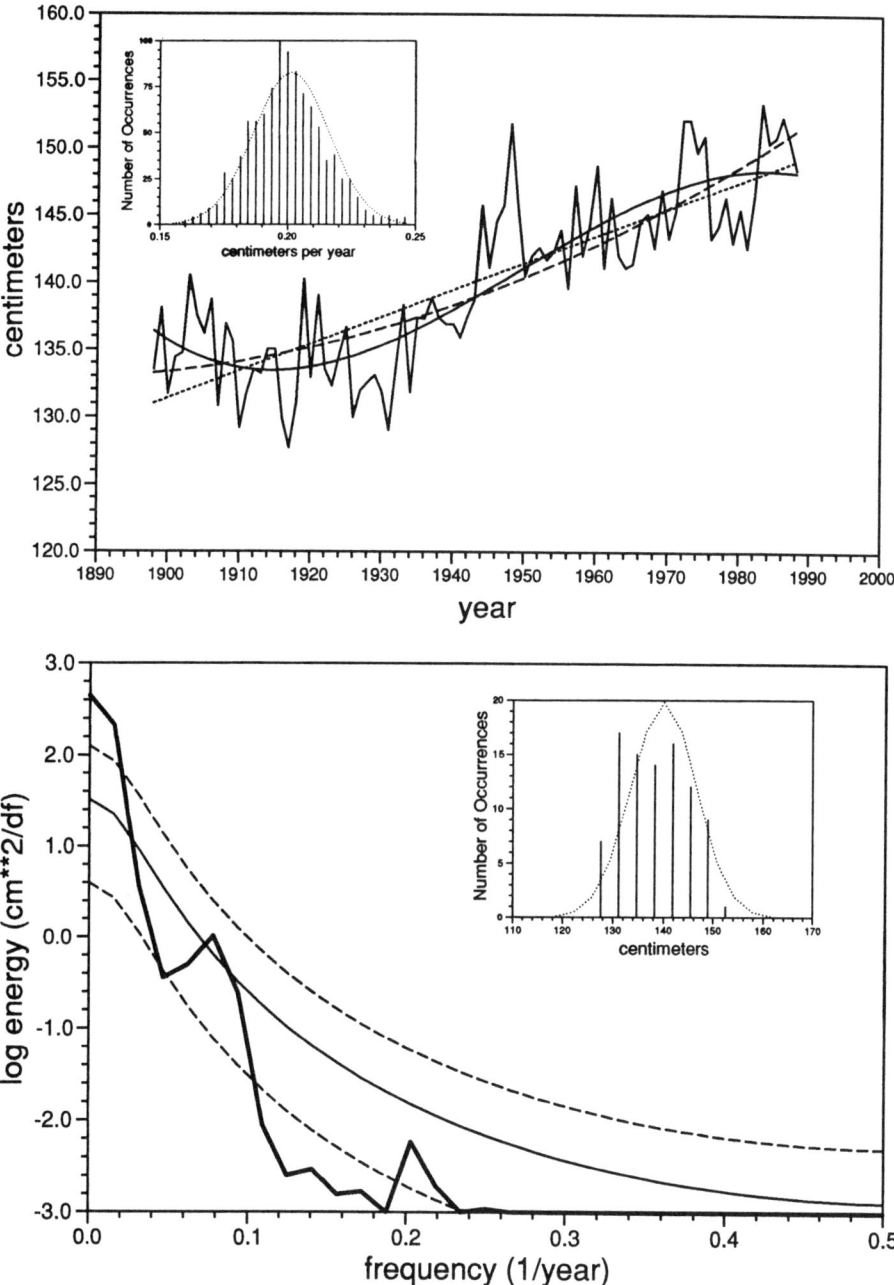

Figure 19. a: (upper panel) Annual mean sea level at Fernandina Beach, Florida, showing first (dotted), second (dashed), and third (solid) order polynomial fits. b: (lower panel) Spectrum of Fernandina RSL, Markov spectrum (±95% confidence limits dashed), and inset, frequency distribution of RSLs.

is calculated, 91 more random sample and replace selections were made, and the calculation repeated 999 more times. The inset shows the frequency distribution of the 1,000 estimates of the linear trend with the Gaussian distribution (dots) shown as an overlay. The ensemble mean trend is 0.201 cm/yr±0.015 cm/yr at the ±1σ level, and the distribution appears to be normal with a small amount of skewness. The bootstrap thus provides a handy graphical assessment of the regression coefficient (or any other parameter) in a geophysical time series, and makes no *a priori* statistical assumptions. Indeed, L1 statistics (Mean Absolute Deviation, MAD) are as readily applied with the bootstrap as these (L2) least squares statistics.

The WMO[15] emphasizes that climate data do not fit standard statistical tests because they are serially correlated; *i.e.*, year N+1 is not independent from year N due to persistence. To study this we compute r_1, the serial correlation coefficient from

$$r_1 \simeq [\{N^2/(N-1)\} \cdot \Sigma x_i\, x_{i+1} - (\Sigma x_i)^2]/[N\Sigma x_i^2 - (\Sigma x_i)^2]$$

which is the approximation to the "non-circular" defined r_1. If $r_2 \simeq r_1^2$ and $r_3 \simeq r_1^3$, etc., then the appropriate null continuum (S_k) is assumed to be Markov "red noise," and is given by

$$S_k = \overline{S}\,[(1-r_1^2)/(1+r_1^2 - 2r_1\cos[\pi k/m])]$$

where $\overline{S}$ is the average of the m+1 raw spectral estimates in the computed spectrum under consideration. The degrees of freedom (ν) for S_k with record length N and a maximum lag of m units is approximately

$$\nu = [2N - m/2]/m.$$

Confidence limits for S_k are computed from ν using the 5 and 95% points of the χ^2/ν distribution from appropriate statistical tables.[15] In the lower panel of Figure 19 is plotted the spectrum for Fernandina sea level, S_k, and the 95% confidence limits (dashed) for the Markov spectrum.

Using Table IV as a guide, results in Figure 19b (the lower panel) are summarized as follows. The histogram inset in the spectrum clearly shows that the distribution of Fernandina RSL is not Gaussian. In fact, there is so much persistence that the autocorrelation function (not shown) never becomes less than 0.7; this is caused by the trend, however, but even when detrended, the e-folding autocorrelation exceeds three years. Only the lowest frequency exceeds the 95% confidence limit of the Markov spectrum, and (*q.v.* Table IV) one concludes there is a real trend. Several peaks in the spectrum of detrended Fernandina RSL (not shown) are larger than the Markov spectrum, notably at ~13 years and at ~50 years. There is no evidence of H_T, the 18.61-year nodal tide, and the ~50 year peak is very close to the second harmonic of this 91-year record. Since detrended Fernandina RSL has ~50-year or longer spectral energy, a C_3 residual could be found that is caused by a short record with low frequency energy. Thus, there is no unequivocal evidence for acceleration in this record, but the physical significance of the 13-year peak needs to be investigated.

The penultimate issue is that of future observations, particularly space-based measurements. The use of satellite altimetry to study sea level variability is probably confined to analysis of interannual and shorter time scales and ocean-basin and shorter length scales.[26] The centimeter level of accuracy required over decadal time scales seems beyond the altimetry state of the art. Space-based geodetic techniques use a combination of Very Long Baseline Interferometry (VLBI) with quasars as a radio source, the Global Positioning System (GPS) for short baseline ties between tide gauges and VLBI sites, and Satellite Laser Ranging (SLR). Superconducting gravimeters and absolute gravimeters at fixed VLBI/GPS/SLR sites are used for improved precision. The combination seems capable of measuring H_G adequately for climate studies.[26,44]

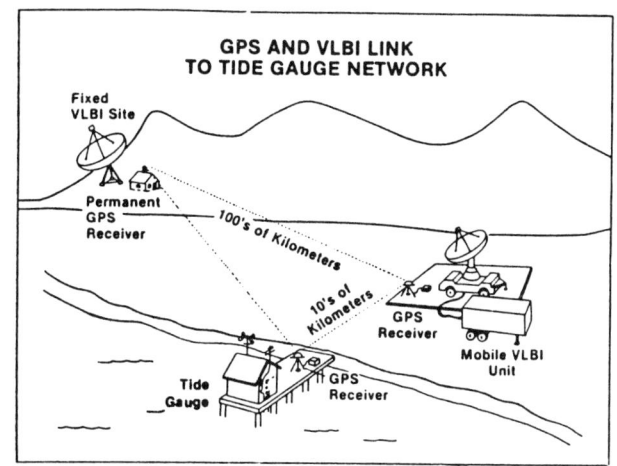

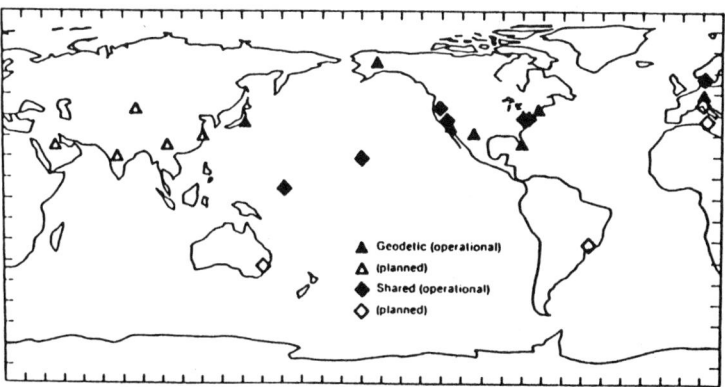

Figure 20. Upper panel: Sketch of the scheme of tying RSL to a global reference frame using very long baseline interferometry and the global positioning system. Bottom Panel: Existing and planned VLBI sites (from Diamonte et al.[30]).

The upper panel of Figure 20 is a sketch from Diamonte et al.[30] showing how a VLBI/GPS tie will be made, and Table V from Bilham[44] is a summary of the errors. If all the errors in Table V are random, then one can improve the accuracy of the VLBI/GPS/SLR/absolute-g tie by $\sqrt{N}$ where N is the number of geodetic ties; an experiment is now being conducted whereby a GPS receiver is "permanently" installed at the Miami tide gauge for correlation with the Richmond, Florida, VLBI/GPS/SLR/absolute-g site 20 km away. One important aspect of measuring absolute sea level is that only a dozen or so tide gauges (see bottom panel of Figure 20) will be within 100 km of the VLBI/GPS/SLR/ absolute-g sites, and only coastal or island sites will be included. It is entirely possible that global sea level will change in a manner not fully observed by such coastal instrumentation. Hence, altimetry will be incorporated in analysis of global change, albeit with increased uncertainty.

Table V. Estimated uncertainties (mm) in relating a coastal sea level measurement to a global datum (after Bilham[44]).

Error Source	Optimistic (mm)	Pessimistic (mm)
VLBI/SLR/absolute-g datum	6.0	15
Ground ties VLBI/SLR	0.5	1
Phase center GPS antenna to ground	0.5	2
GPS/GPS Link	10.0	30
Phase center GPS antenna to tidal datum	0.3	1
Tidal datum to sea level transducer	0.5	2
Local corrections to annual $\overline{RSL}$	5.0	10
TOTAL	22.8	61
σ if all uncertainties are random	±12.7	±35

Tablel VI. Summary of principal conclusions.

1. Measurements of "global" land, marine air, and sea surface temperatures show approximately a 0.3°C–0.5°C increase during the last century.
2. U.S.A. regional land air temperature measurements show no statistically significant trend during the last century.
3. Changing observing practices, changing spatial sampling distributions, and changing technology all add to the uncertainty of temperature determinations.
4. Systematic errors may still exist in temperature records from land and sea.
5. Satellite sea surface temperatures show mixed results but satellite middle tropospheric air temperatures show stable results.
6. The average rate of sea level rise has been 1–2 mm/yr during the last century.
7. Most island and coastal tide gauge sites have significant vertical land motion due to subsidence, post-glacial rebound, and/or plate tectonics.
8. There is no firm evidence of acceleration in sea level rise.
9. Both temperature and sea level records are complicated by considerable interannual, decadal, and lower frequency variability, which are inadequately understood physical processes.

CONCLUSIONS

To facilitate discussions the principal conclusions of this chapter are given in Table VI. In general, these conclusions lean toward the lower values of the many opinions surveyed herein because measurements of LAT, MAT, and SST all have been shown to be systematically biased towards higher values in the later half of the century. For the contiguous U.S.A., which has the most developed climate monitoring network, the data show a warming in the western states which is balanced by a cooling in the eastern states.[45] The net U.S.A. temperature trend is zero.[14,45] Recent studies of natural variability of the climate system[46] suggest that natural trends of up to 0.3°C per century are quite expected, and this value is at the lower limit of the "global" increase Earth is probably experiencing in the 20th century.

Peltier and Tushingham[47] argue that a "global" sea level rise has occurred that is not subject to the same type of temperature errors discussed, and that sea level may be an unequivocal harbinger of greenhouse warming. Douglas[48] determined global sea level rise to be

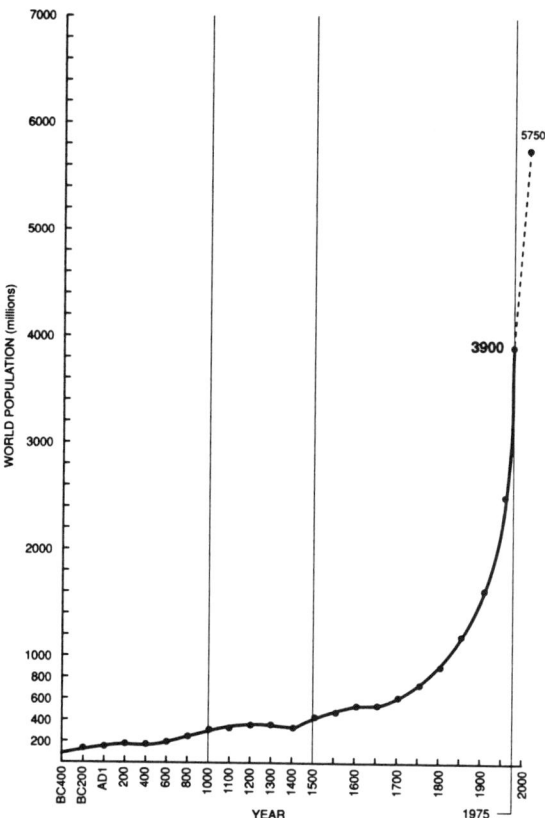

Figure 21. Human population during the last two millennia (from NAS[5]).

1.8 mm/yr ± 0.1 mm/yr using yet another Tushingham and Peltier post glacial rebound model, and by selecting stations that were removed from converging tectonic plates. On balance, however, not all geodynamic models agree.[32,33] Many of the conclusions concerning sea level rise hinge on these models, just as many of the CO_2 doubling forecasts depend upon models that have many disagreements.[4]

It is clear from the summary in Table VI that a consistent argument can be made for global warming. The student of the history of science is also aware that, less than 15 years ago, the popular press was telling us that we were possibly on the brink of another ice age,[49] or at least that Earth was cooling![9] Wyrtki[37] concluded an article on Pacific sea level rise with a graph showing the exponential growth of the human population (cf. Figure 21)[5] and made the point that resource limitations will probably be of more immediate socioeconomic concern than climate change. J. H. Ausubel even asks "Does climate [*change*] still matter?"[50,51] Wisely, however, the United Nations Environment Programme established a series of "Implications of Climatic Change" studies based on the 1985 UNEP/WMO/ICSU scenario of a global 20 cm sea level rise and a 1.5°C temperature rise by 2025 AD. Both the 20 cm and the 1.5°C "global" rises require a nonlinear extrapolation of the observations to reach these levels in the next 35 years, but records (*q.v.* Figures 4, 5a, 7, 12b, 14, 16b, 17b, 19a) are so dominated by interannual variability that detection of accelerations in trends is not statistically significant.

Much of "the great climate debate," as National Academy of Engineering President Robert M. White has written, centers on the question: "Is the rise in global temperatures a natural fluctuation or a result of the increase in greenhouse gases?"[3] White answers the question that our research efforts should be directed toward better modeling and better comparisons between predictions and observations. Coupled ocean–atmosphere models are just becoming available, and they are giving surprising results. Stouffer et al.,[52] for example, report a marked and unexpected interhemispheric asymmetry in air temperature rise due to CO_2 doubling using a coupled model. Other coupled models show that the notion of an eustatic or global sea level rise is also incorrect; there is no reason to expect that the ocean will respond uniformly to CO_2 doubling. Another model by Stocker and Wright[53] warns that fresh water flux in the high latitude North Atlantic Ocean controls the thermohaline circulation; should we be paying more attention to precipitation than temperature or sea level change?

There are many other physical problems to be solved. Temperatures today, for example, have risen back almost to the level reached during the Medieval Warm Period. The Medieval Warm Period preceded the Little Ice Age (q.v. Figure 1), which was a devastating event in human history.[8,9] There is no physical explanation as yet that fully elucidates this 1°C temperature swing over ~700 years; anthropogenic activity cannot be the cause. Predicting future climate before being able to adequately explain such a significant event as the Medieval Warm Period–Little Ice Age oscillation seems discomforting. For physicists, both theoretical and observational challenges abound, but beware: no oceanographer or meteorologist has ever been awarded the Nobel Prize in physics.

ACKNOWLEDGMENTS

Preparation of this chapter has benefited from many conversations with K.J. Hanson, C.G.H. Rooth, D.V. Hansen, and S.R. Baig. I am particularly indebted to T.M.L. Wigley and T.R. Karl who sent me numerous reprints and preprints of their very significant work. Finally, I express my appreciation to B. Levi and D. Hafemeister and the APS Forum on Physics and Society for their invitation to conduct this study and for their editorial efforts, and to G. Derr for word processing, D. Senn for figure layout, and A. Ramsay for photography.

REFERENCES

1. La Brecque, M. Detecting Climate Change, I and II. *Mosaic*, 20(4):2–17, 1989.
2. Schneider, S.H. The Greenhouse Effect: Science and Policy. *Science*, 243:771–781, 1989.
3. White, R.M. Greenhouse Policy and Climate Uncertainty. *Bull. Amer. Meteorol. Soc.*, 70(9):1123–1127, 1989.
4. Houghton, J.T., G.J. Jenkins, and J.J. Ephraums (eds.). *Climate Change, The IPCC Scientific Assessment*. ©Intergovernmental Panel on Climate Change, Cambridge Univ. Press, 356 pages, 1990.
5. NAS (National Academy of Sciences). *Policy Implications of Greenhouse Warming*. National Academy Press, Washington, 127 pages, 1991.
6. Folland, C.K., T.R. Karl, and K.Ya. Vinnikov. Observed Climate Variations and Change. Chapter 7: Houghton, J.T., G.J. Jenkins, and J.J. Ephraums (eds.), *Climate Change, The IPCC Scientific Assessment*. ©Intergovernmental Panel on Climate Change, Cambridge Univ. Press, 195–238, 1990.
7. Fairbanks, R.G. A 17,000-Year Glacio-Eustatic Sea Level Record: Influence of Glacial Melting Rates on the Younger Dryas Event and Deep-Ocean Circulation. *Nature*, 342:637–642, 1989.
8. Ladurie, E.L. *Times of Feast, Times of Famine: A History of Climate since the Year 1000.* ©Noonday Press, New York, 438 pages, 1988.

9. Lamb, H.H. *Climate, History, and the Modern World.* ©Methuen, London and New York, 387 pages, 1982.
10. Hansen, J., and S. Lebedeff. Global Trends of Measured Surface Air Temperature. *J. Geophys. Res.*, 92(D11):13,345-13,372, 1987.
11. Wigley, T.M.L. Measurement and Prediction of Global Warming. Jones, R.R. and T. Wigley (eds.), *Ozone Depletion: Health and Environmental Consequences.* ©John Wiley & Sons, Chichester, 85-97, 1989.
12. Jones, P.D., T.M.L. Wigley, and G. Farmer. Marine and Land Temperature Data Sets: A Comparison and a Look at Recent Trends. M.E. Schlesinger (ed.), *Greenhouse-Gas-Induced Climatic Change: A Critical Appraisal of Simulations and Observations.* ©Elsevier, Amsterdam (in press), 1991.
13. Vinnikov, K.Ya., P.Ya. Groisman, K.M. Lugina, and A.A. Golubev. Variations in Northern Hemisphere Mean Surface Air Temperature over 1841-1985 (in Russian). *Meteorology and Hydrology,* 1:45-53, 1987.
14. Hanson, K., G.A. Maul, and T.R. Karl. Are Atmospheric "Greenhouse" Effects Apparent in the Climatic Record of the Contiguous U.S. (1895-1987)? *Geophys. Res. Lett.,* 16(1):49-52, 1989.
15. WMO (World Meteorological Organization). *Climatic Change,* Report of a Working Group of the Commission for Climatology. WMO Technical Note No. 79, 82 pages, 1966.
16. Solow, A.R. Testing for Climate Change: An Application of the Two-Phase Regression Model. *J. Clim. Appl. Meteorol.,* 26:1401-1405, 1987.
17. Woodruff, S.D., R.J. Slutz, R.L. Jenne, and P.M. Steurer. A Comprehensive Ocean-Atmosphere Data Set. *Bull. Amer. Meteorol. Soc.,* 68(10):1239-1250, 1987.
18. Oort, A.H., Y.H. Pan, R.W. Reynolds, and C.F. Ropelewski. Historical Trends in the Surface Temperature over the Oceans based on the COADS. *Climate Dynamics,* 2:29-38, 1987.
19. Jones, P.D., T.M.L. Wigley, and P.B. Wright, Global Temperature Variations between 1861 and 1984. *Nature,* 322(6078):430-434, 1986.
20. Karl, T.R., and P.D. Jones. Urban Bias in Area-Averaged Surface Air Temperature Trends. *Bull. Amer. Meteorol. Soc.,* 70(3):265-270, 1989.
21. Strong, A.E. Greater Global Warming Revealed by Satellite-Derived Sea Surface Temperature Trends. *Nature,* 338:642-645, 1989.
22. Reynolds, R.W., C.K. Folland, and D.E. Parker. Biases in Satellite-Derived Sea-Surface-Temperature Data. *Nature,* 341:728-731, 1989.
23. Spencer, R.W., and J.R. Christy. Precise Monitoring of Global Temperature Trends from Satellites. *Science,* 247:1558-1562, 1990.
24. Maul, G.A. *Introduction to Satellite Oceanography.* ©Martinus Nijhoff Publishers, Dordrecht/Boston/Lancaster, 606 pages, 1985.
25. Deepak, A. (ed.) *Inversion Methods in Atmospheric Remote Sensing.* ©Academic Press, New York/San Francisco/London, 622 pages, 1977.
26. Eden, H.F. (ed.) *Towards an Integrated System for Measuring Long-Term Changes in Global Sea Level.* ©Joint Oceanographic Institutions, Washington, DC, 178 pages, 1990.
27. Pugh, D.T. *Tides, Surges, and Mean Sea Level.* ©John Wiley & Sons, Chichester, 472 pages, 1987.
28. Emery, K.O., and D.G. Aubrey. Relative Sea-level Change from Tide Gauge Records of Western North America. *J. Geophys. Res.,* 91:13,941-13,953, 1986.
29. Barnett, T.P. Estimation of "Global" Sea Level Change: A Problem of Uniqueness. *J. Geophys. Res.,* 89:7980-7988, 1984.

30. Diamonte, J.M., T.E. Pyle, W.E. Carter, and W. Scherer. Global Change and the Measurement of Absolute Sea Level. *Prog. Oceanogr.*, *18*:1-21, 1987.
31. Carter, W.E., M. Chin, J.R. MacKay, G. Peter, W. Scherer, and J. Diamonte. Global Absolute Sea Level: The Hawaiian Network. *Marine Geodesy*, *12*(4):247-257, 1988.
32. Peltier, W.R. Deglaciation Induced Vertical Motion of the North American Continent. *J. Geophys. Res.*, *91*:9099-9123, 1986.
33. Nakiboglu, S.M., and K. Lambeck. Secular Sea-Level Change. R. Sabadini and K. Lambeck (eds.), *Glacial Isostasy, Sea Level and Mantle Rheology*. ©Kluwer Academic Pub. (in press), 1991.
34. Maul, G.A., and K. Hanson. A Century of Southeastern United States Climate Change Observations: Temperature, Precipitation and Sea Level. *Global Change: A Southern Perspective*. ©Southeast Regional Climate Center, Columbia, SC, 139-155, 1990.
35. Warrick, R., and J. Oerlemans. Sea Level Rise. Chapter 9: Houghton, J.T., G.J. Jenkins, and J.J. Ephraums (eds.), *Climate Change, The IPCC Scientific Assessment*. ©Intergovernmental Panel on Climate Change, Cambridge Univ. Press, 257-281, 1990.
36. Gornitz, V., S. Lebedeff, and J. Hansen. Global Sea Level Trends in the Past Century. *Science*, *215*:1611-1614, 1982.
37. Wyrtki, K. Sea Level Rise: The Facts and the Future. *Pacific Science*, *44*(1):1-16, 1990.
38. Efron, B. *The Jackknife, the Bootstrap, and other Resampling Plans*. Society of Industrial and Applied Mathematics, SIAM Monograph No. 38, 92 pages, 1982.
39. Hanson, K.J. The South Pole Station Temperature Chronology and the Global Warming Problem. *Antarctic J. U.S.*, *25*(5), in press, 1991.
40. Roemmich, D. Sea Level and the Thermal Variability of the Ocean. Chapter 13: *Sea Level Change*, ©National Academy Press, Washington, 208-217, 1990.
41. Schroeder, E., and H. Stommel. How Representative is the Series of PANULIRUS Stations on Monthly Mean Conditions off Bermuda? *Prog. Oceanogr.*, *5*:31-40, 1969.
42. Levitus, S. Interpentadal Variability of Steric Sea Level and Geopotential Thickness of the North Atlantic Ocean, 1970-74 versus 1955-59. *J. Geophys. Res.*, *95*(C4):5233-5238, 1990.
43. Woodworth, P.L. A Search for Accelerations in Records of European Mean Sea Level. *Int'l. J. Climatol.*, *10*:129-143, 1990.
44. Bilham, R. Earthquakes and Sea Level: Space and Terrestrial Metrology on a Changing Planet. *Rev. Geophys.*, *29*:1-29, 1991.
45. Plantico, M.S., T.R. Karl, G. Kukla, and J. Gavin. Is Recent Climate Change Across the United States Related to Rising Levels of Anthropogenic Greenhouse Gases? *J. Geophys. Res.*, *95*(D10):16,617-16,637, 1990.
46. Wigley, T.M.L., and S.C.B. Raper. Natural Variability of the Climate System and Detection of the Greenhouse Effect. *Nature*, *344*(6264):324-327, 1990.
47. Peltier, W.R., and A.M. Tushingham. Global Sea Level and the Greenhouse Effect: Might they be Connected? *Science*, *244*:806-810, 1989.
48. Douglas, B.C. Global Sea Level Rise. *J. Geophys. Res.*, *96*(C4):6981-6992, 1991.
49. Bryson, R.A., and T.J. Murray. *Climates of Hunger: Mankind and the World's Changing Weather*. ©U. Wisconsin Press, Madison, 171 pages, 1977.
50. Ausubel, J.H. Does Climate Still Matter? *Nature*, *350*:649-652, 1991.
51. Ausubel, J.H. A Second Look at the Impacts of Climate Change. *Amer. Scientist*, *79*:210-221, 1991.
52. Stouffer, R.J., S. Manabe, and K. Bryan. Interhemispheric Asymmetry in Climate Response to a Gradual Increase of Atmospheric CO_2. *Nature*, *342*:660-662, 1989.
53. Stocker, T.F, and D.G. Wright. Rapid Transitions of the Ocean's Deep Circulation Induced by Changes in Surface Water Fluxes. *Nature*, *351*:729-732, 1991.

SHORT TERM CLIMATE VARIABILITY AND PREDICTIONS

J. Shukla

CENTER FOR OCEAN–LAND–ATMOSPHERE INTERACTIONS
Department of Meteorology
University of Maryland
College Park, MD 20742

ABSTRACT

This paper first describes the nature of short term variability of the coupled atmosphere–ocean– biosphere system as shown by analysis and diagnosis of observations during the past 100 years. By "short term" we mean those fluctuations of the coupled climate system whose time scales range from 10 days to 1000 days. We have deliberately excluded any discussion of short range weather forecasting (less than 10 days) and decadal changes (more than 1000 days).

We next present a discussion of the present status of our knowledge of the predictability of short term fluctuations of the coupled climate system. Based on a large number of observational and modeling studies using complex models of this system, we suggest that most of the major short term climate fluctuations observed during the past 100 years of reliable data are consequences of the interactions among the different components of the climate system. For example, interactions between the atmosphere and the biosphere play an important role in the maintenance of prolonged drought conditions over the land areas. Interactions between the oceans and the atmosphere produce large and significant changes in the locations and the intensities of the large scale rain belts and also produce large changes in the global atmospheric circulation patterns.

We then present a brief description of TOGA (Tropical Oceans and Global Atmosphere), which is a 10 year program (1985–1995) launched by the World Climate Research Programme (WCRP) to monitor and model the interactions between the tropical oceans and the global atmosphere. Its ultimate objective is the design and development of an ocean–atmosphere observing system for operational climate prediction using advanced models of the coupled climate system.

We also point out that the natural variability of regional climate is so large that the uncertainty of predicted climate change due to such factors as increase of greenhouse gases would be significant, unless the climate models can realistically simulate the interannual variability of the coupled climate system. For example, decadal mean global temperature can be significantly affected by the number and intensity of El–Niño events, which can be produced by interactions between the atmosphere and the oceans. Better simulation and prediction of short–term climate variability will increase our confidence in climate models used to predict climate change.

Finally, we make a recommendation that as a natural extension of the earlier and ongoing programs like the Global Atmospheric Research Program and TOGA, WCRP should now initiate a comprehensive Global Climate Prediction Program to investigate the feasibility of operational prediction of monthly, seasonal and interannual variability of regional climatic anomalies over the globe. Such a Global Climate Prediction Program would utilize realistic models of the atmosphere, oceans and biosphere including snow and sea ice. Most of the important ingredients of a global climate prediction program are already in place. The ongoing program of TOGA would need to be expanded to cover global oceans; the ongoing WCRP project of hydrological and atmospheric pilot experiments (HAPEX) would provide better treatment of the biosphere; the ongoing WCRP radiation projects would help improve the treatment of clouds and radiative processes; and the ongoing projects on sea ice research would help improve the atmosphere–ocean–sea ice interactions in the climate system model.

1. Introduction

Before coming to the topic of short–term climate variability and prediction, let us begin by asking a more fundamental question. What determines the mean climate of any region of earth's atmosphere–ocean–biosphere system?

The primary energy source for atmospheric motions is the radiation heating of the warm equatorial regions and the cooling of the cold polar regions. The actual rates of heating and cooling are determined by astronomical parameters (the earth's distance from the sun, the periods of rotation of the earth around its axis and of the earth's orbit around the sun) and the planetary parameters (size, shape and mass of the earth; chemical composition of the atmosphere; ocean and biosphere; and distribution of land, ocean, mountains, and vegetation). In addition to these fixed parameters, the amount of heat transported around by atmospheric and oceanic currents, which we would refer to as the dynamical parameters, also plays an important role in determining the mean climate of any region on the earth.

The mean climate of the earth, therefore, is an equilibrium resulting from the various factors described above. This mean climate contains strong spatial and temporal gradients of pressure, temperature, salinity, velocity and water vapor. These gradients combined with the rotation of the earth give rise to day–to–day fluctuations of "weather" in the atmosphere and oceans which are routinely measured and diagnosed in order to predict their future evolution. It is the statistical average of these day to day weather fluctuations which gives rise to the weekly, monthly, seasonal and annual average climates, whose variability from one year to another is referred to as the interannual variability. It is no surprise, therefore, that superimposed on a well defined seasonal cycle − which itself has a rich space–time structure − there are large weekly, monthly, seasonal and interannual variations in the earth's climate system. For convenience of discussion in this paper, we shall make the following somewhat arbitrary classification of various time scales:

Time/Scale	Qualitative Description
0–10 days	Hourly & Daily Changes
10–100 days	Monthly, Intraseasonal & Seasonal Changes
100–1,000 days	Annual & Interannual Changes
1,000–10,000 days	Decadal & Interdecadal Changes
10,000–100,000 days	Centennial & Beyond

For discussion in this paper I have chosen to include the variations in the range of 10 days to 1000 days to define short–term climate variability. This means that we will not discuss weather prediction, and we will not discuss decadal and longer climate change. As a partial justification for this choice of time scales, it should be noted that the needs of water, energy and agriculture, and, in fact the entire socio–economic fabric of the global community are affected significantly by climatic fluctuations on these time scales.

Understanding and predicting short–term climate variability are also important, for such variability can be helpful in verifying climate models which are used for the prediction of climate change. Just as the numerical weather prediction models were useful for simulation and prediction of short–term climate variability, likewise, models with realistic simulation of short term climate variability will enhance our confidence in predictions of long–term climate change. It should be noted that most of the climate models that have been used so far to predict the climate change due to the increase in greenhouse gases have not been sufficiently validated in terms of the simulation and prediction of short–term climate variability.

In addition to the two factors of societal importance and validation of climate models, there is yet another important reason for our discussion of short–term climate variability. A large body of modeling and empirical studies, and our current understanding of the mechanisms that govern interannual changes, suggest that there is a scientific basis, and indeed some hope, for making useful predictions of climate variations on seasonal and interannual time scales.

It is, of course, true that day to day atmospheric fluctuations are not predictable after a few weeks because of the chaotic nature of atmospheric motions; however, it is now recognized that there is predictability in the midst of chaos. The interactions between the atmosphere and ocean, and atmosphere and biosphere produce long–period variations in the coupled system which enhance the predictability of the coupled system for months to years. These interactions are found to be much stronger in the tropics than in the extratropics, and therefore, the predictability of the short–term climate variability is also much higher for the tropics. We will come to this point in a later section.

In this paper, we shall address the following aspects of the short–term climate variability.

- Examples of short–term climate variability.

- Mechanisms of short–term climate variability.

- Predictability of short–term climate variability.

- Tropical Oceans Global Atmosphere (TOGA).

- A proposal to initiate an international program for prediction of global short–term climate variability.

2. Examples of Short–term Climate Variability

During the past 100 years of global observations of the earth's climate, there are many examples of significant short–term climate variability, such as the El–Niño–Southern Oscillation (ENSO), the monsoons, the tropical droughts and heat waves/severe cold winters in the extra–tropical regions. It will be pointed out in the next section that these regional short–term climate anomalies are manifestations of regional and global scale atmosphere–biosphere and atmosphere–ocean interactions. The interrelationships among El Niño, Southern Oscillation and monsoons are the most remarkable examples of interannual changes in the coupled climate system which affect global circulation and rainfall. A comprehensive summary of mechanisms of air–sea interaction and worldwide climate anomalies associated with the 1982–82 El Niño has been presented by Rasmusson and Wallace (1983).

It was noted by Walker (1924) that "when pressure is high in the Pacific Ocean, it tends to be low in the Indian Ocean from Africa to Australia", and for this recurrent pattern of planetary scale atmosphere fluctuations he coined the term "Southern Oscillation." It was later suggested by Bjerknes (1966) that the Walker's Southern Oscillation is but one component of a coupled ocean–atmosphere climate system, the other being the sea surface temperature fluctuations in the tropical Pacific Ocean. It is the interaction between the ocean (ocean warming off the South American coast being referred to as El Niño) and the atmosphere that is responsible for such fluctuations and produces short–term climate anomalies in different regions of the globe. Figure 1 shows fluctuations of surface pressure over Darwin, Australia (dashed line). The anomalies (departures from climatological mean) are first smoothed by a 12 month running mean and then divided by the standard deviation, and then smoothed again by a 12 month running mean. Darwin pressure has been chosen to illustrate the fluctuations in southern oscillation. The solid curve in this figure represents the sea surface temperature (SST) anomalies for the eastern equatorial Pacific. It can be seen that both variables, surface pressure and SST, show long–period (2–5 years) fluctuations, and it is also remarkable that the two are highly correlated. These coupled fluctuations of tropical ocean and atmosphere are referred to as ENSO (El Niño–Southern Oscillation).

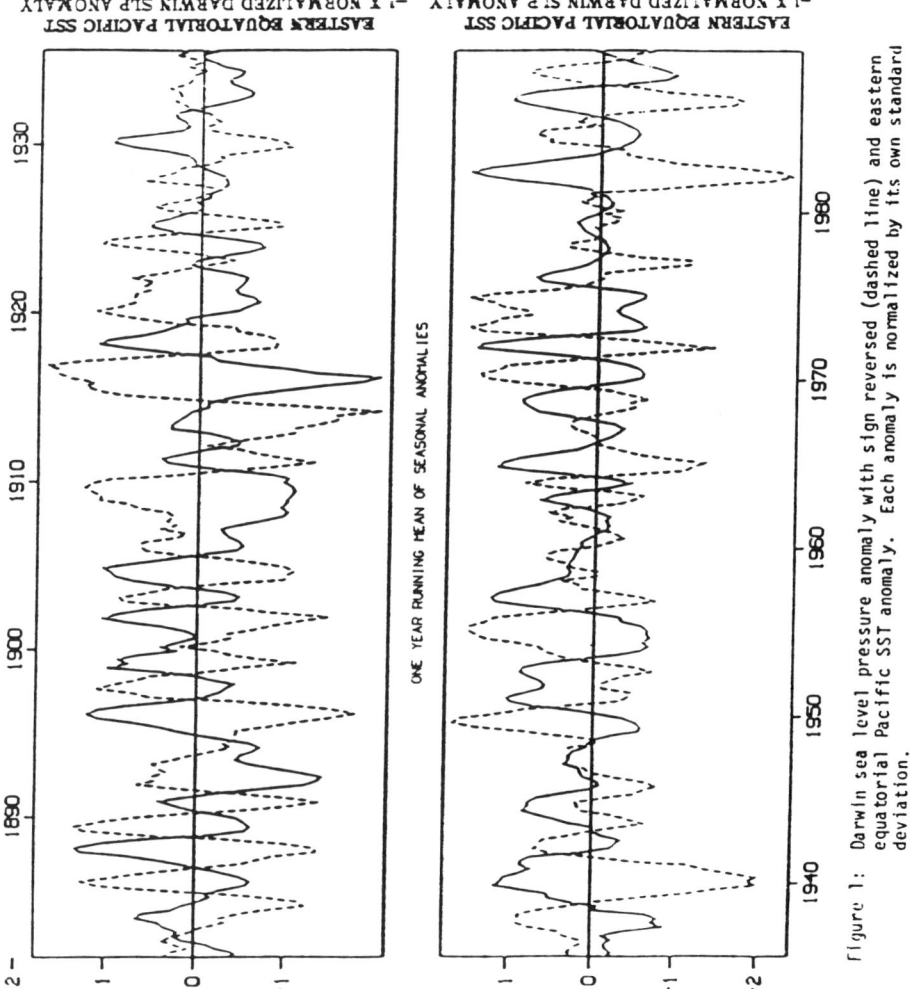

Figure 1: Darwin sea level pressure anomaly with sign reversed (dashed line) and eastern equatorial Pacific SST anomaly. Each anomaly is normalized by its own standard deviation.

Interannual variations in seasonal mean rainfall averaged over the whole of India also show very large interannual variability. Figure 2 shows the summer (June, July, August, September) monsoon rainfall averaged for more than 300 stations over whole India except the northern and the eastern hilly regions. It can be seen that even after averaging over a large spatial region and the whole monsoon season, there are significant fluctuations in monsoon rainfall from year to year, and from decade to decade. The solid line shows the 30 year running mean of the seasonal mean rainfall and the dashed line shows the standard deviation of rainfall for each 30 year period. It is remarkable that the 30 year mean as well as the variability within a thirty year mean show such large changes from one thirty year to the other thirty year period. It is unlikely that such large-scale, long-period fluctuations could be explained as a mere consequence of different sampling of high-frequency small scale rain-producing disturbances. It is more likely that such fluctuations are produced by planetary scale, long-period fluctuations of the coupled ocean-land-atmosphere system.

These selected examples of short-term climate variability suggest that quite large changes in our climate system can, and do, occur which are not necessarily either due to external or anthropogenic factors. We will refer to such fluctuations at the natural variability of our climate system. For lack of a better definition of natural variability, we would define it as the climate variability that would occur if the planet were never inhabited by the human species. This provides a baseline for detecting and predicting changes in climate and climate variability due to human influences. According to this definition, the examples described above will be categorized as being part of the natural variability of the climate system.

3. Mechanisms of short-term climate variability

The mechanisms responsible for the short-term climate variability can be described conceptually in two categories (Shukla, 1981).

- Internal dynamics of the individual components of the climate system.
- Interactions among the various components of the climate system.

Internal Dynamics: Even if the external forcings of the solar radiation and boundary conditions at the earth's surface were constant in time, the regional atmospheric circulation will exhibit short-term climate variability due to the combined effects of the hydrodynamical instabilities of the climate system and nonlinear interactions among various scales of motion. Although the distribution of oceans, continents, and mountains are fixed with time, their interactions with fluctuating winds can produce short-term climate variability. The occurrence of nearly zonal or persistent non-zonal regional circulation regimes ("blocking") are possible examples of anomalies due to the internal dynamics of the atmosphere. Likewise, the internal dynamical instabilities of ocean currents can produce variability in the ocean circulation and possibly the overlying atmosphere. We know that the spectrum of the

J. Shukla 119

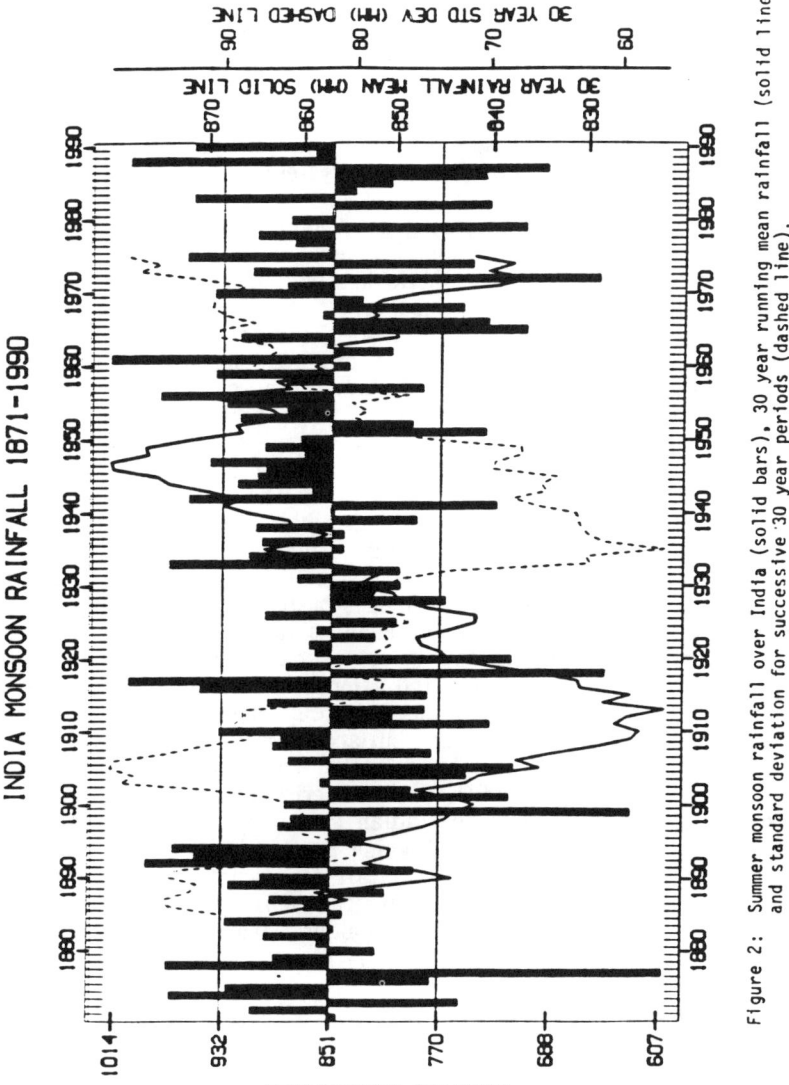

Figure 2: Summer monsoon rainfall over India (solid bars), 30 year running mean rainfall (solid line) and standard deviation for successive 30 year periods (dashed line).

atmospheric observations is red. Certain amount of interannual variability will be produced solely due to the unpredictable weather and therefore, that will remain unpredictable too.

Interactions: (Atmosphere–Ocean; Atmosphere–Biosphere; Atmosphere– Cyrosphere): We suggest that all the major events of short–term climate variability observed during the past 100 years – a period for which reasonably reliable instrumental measurements of climate variables exists – are due to the interactions among the atmosphere, ocean, biosphere and cyrosphere components of the climate system.

Atmosphere–Ocean Interactions: Changes in SST produce changes in evaporation, sensible heat flux and low level moisture convergence which in turn produce changes in atmospheric heating. The anomalous atmospheric heating produces changes in atmospheric circulation which in turn produce changes in wind stress and heat flux at the ocean surface. If this air–sea coupling has a positive feedback, it can produce long–lived anomalies of SST and the associated atmospheric circulation. Because of the differing rotational forces in the tropics and the extra–tropics, and because tropical ocean temperatures are warmer, even a small change in SST in the tropics can produce much larger changes in moisture convergence and heating than similar SST changes in the extratropics. This is the main mechanism for the occurrence of tropical droughts and floods which are manifestations of spatial and temporal shifts of mean climatological maxima of rainfall. The tropical atmosphere and oceans also do not have strong dynamical advections (as they do in midlatitudes) and therefore changes in atmospheric heating and surface wind stress can produce significant changes in atmospheric circulation and SST respectively. This is the primary reason why short–term climate variability of the tropical oceans and atmosphere are so strongly linked, and also why there is hope for predicting this coupled variability. In Figure 1 it was seen that the tropical SST and surface pressure fluctuations were highly correlated. Many researchers have further shown that in association with ENSO, several large regions of the globe experience droughts and heavy floods. For example, Figure 3 shows the Indian monsoon rainfall fluctuations (as in Figure 2) except that the years in which the tropical Pacific SST was rising and falling during the monsoon season are represented by black and hatched bars respectively. It is again remarkable that most of the severe droughts and floods over India occur during the anomalous warming (El Niño) and cooling of the equatorial Pacific ocean respectively.

Atmosphere–Biosphere Interaction: Changes in vegetation produce changes in albedo, surface roughness and soil moisture. These changes in turn produce changes in ground temperature, evaporation and sensible heat flux. Changes in horizontal gradients of ground temperature produce changes in convergence of moisture, and changes in vertical gradients of temperature along with moisture convergence produce changes in convection and rainfall, which in turn changes the soil moisture. The nature and degree of this interaction again depends on the character of the dynamical circulation regime where the land surface changes are taking place. The occurrence of prolonged

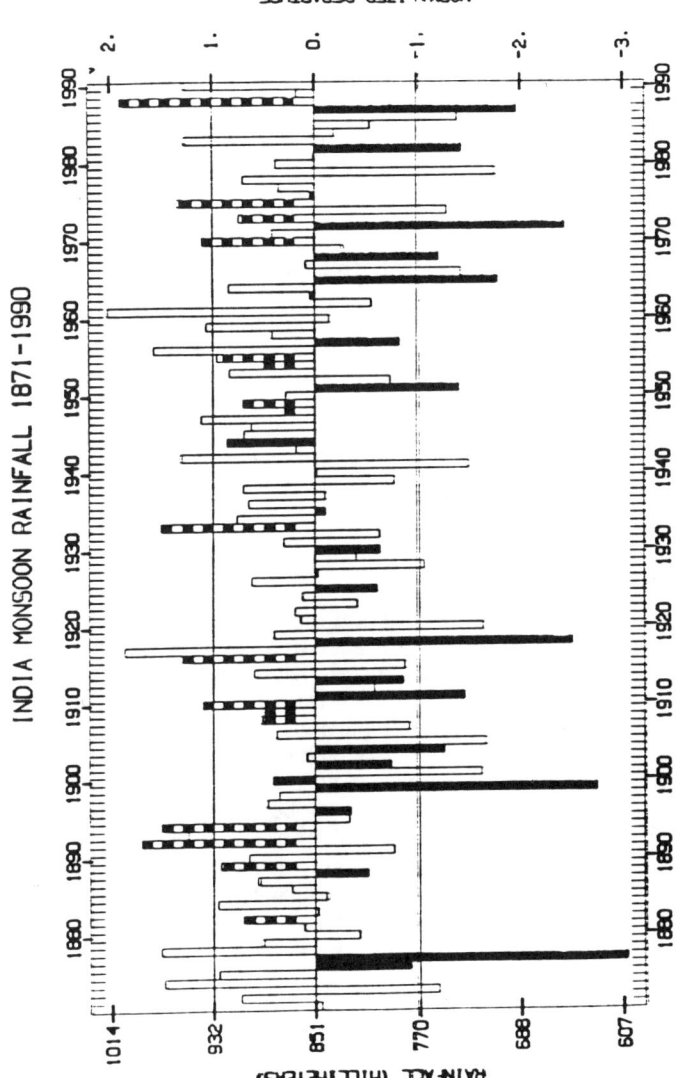

Figure 3: Summer monsoon rainfall over India. Solid and hatched bars denote the years when the eastern equatorial Pacific SST anomaly was rising and falling respectively.

droughts in sub–tropical regions (where the atmospheric dynamics are relatively weak) and even the tendency of heat waves to persist in the extra–tropical regions can be explained, at least in part, by such atmosphere–biosphere interactions. The West African Sahel has experienced persistent drought conditions for more than 20 years with significant interannual changes during these 20 years. Figure 4 shows fluctuations of rainfall over west (Africa) Sahel during the period 1940–1990. It can be seen that Sahel rainfall, like the Indian monsoon rainfall shown in Figure 2, also displays large year to year changes in seasonal mean rainfall. However, in addition there is a significant shift in the mean rainfall after 1968. While such shifts in seasonal and annual mean values are not uncommon for regionally averaged climatic parameters, this is a rather unique case because there is not a single year during the past 20 years when the seasonal rainfall was significantly above the climatological normal. Atmosphere–ocean interactions over the global oceans, as well as local atmosphere–biosphere interactions have been suggested as possible causes for these changes. Since monsoon rainfall over India as well as China also showed a notable shift (reduction) from the decade of 1950s to the decade of 1970s, it is reasonable to assume that this decadal shift in rainfall was perhaps due to some planetary scale circulation changes. However, it is quite likely that the local atmosphere–biosphere interactions, exacerbated by possible human activities leading to changes in the land–surface properties, could contribute towards the continuation of the reduced rainfall regime.

4. Predictability of short–term climate variability

It is well known that the instantaneous weather conditions are not predictable beyond a few days. It is also well understood that this lack of predictability is due to dynamical instabilities and non–linear interactions which amplify even very small initial uncertainties which may be either due to inadequate observations or imperfect equations for physical principles (Lorenz, 1965). However, it should be noted that lack of deterministic predictability of day–to–day weather beyond a few weeks does not necessarily mean that space and time averages for a month or season or beyond are also not predictable. In fact, we would like to propose that the large body of observational, theoretical and modeling results collectively suggest that there is indeed a scientific basis for the predictability of space–time averaged short–term climate variability. The primary scientific reasons for such an optimistic view on the predictability of short–term climate can be summarized as follows:

The space time spectra of atmospheric observations show that most of the variance in the interannual variability is accounted for by long–period, large scale fluctuations which are intrinsically more predictable than the day–to–day small–scale weather systems, and it is these relatively longer period large scale variations which are important for the prediction of short–term climate variability. In addition, atmosphere–ocean and atmosphere–biosphere interactions produce predictable changes in the coupled climate system. The atmospheric circulation anomalies are likely to be more predictable for those time scales for which the boundary forcings due to the anomalies in SST and soil moisture can also be predicted. For time periods

J. Shukla 123

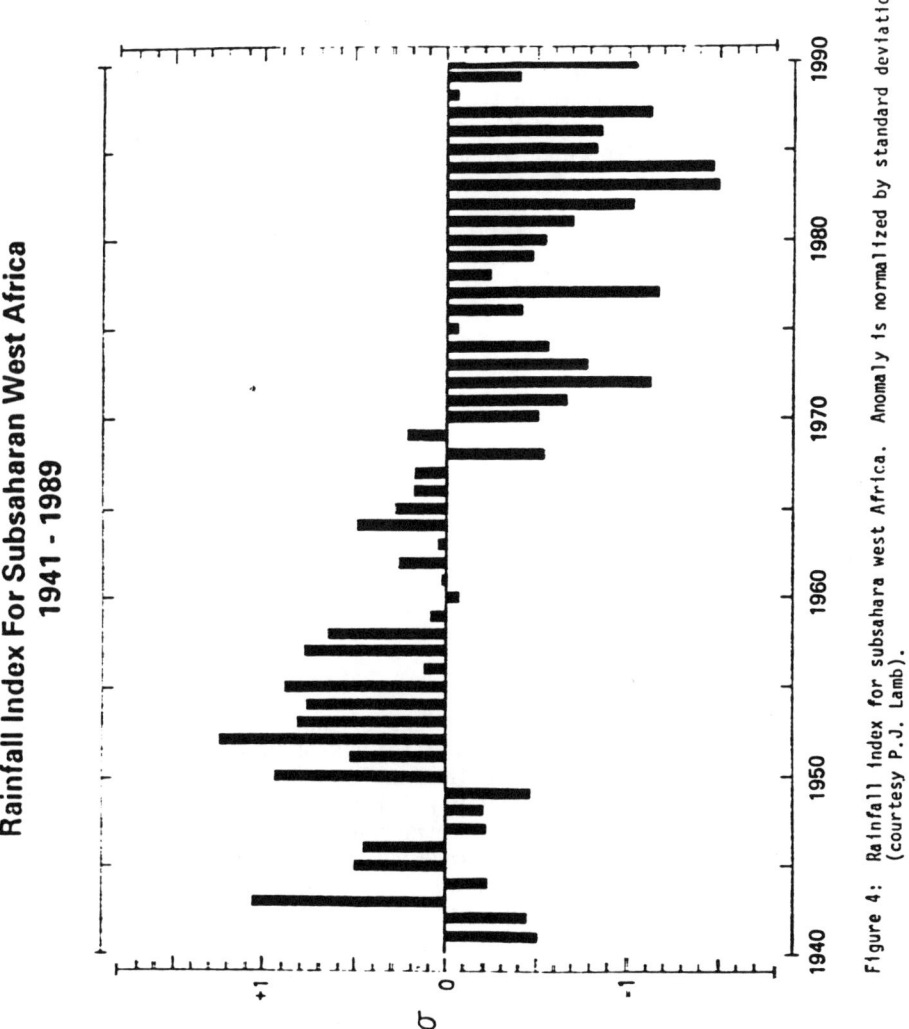

Figure 4: Rainfall index for subsahara west Africa. Anomaly is normalized by standard deviation (courtesy P.J. Lamb).

beyond the persistence of boundary forcings, we must be able to predict the evolution of the boundary forcings themselves. Recent developments in the modeling of the coupled system suggests that for the particular example of ENSO, the coupled tropical ocean–atmosphere climate system is theoretically predictable up to 1–2 years (Goswami and Shukla, 1990).

However, while considering the predictability of any climate signal, we must also consider the possible sources of climate noise which would tend to introduce a lack of confidence in the predictions. The following table gives some simple examples of possible signals we may wish to predict and important sources of noise which will make the predictions unreliable.

SIGNAL	NOISE
Daily Weather	Thunderstorms
1–10 Day Mean	Cyclones
Monthly Mean Climate	Blocking
Seasonal Mean Climate	30–60 Day Oscillations
Interannual (1–3 Year) Climate	Coupled Air–Sea Instabilities (ENSO)
Decadal Climate Change	ENSO

It should be noted that predictions of monthly and seasonal mean climate anomalies (from a given initial state) will be affected by the presence or absence of blocking regime, and the amplitude and phase of 30–60 day oscillations. Similarly, the predictability of ENSO will be determined by the instabilities of the coupled ocean–atmosphere system much like the predictability of short and medium range weather is influenced by baroclinic instability in the atmosphere. It should also be noted that the predictability of decadal climate changes (e.g., greenhouse warming) will be strongly influenced by the intensity and frequency of ENSO events. Pan and Oort (1983) have shown that interannual changes in the global mean temperature of the entire atmospheric column are highly correlated with SST anomalies in the eastern equatorial Pacific.

This is particularly relevant for the ongoing controversy over the detection and prediction of greenhouse warming. Based on the observational record of the global mean temperature, it is not possible to conclude that the climate change has been detected. The observed changes are entirely within the range of the natural variability of the coupled climate system. Likewise, the present climate models – the ones used for predicting effects of increased greenhouse gases – have not been adequately validated against the actual observed climate variability during the past. Therefore, there is no strong basis to accept the model predicted changes in global climate. The present climate models show large systematic errors in simulating the mean climate. It is assumed that although the simulated mean climate of the models is wrong, the differences between the simulated climate for the future (with increased greenhouse gases) and the simulated climate for the present can be accepted because the errors in simulating the present climate get removed when we substract one model simulation from the other. This assumption is generally correct if the models were used to simulate the direct response of a strong external forcing. The simulation of the effects of increased greenhouse gases falls in an altogether different category. The direct radiative forcing due to increased greenhouse gases is quite small (3–4 watts m^{-2} compared to the mean value of about 300 watts m^{-2}), and the corresponding increase in surface temperature will also be quite small. Thus, the model predicted climate changes due to greenhouse gases are entirely due to a number of positive feedbacks as formulated in the model parameterizations. Therefore, in order to accept the model results, it is quite important that the models are validated against some known climate variations in the past before we can have confidence in the model predictions for the future. It is in this context that it is considered particularly important that the climate models are validated for their ability to simulate and predict the short–term climate variability including the frequency and intensity of El Niño events.

5. Tropical Ocean Global Atmosphere (TOGA)

Recognizing the role of tropical SST anomalies in forcing global scale atmospheric circulation anomalies, and that the tropical SST anomalies are deterministically forced by atmospheric circulation anomalies, the World Meteorological Organization (WMO) and the International Council of Scientific Unions (ICSU) established a scientific steering group for the international TOGA program which is an element of the World Climate Research Programme (WCRP).

The scientific objectives of TOGA are (WMO, 1985):

1. To gain a description of the tropical oceans and the global atmosphere as a time–dependent system, in order to determine the extent to which this system is predictable on time scales of months to years, and to understand the mechanisms and processes underlying that predictability.

2. To study the feasibility of modeling the coupled ocean–atmosphere system for the purpose of predicting its variations on time scales of months to years.

3. To provide the scientific background for designing an observing and data transmission system for operational prediction if this capability is demonstrated by coupled ocean–atmosphere models.

TOGA was conceived as a ten year program (1985–1995) of data collection and modeling research. It is hoped that by use of advanced four–dimensional data assimilation techinques, all the surface and sub–surface ocean observations can be synthesized to produce an internally consistent basin–wide synoptic description of tropical oceans. Likewise, an internally consistent homogeneous data set for the four–dimensional structure of the global atmosphere can also be produced. Current efforts of tropical ocean data assimilation in USA and France have already begun to produce basin scale synoptic maps of ocean circulation. This is a major breakthrough for dynamical oceanography.

The ongoing modeling efforts using atmospheric models with prescribed SST, ocean models with prescribed wind stress and heat flux, and coupled ocean–atmosphere models have produced the highly promising result that interactions between the tropical oceans and the global atmosphere enhance the predictability of short–term climate variability. A description of the accomplishments in the first five years of the U.S. TOGA program and challenges for the future are summarized in a recent report (TOGA, 1990).

It should be recognized, however, that although the largest predictable part of the interannual variability of climate arises from ENSO, and more generally TOGA phenomena, to successfully carry out the prediction of short–term climate variability on seasonal and longer time scales will require adequate treatment of the other important interactive components of the climate system, including the extra–tropical oceans, land surface processes (biosphere) and variations in snow cover and sea ice.

6. A Proposal

In order to exploit the scientific advances in understanding the dynamics of the coupled Tropical Ocean/Global Atmosphere system as well as relevant results of other studies by WCRP and Climate and Global Change Programs, serious consideration should be given to the initiation of an international program for the prediction of global short–term climate variability.

The overall objectives of this program might be to:

• Provide real–time predictions of variations of the earth climate system on time scales of seasons to several years

• Validate predictive models of global climate change by demonstrating skillful forecasts of short–term variations of the coupled ocean–land–atmosphere climate system

A transition form TOGA and other ongoing WCRP programs to this project will require a transition from ocean models that focus entirely on tropical oceans to global atmospheric models with fully interactive land–surface processes including snow cover, and global oceans including sea ice. It is likely that in the initial phase, such a program may need to take into account only the thermodynamics of the global upper ocean and the fast dynamics of tropical ocean basins. The information base and the results from this program will be highly valuable in putting quantitative confidence limits on predictions of climate change at decadal time scales.

The program might include three main components, in addition to development of global observing systems foreseen in the framework of the World Weather Watch (WWW), World Ocean Circulation Experiment (WOCE) and Global Energy and Water Cycle Experiment (GEWEX). These components are:

i) A global atmosphere–ocean–land climate data analysis and prediction component, based on one or several dedicated central facilities for data acquisition, analysis, quality control and climate forecasts.

ii) An operational global observing system to provide the required data inputs for atmosphere, surface and upper ocean, sea ice, snow cover and soil moisture.

iii) A research component to address outstanding problems and new scientific issues which may arise in the course of the program.

It is recognized that for entirely fundamental scientific reasons (differences in the rotational force, dynamical instabilities and non–adiabatic heat sources), the potential predictability of the tropical atmosphere and oceans is much higher than that of the extratropics. Therefore, initially, the greatest beneficiaries of any organized, internationally coordinated effort in short–term climate prediction will be the tropical countries. However, the tropical countries do not have, at this time, the required resources of trained scientific personnel and computation–communication facilities to exploit this gift from nature for the well being of their respective societies. Therefore, we conclude with a suggestion that the nations of the world join together to exploit the recent scientific advances by initiating an international program on the prediction of short–term climate variability.

REFERENCES

Bjerknes, J.: 1966, A possible response of the atmospheric circulation to equatorial anomalies of ocean temperature. Tellus, 18, 820–829.

Goswami, B.N. and Shukla, J.: 1990, Predictability of a Coupled Ocean–Atmosphere Model. J. of Climate, 3, 2–23.

Lorenz, E. N.: 1965, A study of the predictability of a 28–variable atmospheric model. Tellus, 17, 321–333.

Pan, Y. H. and Oort, A. H.: 1983, Global climate variations connected with sea surface temperature anomalies in the eastern equatorial Pacific Ocean for the 1958–73 period. Mon. Wea. Rev., 111, 1244–1258.

Rasmusson, E. M. and Wallace, J. M.: 1983, Meteorological aspects of the El Niño/Southern Oscillation. Science, 222, 1195–1202.

Shukla, J.: 1981, Dynamical predictability of monthly means. J. Atmos. Sci., 38, 2547–2572.

Tropical Ocean Global Atmosphere (TOGA): 1990, A review of progress and future opportunities. National Research Council, National Academy Press, Washington, D.C.

Walker, G. T.: 1924, Correlations in seasonal variation of weather. X. Mem. India Met. Dept., 24, 333–345.

World Meteorological Organization (WMO): 1985, Scientific plan for the Tropical Ocean Global Atmosphere Programmee. WCRP Publication Series No. 3, WMO/TD 64, Geneva.

The Great Ocean Conveyor

Wallace S. Broecker
Lamont-Doherty Geological Observatory of
Columbia University
Palisades, New York 10964

INTRODUCTION

A diagram depicting the ocean's "conveyor belt" has been widely adopted as a logo for the Global Change research initiative. This diagram (see figure 1) first appeared as an illustration in an article about the Younger Dryas event which appeared in the November 1987 issue of Natural History. It was designed as a cartoon to help the largely lay readership of this magazine to comprehend one of the elements of the deep sea's circulation system. Had I suspected that it would be widely adopted as a logo, I would have tried to "improve" its accuracy. In hindsight such repairs would likely have ruined the diagram both for the readers of Natural History and for use as a logo.

The lure of this logo is that it symbolizes the importance of linkages between realms of the Earth's climate system. The ocean's conveyor is driven by the salt left behind as the result of the transport of water vapor through the atmosphere from the Atlantic Basin to the Pacific Basin. A byproduct of its operation is the heat which maintains the anomalous warm winter air temperatures enjoyed by northern Europe. A millinneum of very cold conditions known as the Younger Dryas appears to have ben the result of a temporary shutdown of the conveyor. Thus the conveyor logo portrays the concern which led to the launching of the Global Change research initiatives; namely devilishly complex interconnections among the elements of our Earth's climate system will greatly complicate our task of predicting the consequences of global pollution.

The objective of this paper is to provide a readable summary, from my perspective, of the conveyor's operation present and past.

ITS PATH

The main problem with the logo is that it implies that if one were to inject a tracer substance into one of the conveyor's segments it would travel around the loop as a neat package eventually returning to its starting point. As we all know this is hardly the case. Other circulation "loops" exist in the ocean and mixing occurs

130   The Great Ocean Conveyor

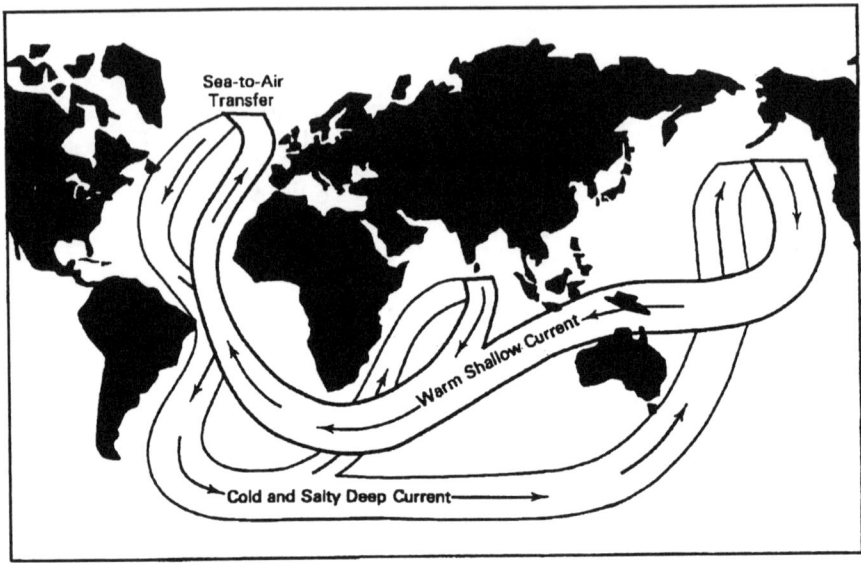

Figure 1. The great ocean conveyor logo (Broecker, 1987) (Illustration by Joe Le Monnier, Natural History Magazine.)

among the waters traveling along these intersecting pathways. The logo symbolizes a far more complex situation.

To understand the logo's message, let's start at the point of origin of its lower limb and work our way around the ocean. Waters in the vicinity of Iceland are cooled through contact with the cold winter air masses which sweep in from the Canadian Arctic. The cooling densifies the surface water to the point where it can sink to the abyss and flow southward forming the conveyor's lower limb. In the logo this flow is depicted as a ribbon of water which jets its way through the deep Atlantic from the vicinity of Iceland to the tip of Africa. In reality it is a sluggish mass which fills most of the deep Atlantic. Its flow is more akin to that in a slow moving river than to that in a fast moving mountain stream. This water mass, known to oceanographers as the North Atlantic Deep Water (NADW), stands out as a tongue of high salinity, low nutrient content and high $^{14}C/^{12}C$ ratio water in sections drawn along the Atlantic's length (see Figure 2). Its only competitor for space in the deep Atlantic is a wedge of Antarctic Bottom Water (AABW) which underrides the NADW mass. This intruding water is mixed upward into the southward flowing NADW, increasing the transport by the conveyor's lower limb.

Southward of 30°S the lower limb of the conveyor joins a rapidly moving deep current which encircles the Antarctic continent. This current serves as the great mix-master of the world ocean. It blends the NADW exiting the Atlantic with new deep water generated back into the Antarctic from the deep Pacific and Indian Oceans. So efficient is this blending that the NADW entering from the Atlantic loses its

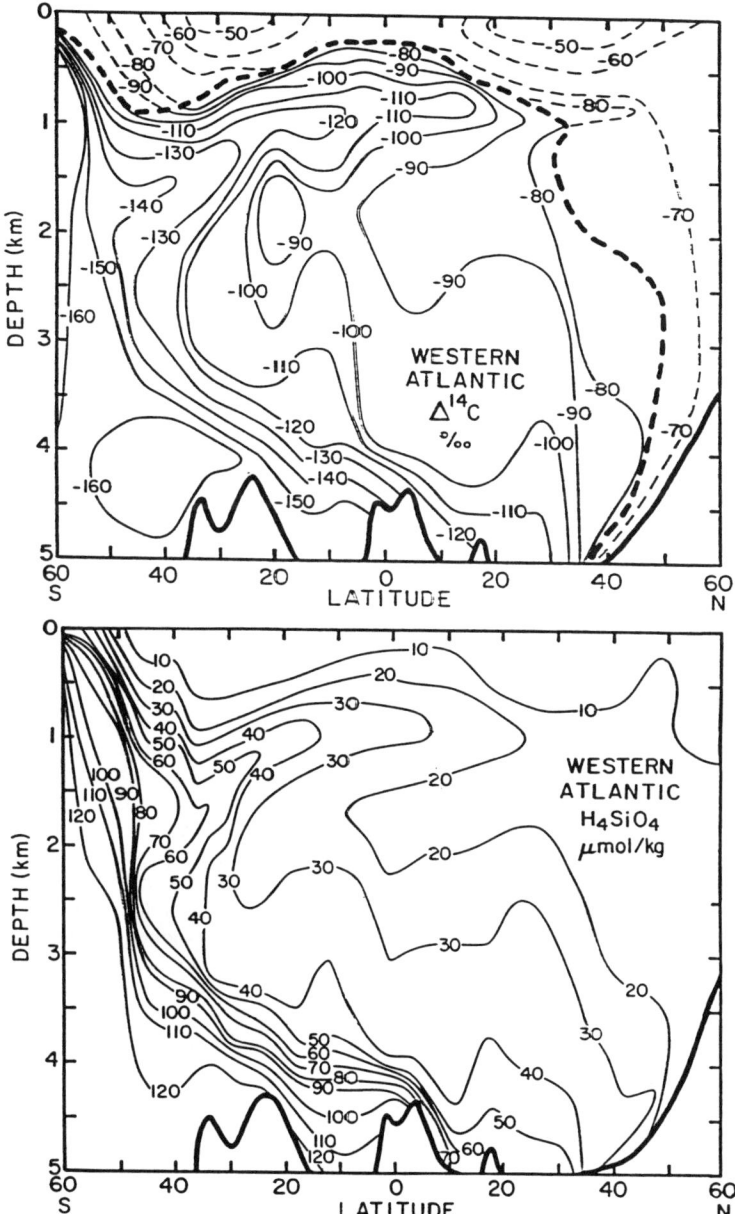

Figure 2. Sections of radiocarbon ratios and of dissolved silicate in the western Atlantic based on a few prenuclear measurements made as part of the GEOSECS program (from Broecker and Peng 1982). The dark dashed line indicates the extent of waters into which bomb testing ^{14}C has penetrated. Tritium (all from bombs) has been used to define this depth. In both, the North Atlantic Deep Water clearly stands out from the over and underlying waters of Antarctic origin. The intermediate and bottom waters which enter the Atlantic from the Antarctic have higher in silica concentrations and lower $^{14}C/^{12}C$ ratios than the NADW which constitutes the conveyor's lower limb.

identity before it passes even one half of a revolution around the Antarctic!

A rough quantification of the contribution of NADW to the deep waters of world ocean is provided by a property called PO_4 (Broecker, et al., 1991) which is defined as follows:

$$PO_4^* = PO_4 + \frac{O_2}{175} - 1.95 \; \mu m/kg$$

where PO_4 and O_2 are the measured phosphate and dissolved oxygen gas concentrations in a given water sample. The coefficient, 175, is the global Redfield coefficient relating O_2 consumption to PO_4 release during respiration of organic material (Broecker, et al., 1985), and the coefficient 1.95 is arbitrarily introduced in order to bring the values of PO_4^* into the range of deep water PO_4 concentrations. To the extent that the Redfield coefficient is a constant, PO_4^* constitutes a conservative property of any given deep water parcel; the increase in PO_4 due to the oxidation of organic material is exactly balanced by the decrease in $O_2/175$. PO_4^* is attractive as an indicator of the contribution of NADW because deep waters formed in the northern Atlantic have much lower PO_4^* values than those formed in the southern Ocean. Further the range of PO_4^* values for the northern source waters (0.73 ± 0.03) and for southern source waters (1.67 ± 0.10) is small compared to the difference between the means for these end member values (1.67 - 0.73 = 0.94). The nomogram in figure 3 permits the conversion of deep sea PO_4^* values into percentage contribution of NADW.

As can be seen in the map in figure 4, at a depth of 3 kilometers the contribution of NADW to the deep water mix remains strong throughout the Atlantic, but after the conveyor's lower limb passes around the southern tip of Africa into the Antarctic it rapidly becomes blended with the high PO_4^* deep water generated along the edge of the Antarctic continent. In this way an ambient deep water mix with a PO_4^* value of 1.37 is produced (see histograms in figure 4). This blend consists of 1 part deep water produced in the northern Atlantic with about 2 parts of deep water produced in the Antarctic floods the deep Pacific and Indian Oceans.

A more complete picture of the geometry of the blending process is given by the PO_4^* sections in figure 5. As can be seen, the more dense and higher PO_4^* waters produced in the Antarctic mix upward and northward into the less dense and lower PO_4^* waters entering from the Atlantic. By the time the water reaches the deep Indian, the deep Pacific and the Drake Passage, these endmember waters have been well blended.

As depicted in the logo the lower limb water returns to the surface in the northern Indian and Pacific Oceans. In reality this upwelling is widely spread with a large amount taking place in the Antarctic. The logo also suggests that the major route for return flow to the Atlantic (i.e. the conveyor's upper limb) is through the Indonesian archipelago and around the tip of Africa. This view was impressed on my brain by enthusiastic presentations by my colleague Arnold Gordon who stressed the role of the Agulhas current in global circulation (Gordon and Piola, 1983; Gordon, 1985 and 1986). As discussed below, this route probably accounts for only about one quarter of the return flow. A more important pathway is that through the Antarctic via the Drake Passage into the South Atlantic. These additional upwelling and return flow pathways are portrayed in figure 6.

ITS FLUX

It is my view that the magnitude of transport by the conveyor is best constrained by radiocarbon measurements on samples of deep water from the Atlantic

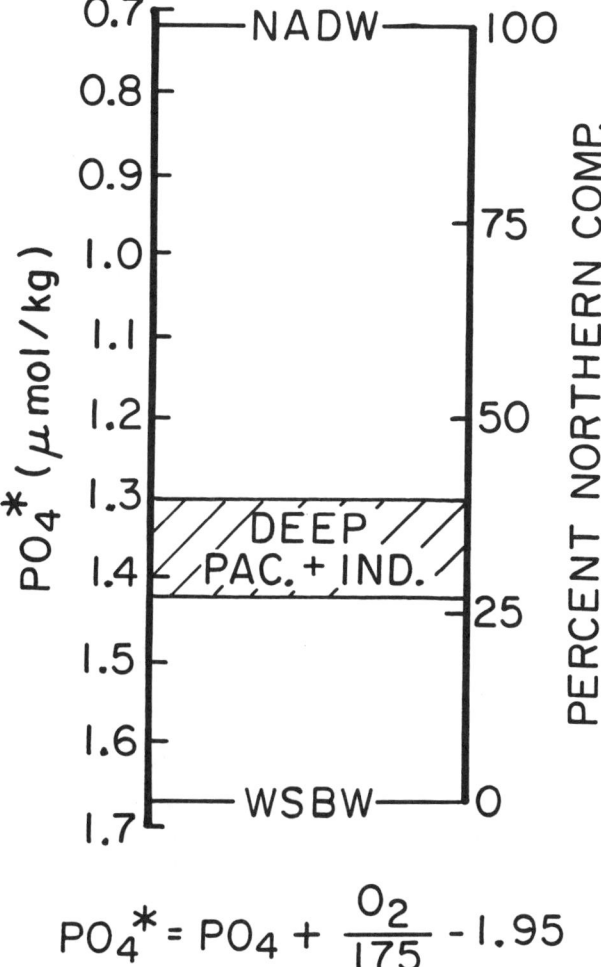

Figure 3. Nomogram showing the relationship between PO_4^* value (Broecker, et al., 1991a) and the percentage contribution of NADW to waters in the deep sea mix. Deep waters formed in the northern Atlantic have nearly uniform low PO_4^* values (0.73 ± 0.03 μm/kg). Those which form around the Antarctic continent have high values (between 1.6 and 1.7 μm/kg. The deep Pacific and Indian Oceans are flooded with a nearly uniform mix of these two end members (33% NADW + 67% WSBW).

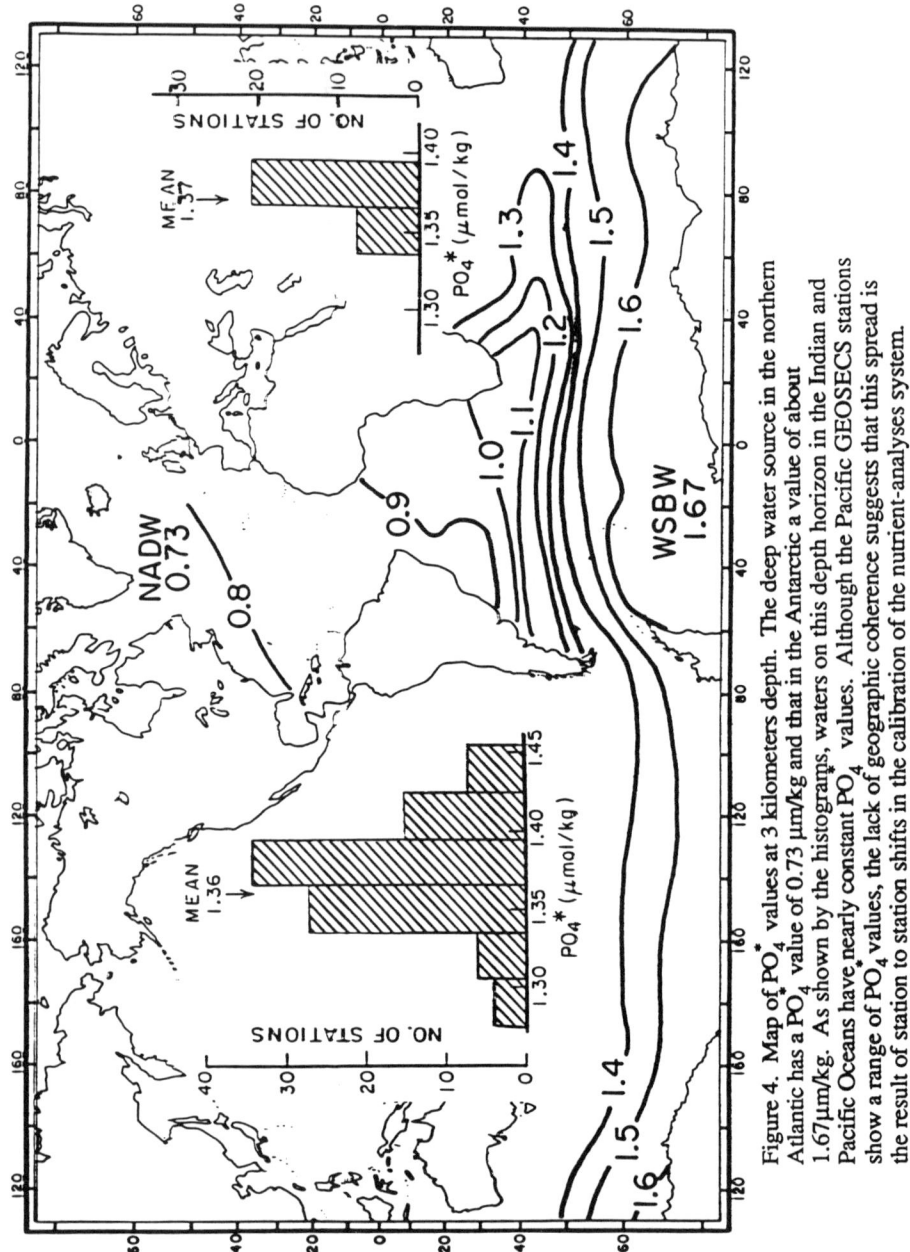

Figure 4. Map of PO_4^* values at 3 kilometers depth. The deep water source in the northern Atlantic has a PO_4^* value of 0.73 μm/kg and that in the Antarctic a value of about 1.67 μm/kg. As shown by the histograms, waters on this depth horizon in the Indian and Pacific Oceans have nearly constant PO_4^* values. Although the Pacific GEOSECS stations show a range of PO_4^* values, the lack of geographic coherence suggests that this spread is the result of station to station shifts in the calibration of the nutrient-analyses system.

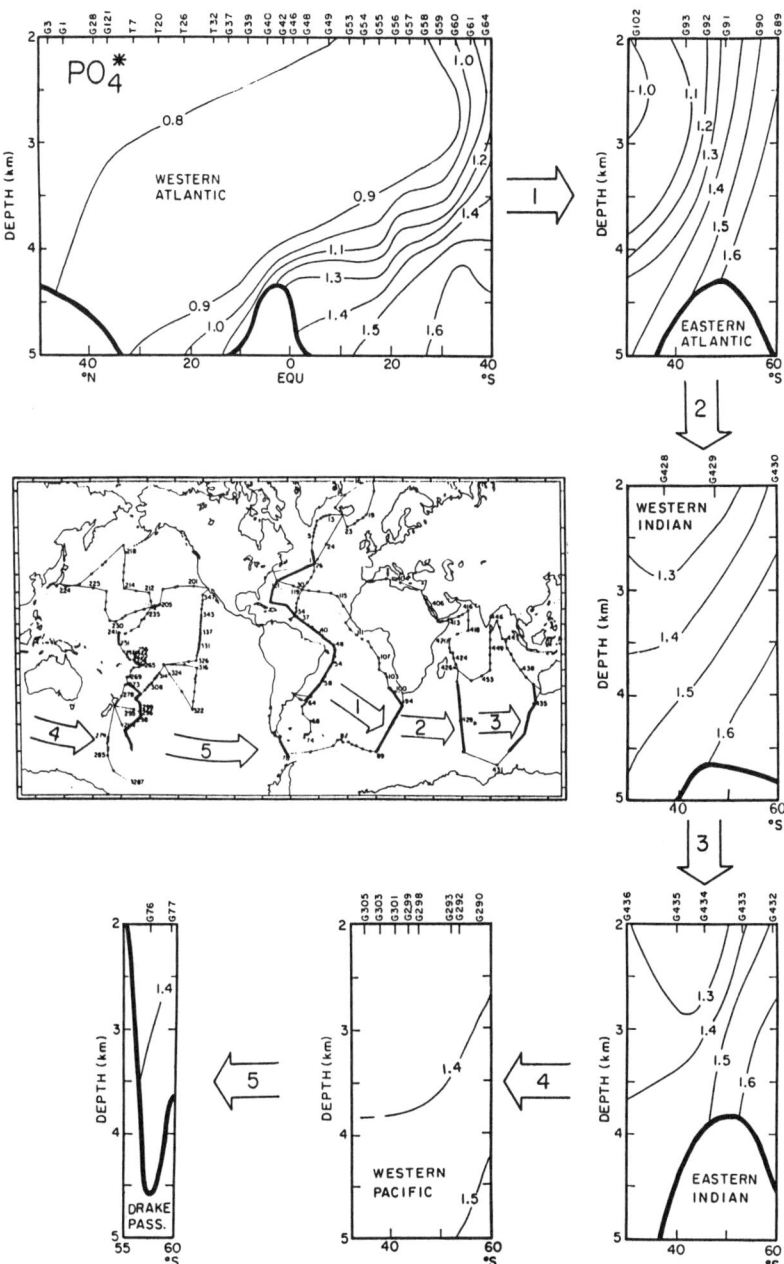

Figure 5. Sections of PO_4^* down the western basin of the Atlantic and around the Southern Ocean. The map shows the location of the GEOSECS stations and of the sections. Note how the <1.3 μm/kg and the .1.6 μm/kg waters blend as they move around the Antarctic continent eventually producing a nearly uniform mix with a PO_4^* value close to 1.4 μm/kg.

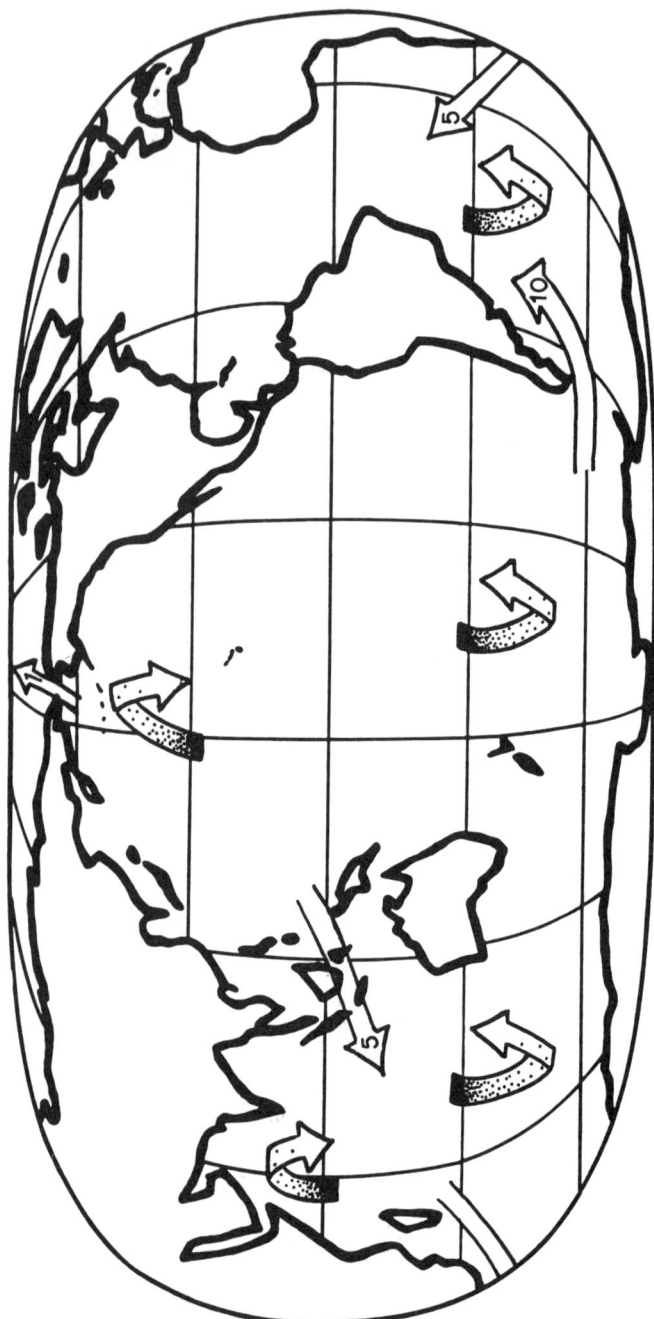

Figure 6. The dotted arrows show the location where intense upwelling of deep water occurs. The blue arrows show the upper ocean routes of return flow which balance the outflow from the Atlantic of lower-limb water. The numbers in the arrowheads represent the rough estimates of the magnitude of the fluxes (in Sverdrups). In addition to the features portrayed by the logo, this diagram shows that much of the upwelling occurs in the Antarctic and that much of the return flow occurs through Drake Passage at the tip of South America..

Ocean. Some physical oceanographers might dispute this claim and opt instead for estimates derived from a combination of current meter measurements and geostrophic flow calculations. Fortunately the two approaches yield similar answers.

The radiocarbon based estimate of the flux of NADW into the deep Atlantic is obtained by dividing the volume of water contained in the deep Atlantic by the radiodecay based mean residence time. This estimate must be corrected for the contribution made by AABW to the conveyor's lower limb. It must also be corrected for the impact of temporal changes in the ^{14}C/C ratio for atmospheric CO_2.

The major obstacle to calculation from radiocarbon measurements of residence times for water in the deep Atlantic is the determination of the initial ^{14}C/C ratio for each parcel. The reason is that all waters in the deep Atlantic are mixtures of northern component water with a comparatively high ^{14}C/C ratio (Δ^{14}C = -68‰) and of southern component water with a comparatively low ^{14}C/C ratio (Δ^{14}C = -158‰). Because of the large difference in the Δ^{14}C values* for these end members much of the variation in ^{14}C/C ratio within the deep Atlantic is created by differences in the end member blend. As shown by Broecker et al. (1991a), PO_4 provides quite an accurate means of establishing the proportions of northern and southern component water in the sample analyzed for radiocarbon. The measured radiocarbon concentration is then subtracted from the initial concentration calculated for the mixture, yielding the deficiency attributable to radiodecay. An example of this calculation is shown in Table 1. The radiocarbon measurements used in this analysis were made in the laboratories of Gote Ostlund at the University of Miami and Minze Stuiver at the University of Washington.

Water column averages for radiodecay deficiencies are shown in figure 7 for all the stations occupied during the GEOSECS, TTO, TAS and SAVE expeditions. Little information is lost by this vertical averaging because significant trends with depth are not found for any of the stations. The vertically averaged deficiencies do however show a pronounced geographic trend. The lowest values (<10‰) are found along the western margin of the Atlantic and the highest values (>30‰) along the eastern margin. As radiocarbon decays by 1‰ in 8.27 years, the isolation times

*Footnote

The convention for radiocarbon results on contemporary samples is to express them as per mil differences from that in a universally used NBS standard using the following equation

$$\Delta^{14}C = \delta^{14}C - 2(\delta^{13}C + 25)(1 + \frac{\delta^{14}C}{1000})$$

where

$$\delta^{13}C = [\frac{^{13}C/^{12}C_{sample} - ^{13}C/^{12}C_{standard}}{^{13}C/^{12}C_{standard}}] 1000$$

$$\delta^{14}C = [\frac{^{14}C/C_{sample} - ^{14}C/C_{standard}}{^{14}C/C_{standard}}] 1000$$

The inclusion of the ^{13}C term corrects for the influence of isotope fractionation. Radiocarbon is lower in the sea than in the atmosphere because of the time required to mix newly formed radiocarbon atoms into and through the sea.

Table 1. Example of radiocarbon deficiency calculation

GEOSECS STATION 113 (11°N, 20°W, 4741m)

O_2 μm/kg	PO_4 μm/kg	$\Delta^{14}C$ ‰
239	0.91	-120

$$PO_4^* = \frac{O_2}{175} + PO_4 - 1.95$$
$$= \frac{239}{175} + 0.91 - 1.95$$
$$= 0.91 \text{ μm/kg}$$

$$\text{fraction of northern component} = \frac{1.67 - PO_4^*}{1.67 - 0.73} = \frac{1.67 - 0.91}{1.67 - 0.73}$$
$$= 0.81$$

fraction of southern component = 1 - 0.81
= 0.19

$\Delta^{14}C_{initial} = 0.81(-68) + 0.19(-158)$
$= -84‰$

$\Delta\Delta^{14}C = \Delta^{14}C_{initial} - \Delta^{14}C_{measured}$
$= (-84) - (-120)$
$= 36‰$

$$\text{apparent age} = 8270 \ln\left(\frac{1 - .084}{1 - .120}\right)$$
$$= 331 \text{ y}$$

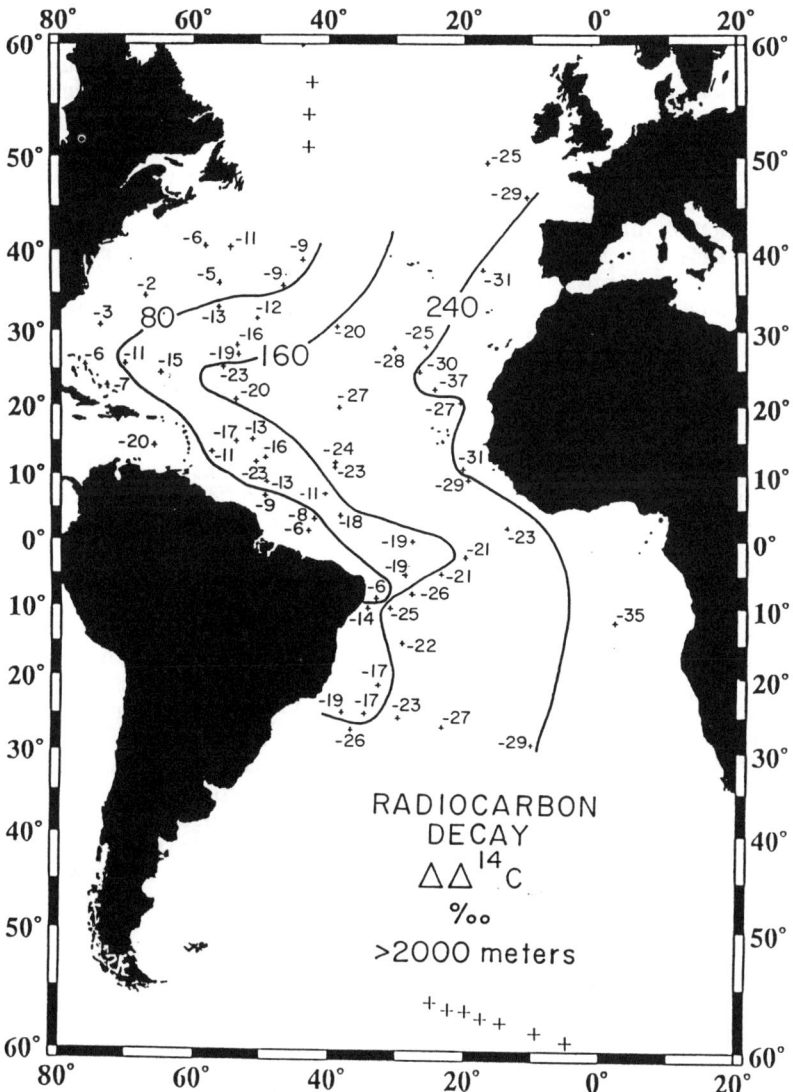

Figure 7. Water-column averages for the radiodecay-induced ^{14}C deficiency (Broecker, et al., 1991a) at stations occupied during the GEOSECS, TTO, TAS and SAVE expeditions. The major gradient is from low deficiencies along the western margin to high deficiencies along the eastern margin. The residence times corresponding to the 10, 20 and 30‰ deficiency contours are respectively 83, 166 and 249 years. The mean deficiency for this entire region of the Atlantic is estimated to be 22‰ which corresponds to an isolation time of about 180 years.

corresponding to these radiocarbon deficiencies range from near zero for the western boundary in North Atlantic to as high as 300 years along the eastern boundary. This suggests rapid ventilation from both ends of the Atlantic along the western boundary coupled with more leisurely dispersion into the interior. The radiocarbon deficiency for the entire deep Atlantic averages about 22‰. This corresponds to a residence time of about 180 years.

The volume of the deep Atlantic reservoir is $1.55 \times 10^{17} m^3$ (i.e. 2500m mean thickness with an area of $6.2 \times 10^{13} m^2$). Hence to achieve this residence time requires a ventilation flux of $8.6 \times 10^{14} m^3/yr$ or 27 Sverdrups (1 Sverdrup = $10^6 m^3/sec$). As the flux of AABW is about 4 Sverdrups, the flux of NADW is estimated to be 23 Sverdrups.

This calculation assumes the system to be at steady state. While we have no way to know whether this is true for the water fluxes, we do know that the atmosphere's $^{14}C/C$ ratio has changed with time. When these changes are taken into account, the flux has to be reduced by a factor of about 0.88 (Broecker et al. 1991b). Hence, we get a flux of close to 20 Sverdrups for the northern component (i.e., NADW). It is difficult to assess the error in this estimate but it is probably on the order of 25% (i.e. ± 5 Sverdrups). To appreciate the immense magnitude of this flux it is important to be reminded that it is 20 times the combined flow of all the world's river and somewhat larger than the rainfall over the entire globe!

ITS DRIVE

My contention is that the conveyor is driven by the excess salt left behind in the Atlantic as the result of vapor export (Broecker, et al., 1985). As can be seen from the map in figure 8, the surface waters of the Atlantic are on the average 1 gm/liter higher in salt content than those in the Pacific. For sea water with temperatures in the range of those constituting the NADW mass (i.e. 2 to 4°C) one gram per liter extra salt has the same impact on the water's density as a cooling of 2 to 3°C. The salinity contrast between surface waters in the northern Atlantic and those at comparable latitudes in the northern Pacific is even larger, ranging from 2 to 3 grams per liter. This difference is so large that surface waters in the northern Pacific even when cooled to their freezing point (i.e. - 1.8°C), sink to a depth of only a few hundred meters before reaching their buoyancy limit. Hence no deep water can form in the northern Pacific.

Three means are available by which the magnitude of the vapor export flux can be estimated. The first approach is based on the water budget for the Atlantic Ocean and its continental drainage basin (see figure 9). Baumgartner and Reichel (1975) have constructed a water budget based on estimates of rainfall and evaporation over the Atlantic Ocean and runoff from its drainage basin. Their result is that vapor is being lost from the Atlantic basin at a rate averaging 0.45 Sverdrups.

A second approach is to estimate the vapor export necessary to maintain the salinity differences in the sea against mixing among the ocean's water masses which tends to homogenize the sea's salt. If the mixing rates within can be determined and incorporated into an ocean mixing model, the fresh water budget for any region of the ocean can be determined. Broecker et al., (1990a) adopted this approach. Using a radiocarbon calibrated ocean box model they obtain a flux of 0.25 Sverdrups. The ocean general circulation model run by Princeton's Geophysical Fluid Dynamics Laboratory (GFDL) yields a water vapor loss of 0.45 Sverdrups (Manabe and Stauffer, 1988) and the Hamburg ocean GCM gives 0.20 Sverdrups (Ernst Maier-Reimer, personal communication).

The third approach involves the determination of the net fluxes of water vapor across segments of boundary separating the Atlantic's drainage basin from the

Figure 8. Map of surface ocean salinity (Levitus, 1982). Regions of low salinity (i.e. S<34‰ are cross hatched. Regions of high salinity (i.e. S> 36‰) are dotted. Note that on the average surface waters in the Atlantic are about 1‰ higher in salinity than their Pacific counterparts. For the high northern latitudes the difference is even larger averaging about 2‰.

142 The Great Ocean Conveyor

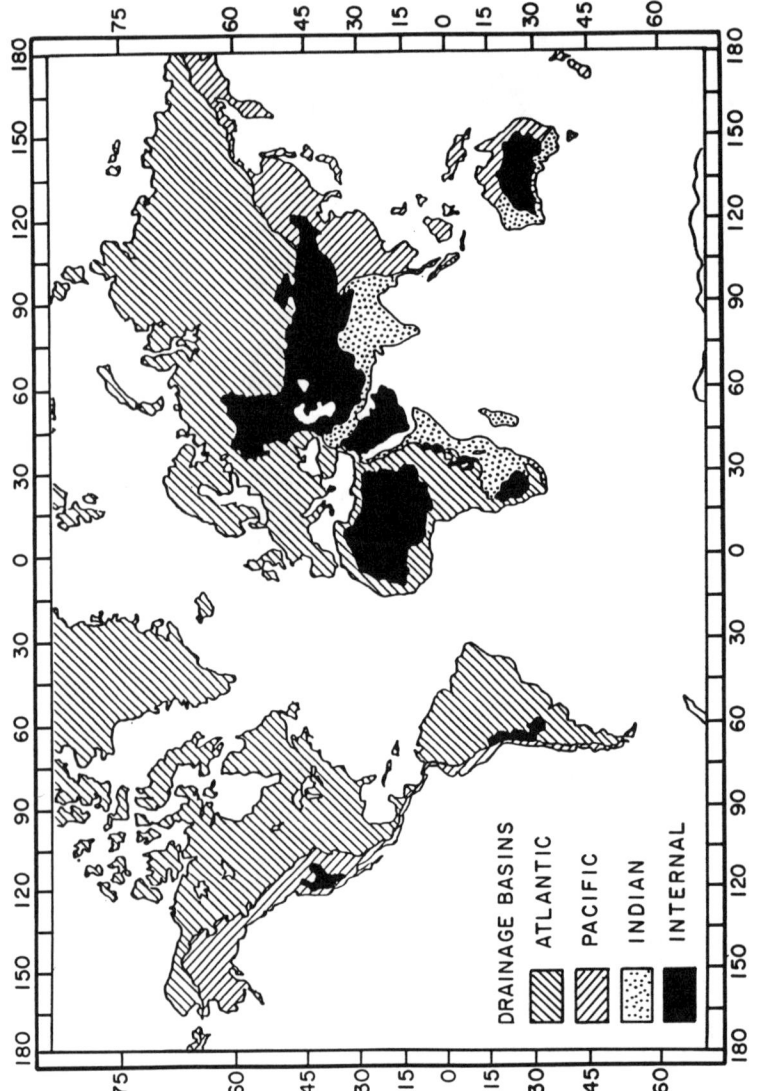

Figure 9. Map showing continental areas whose drainage reaches the Atlantic, Pacific and Indian Oceans. The black areas are deserts from which no drainage to the sea occurs. About 0.35 Sverdrups of fresh water are transported from the Atlantic to the Pacific Ocean moving as vapor in part across Central America and in part across Africa.

remainder of the world. To do this, Zaucker and Broecker (in press) have used wind and humidity data summarized by Oort (1983) to calculate vertically integrated and annually averaged vapor transports for all positions on the globe (using the 4° x 5° grid 11 level eggcrate geometry employed by the Goddard Institute for Space Science (GISS). They obtain in this way a net vapor loss from the Atlantic of 0.32 Sverdrups. In addition to providing an estimate of the magnitude of the vapor loss, this approach also yields the routes for this loss. As is shown in figure 10, vapor loss occurs both in the belt of northern hemisphere westerlies and in the belt of tropical easterlies. For the westerlies substantially more vapor is exported across Eurasia than is imported across the North American cordillera. For the easterlies substantially more vapor is exported across Central America than is imported across Africa.

Based on these results we estimate that the rate of vapor loss from the Atlantic basin is 0.35 ± 0.12 Sverdrups. To put this flux in context it helps to be reminded that it is about twice that for the Amazon River. Over the course of a year vapor export removes an amount of water equal to that in a 15 cm. thick layer covering the entire Atlantic!

ITS SALT BUDGET

Adopting a 20 Sverdrup export rate for lower limb water and an o.35 Sverdrup fresh water loss from the Atlantic, it is of interest to see what combination of return flow water could balance the Atlantic's salt budget. As the outgoing lower limb water has a salinity of about 34.9‰ and the outgoing water vapor a salinity of 0.0°, the average aggregate salinity of the return flow water must be about 34.3‰ (see figure 11) or 0.6‰ lower than that of the outflowing lower limb water. Thus the salinity contrast between sea water and water vapor is about 60 times larger than the salinity contrast between the waters being traded between the Atlantic and the remainder of the ocean! It is for this reason that a measly 0.35 Sverdrup vapor loss can drive a mighty 20 Sverdrup ocean current! It should be kept in mind in this regard that were the salt buildup to go uncompensated, the salinity of the entire Atlantic would increase at the rate of about 1.4 gm/liter per millennium. As we shall see below, the conveyor appears to have been running more or less as it does today for the last 9000 years. Had the salt buildup not been compensated, the Atlantic's salinity would have increased during that time be a staggering 13‰. Clearly this can not have been the case. Rather, on the average over this period of time the export of salt via the conveyor's lower limb must have balanced the enrichment of salt by vapor loss.

The "remainder water," which feeds the upper limb of the Atlantic, has three components; Antarctic surface waters passing through the Drake Passage (S≅33.8‰) and intermediate waters formed at the northern perimeter of the Atlantic segment of the Antarctic (S ≅ 34.3‰). The salinity of the intermediate water matches that required to achieve salt balance. However this salinity could also be achieved by mixing 1.6 parts Drake Passage surface water with 1 part Agulhas water. So based on salinity alone it is not possible to say how much of the remainder water enters the South Atlantic at intermediate depth and how much enters at the surface.

A rough idea of the proportions of intermediate water on one hand and the Drake-Agulhas mix on the other can be obtained by invoking another constraint, namely, that the element phosphorus be conserved within the Atlantic. The reason is that the residence time of phosphorus in the ocean (several tens of thousands of years) is roughly 100 times the residence time of water in the Atlantic (several hundreds of years). Hence, there can be no significant gain or loss of phosphorus during a single ventilation cycle; in other words, the phosphorus contained in the exported water must match that contained in the imported water. Before employing

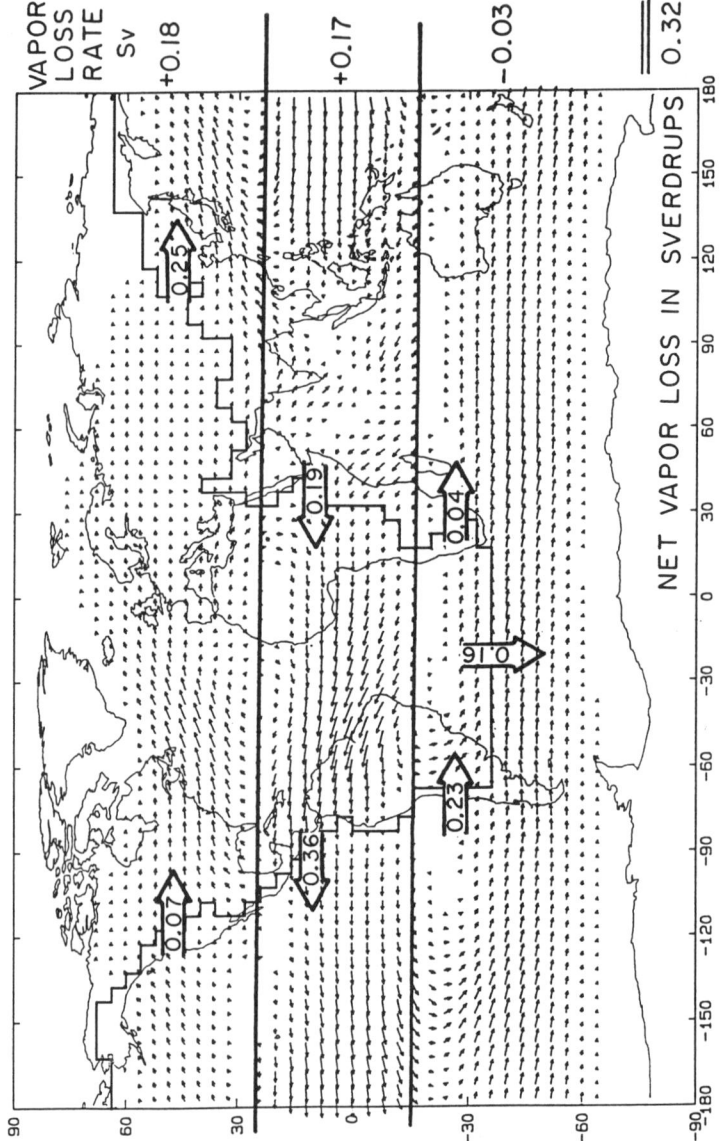

Figure 10. Map showing vertically integrated and annually averaged water-vapor flux vectors compiled by Oort (1983). Also shown is the boundary of the Atlantic's drainage basin and net fluxes across segments of this boundary. The tropical easterlies carry out more water vapor from the Atlantic basin across Central America than they bring in via Africa. So also do the northern westerlies; more water escapes across Asia than enters across the American cordilera. For the entire basin the rate of water vapor loss is 0.32 Sverdrups (see Zaucker and Broecker, in press).

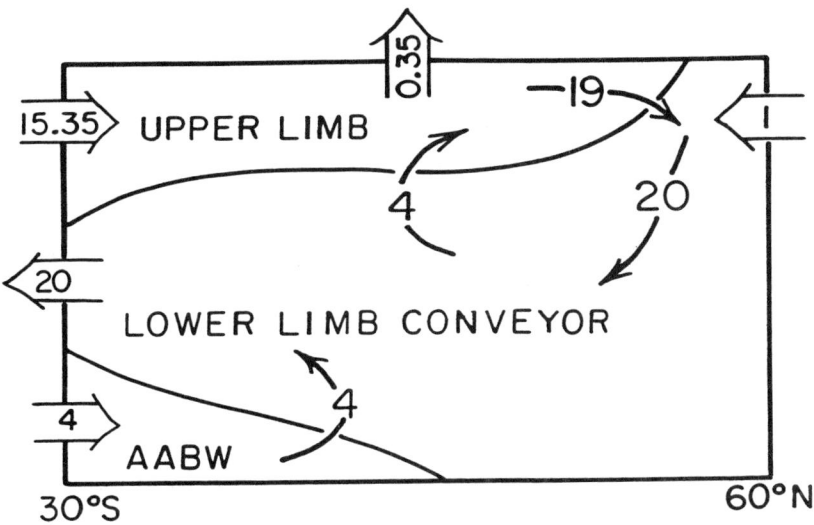

ATLANTIC'S SALT BUDGET

		SALINITY ‰	FLUX Sv
OUT			
	NADW	34.9	20.0
	VAPOR	0.0	0.35
		AVG. 34.3	TOTAL 20.35
IN			
	BERING STS	33.0	1.0
	AABW	34.7	4.0
	REMAINDER	34.3	15.35
		AVG. 34.3	TOTAL 20.35

Figure 11. Approximate water and salt budget for the Atlantic Ocean. The fluxes of Bering Sea surface water and of AABW are independently determined. The remainder water must then have a flux of 19 minus the amount of NADW recirculated within the Atlantic. With the guess of 4 Sverdrups recirculation adopted here the remainder water flux becomes 15.35 Sverdrups. The aggregate return water must have a salinity of 34.3‰ to balance the Atlantic's salt budget. Since the combined Bering Sea (Coachman and Aagaard, 1988) and AABW inputs have a salinity of 34.3‰, the remainder water must also have this salinity.

this constraint I must point out that it has a possible flaw. While we assume in our calculations that phosphorus is transported only in inorganic form (i.e., PO_4), the finding by Suzuki, et al., 1985, that sizable amounts of organic nitrogen are contained in low nutrient surface waters, opens the possibility that phosphorous is contained in the large molecules or fine particles thought to carry the organic nitrogen. If phosphorus is being shuttled back and forth between organic and inorganic forms, then the constraint suggested here becomes invalid.

The phosphorus constraint stems from the fact that only Agulhas water carries less PO_4 than outgoing lower limb water; the other contributors to the return flow have 20 to 50% higher phosphate concentrations. Agulhas water carries no PO_4 and Drake Passage water carries about 1.5 µm/kg PO_4; hence the Agulhas-Drake Passage mix required to yield the right salinity (1 part to 1.6 parts) carries about 0.9 µm/kg. Intermediate water carries about 1.8 µm/kg of PO_4> Hence to match the 1.3 µm/kg of PO_4 carried by the outgoing lower limb water requires a mix of about 56% Agulhas-Drake Passage component and about 44% intermediate water. This then suggests that the 15.35 Sverdrups of remainder water is made up of 6.7 Sverdrups intermediate water, 5.3 Sverdrups of Drake Passage water and 3.3 Sverdrups of Agulhas water. However because of the many simplifications required in order to make this calculation, these fluxes must be taken with a grain of salt (and perhaps also a dram of phosphorus).

ITS BENEFITS

The benefit provided by the conveyor is the heat it releases to the atmosphere over the northern Atlantic. This heat is responsible for Europe's surprisingly mild winters. The amount of heat released to the atmosphere is given by the product of the conveyor's flux and the temperature change required to convert upper limb water to lower limb water (i.e., to create NADW). The temperature of NADW averages about 3°C. As will be shown below, the temperature of the upper limb water averages about 10°C. Thus each cubic centimeter of upper limb water releases 7 calories of heat to the atmosphere during its conversion to deep water. At an average flux of 20 Sverdrups this totals 4×10^{21} calories each year, an amount of heat equal to 35% of that received from Sun by the Atlantic north of 40° latitude!

Manabe and Stauffer (1988) have shown that indeed the thermohaline circulation of the Atlantic maintains high surface water temperatures in the northern Atlantic. Using the GFDL ocean GCM, they demonstrate that circulation in the Atlantic can assume two quite different modes; one with a strong thermohaline component akin to the conveyor and one with no thermohaline circulation. The difference between surface water temperatures maintained by these modes is shown in figure 12. As can be seen when the conveyor is operative, the temperature of surface waters in the northern Atlantic average 5°C warmer than when it is off. Considering that strength of the thermohaline circulation in the Manabe and Stauffers' conveyor-on mode is only 12 Sverdrups, this warming should be even greater in the real ocean with its 20 Sverdrup thermohaline circulation.

Rind et al., (1986) have used the atmospheric GCM of the Goddard Institute for Space Studies to estimate the geographical patter of the winter air temperature change supported by the conveyor's heat output. They adopted for the surface water temperature difference between the conveyor-on and conveyor-off modes that reconstructed by the CLIMAP group for glacial surface water relative to today's. As shown in figure 13, the air temperature anomaly obtained in this way extends across Europe into Siberia. As we will discuss below this geographic pattern matches very nicely that for the Younger Dryas cooling.

Now let us return to the problem of estimating the temperature of the upper limb water supplying deep water formation in the northern Atlantic. The major

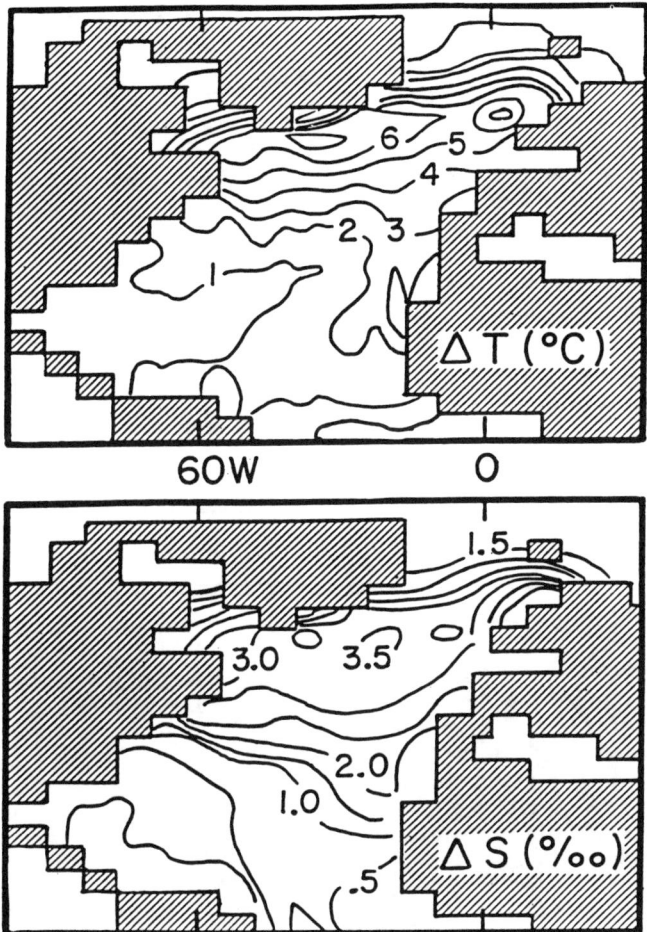

Figure 12. Maps showing the difference in North Atlantic surface water temperature (upper) and surface water salinity (lower) between the Manabe and Stauffer (1988) model circulation schemes with and without thermohaline circulation. As can be seen, a turnoff of thermohaline circulation leads to a pronounced cooling and freshening of the waters in the northern Atlantic.

148 The Great Ocean Conveyor

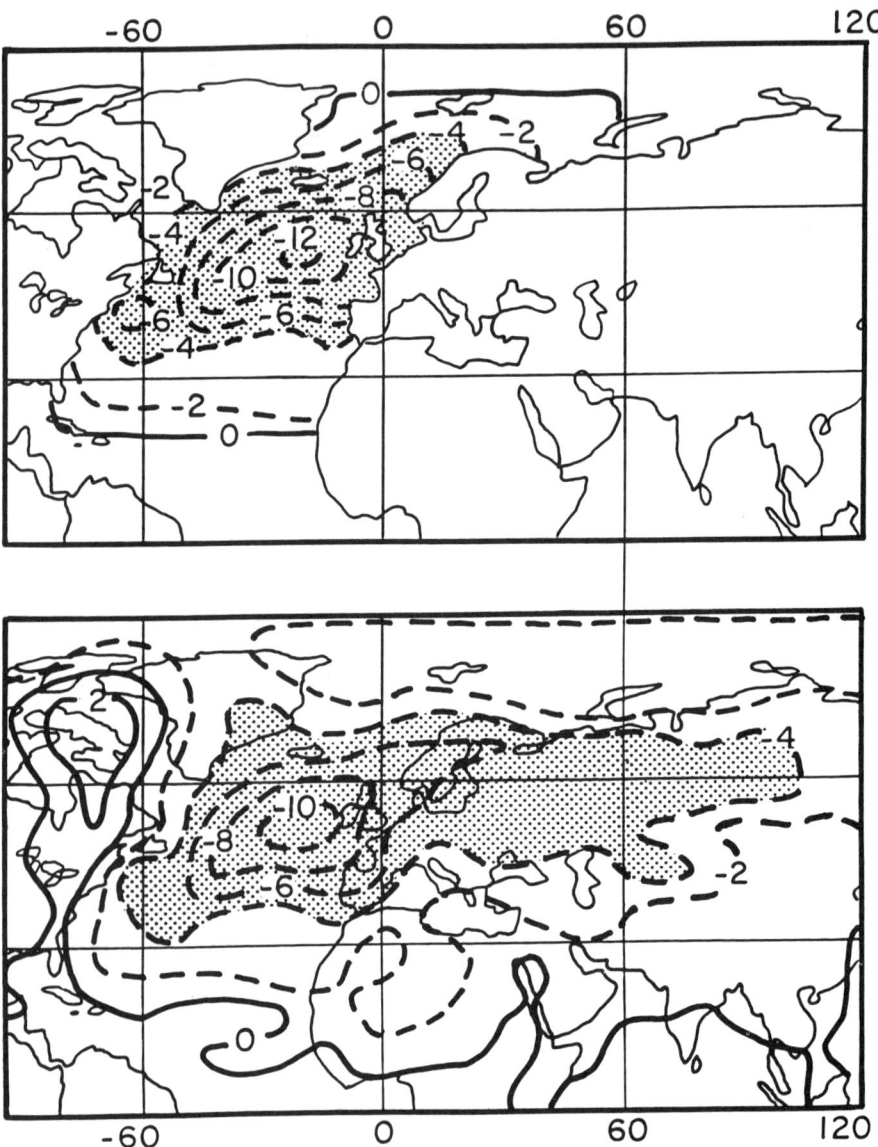

Figure 13. Results of an experiment carried out using the GISS atmospheric GCM (Rind et. al, 1986). Two runs were made which differed only in the surface ocean temperature assigned to the northern Atlantic (see upper panel). The resulting winter air-temperature change produced by this ocean temperature change is shown in the lower panel.

supplier of water to the region of deep water formation is the Gulf Stream. As the roots of this great current extend to a depth of 1500 or so meters, the properties of the water supplied to the NADW source region depend strongly on the depth at which the major transport takes place. A large contrast exists between the water from depths of less than 500 meters and water from depths of greater 800 meters (see figure 14). For example, the upper waters have too high a salinity and the deep waters too low a salinity to match that needed to generate NADW. As summarized in figure 15, when the inputs into the northern region of low salinity Bering Sea water and of fresh water (through river runoff and the excess of precipitation over evaporation) are taken into account the salinity of the aggregate upper limb supply water must be about 35.6‰. This value is intermediate between those for upper and lower Gulf Stream water. A similar situation exists for the nutrient constituents phosphate and silica. The deeper portion of the Gulf Stream carries too high concentrations and the upper portion too low concentrations to match that in outgoing NADW. Thus it appears that a mixture of upper and lower Gulf Stream water is required to create the salinity and nutrient content of new NADW.

Radiocarbon provides a crosscheck on this conclusion (see Broecker, in press a). The prenuclear $\Delta^{14}C$ value by 11‰ while radiodecay lowers it by only 5‰. The prenuclear $\Delta^{14}C$ value for Gulf Stream water was about -50‰, and that for lower Gulf stream water about -80‰. In order to obtain the desired input value of -72‰ requires that about 1 part upper Gulf Stream water be mixed with 2 parts thermocline water carried beneath the Gulf Stream. These proportions match reasonably well those required to create the phosphate and silicate contents of new NADW.

While the proportions of upper and lower Gulf Stream water necessary to achieve the salinity nutrient and radiocarbon contents of new NADW are not well defined, the temperature of the input water should lie between that of about 18°C for the upper water and that of about 6°C for the lower water. A 1 parts upper - 2 parts lower water mixture would have a temperature of 10°C.

ITS ACHILLES HEEL

The addition of fresh water to the northern Atlantic poses a constant threat to the conveyor. Northward of 40°N in the Atlantic, precipitation and continental runoff exceed evaporation by about 0.30 Sverdrups (Baumgartner and Reichel, 1975). In addition, the 1 Sverdrup of low salinity (S≅33.0‰) water entering the Arctic arm of the Atlantic through the Bering Straits contributes the equivalent of 0.06 Sverdrups of fresh water, bringing the total to 0.36 Sverdrups. When the conveyor is running at its current strength of 20 Sverdrups, this fresh water is efficiently swept away causing only an 0.63‰ reduction in the conveyor water's salinity as it passes through the northern Atlantic. If the conveyor were to progressively weaken, this salinity reduction would grow. It would be 0.94‰ at 15 Sverdrups, 1.26‰ at 10 Sverdrups... At some point the salinity reduction would become so large that deep water could no longer form. The conveyor would shut down. Were this to happen, fresh water would pool at the surface of the northern Atlantic much as it currently does in the northern Pacific creating a severe barrier to deep water formation.

Ocean GCM simulations by Manabe and Stauffer (1988) clearly demonstrate the role of this fresh water input. As shown in figure 12, for the conveyor-off mode pooling of fresh waters reduces the salinity surface waters in the northern Atlantic by about 3‰. Maier-Reimer and Mikolajewicz, 1989, using the Hamburg ocean GCM show that a modest dose of excess fresh water to the source region of NADW can kill the model's thermohaline circulation. Furthermore, the demise is abrupt, occurring on the time scale of a few decades. The rapidity of this response is not surprising for it depends on the residence time of water in the source region. At a flushing rate of 20 Sverdrups the entire volume of water contained in the Atlantic north of 45°N can

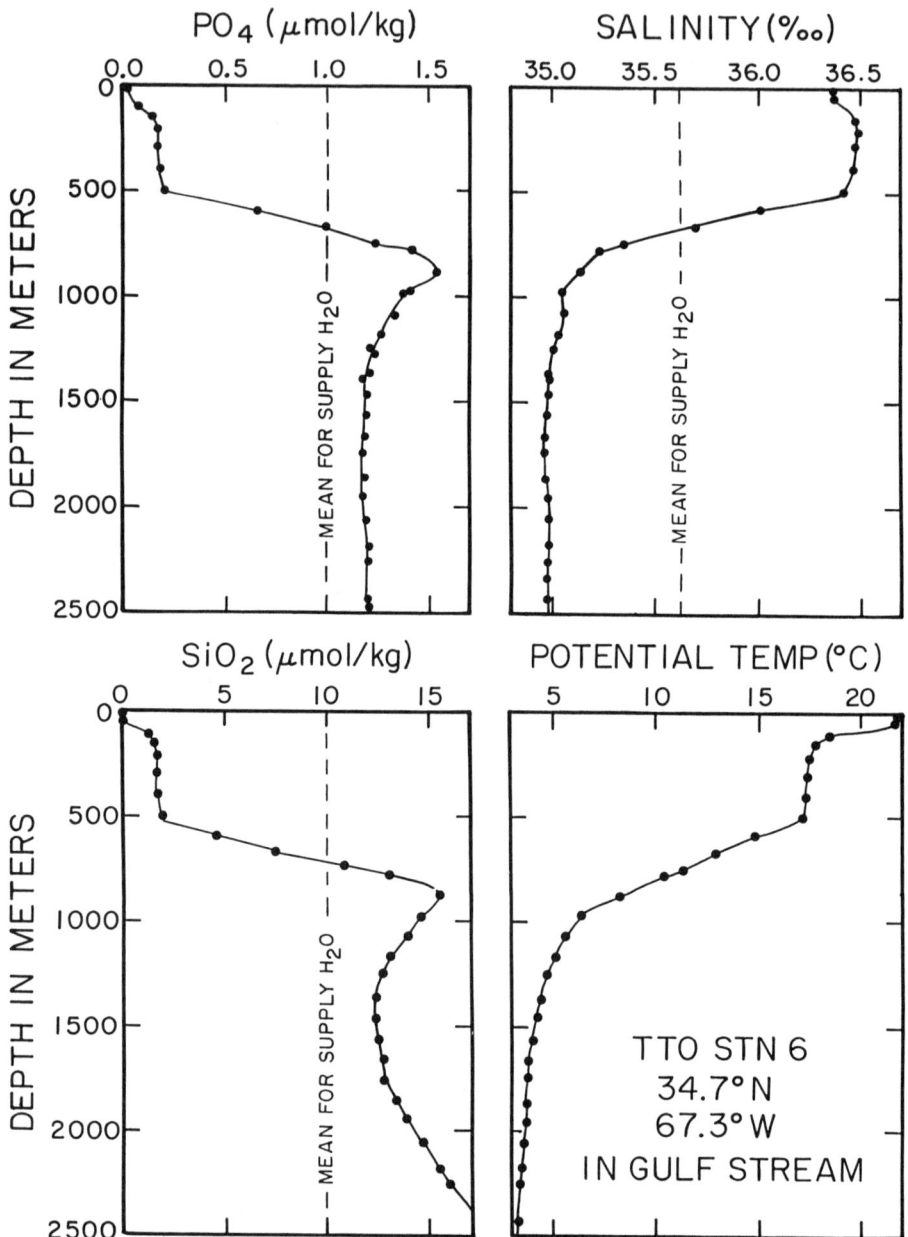

Figure 14. Phosphate, silica, salinity and potential temperature profiles at Transient Tracers in the Ocean survey station 6 in the Gulf Stream. Shown for comparison are the average values for these constituents required for the water constituting the upper limb source for NADW production (vertical dashed line).

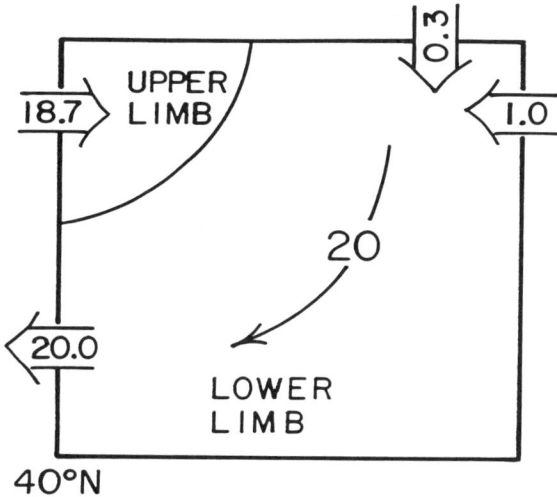

NORTHERN ATLANTIC'S SALT BUDGET

	SALINITY ‰	FLUX Sv
OUT		
NADW	34.95	20.0
	AVG. 34.95	TOTAL 20.0
IN		
BERING STS	33.0	1.0
VAPOR	0.0	0.3
UPPER LIMB	35.6	18.7
	AVG. 34.95	TOTAL 20.0

Figure 15. Water and salt budgets for the region in the northern Atlantic where NADW forms. The fluxes and salinities of new NADW, Bering Straits input and fresh water input are known. The water flux and salinity for the upper limb are obtained by difference.

be replaced in two decades!

We know of no ocean GCM experiment which shows how the conveyor circulation might be restarted. Because of the strong barrier created by the pooling of fresh water, this may prove to ba a tricky task. Microprocesses such as densification through brine formation beneath sea ice may have to be invoked.

ITS HISTORY

The best indicator of the past operation of the conveyor is the air temperature in the northern Atlantic basin. The reason is that, as we have already shown, turning on and off the conveyor causes 5 to 8°C changes in the air temperatures over Greenland and Europe. The most detailed record of air temperature over Greenland and Europe. The most detailed record of air temperature in the northern Atlantic basin is the isotope record preserved in the Greenland ice cap. This record (see figures 16 and 17) had different character during the last period of glaciation than during the present period of interglaciation (Dansgaard et al., 1971, Hammer et al., 1985). In figures 16 and 17 less negative values of ^{18}O concentration correspond to higher temperatures. The deeper the core, the farther back in time. During glacial time air temperatures over Greenland underwent excursions of the magnitude and abruptness expected if the conveyor were turning on and off, on an millennial time scale. By contrast, during the 9 or so thousand years since the period of glaciation came to a close, Greenland's air temperature has remained nearly constant. The impression I get from this is that a frenetic glacial conveyor became firmly locked in the on-position at the beginning of post glacial time and has remained so ever since.

A possible explanation for the behavior of the conveyor during glacial time is that when the northern end of the Atlantic Basin is surrounded by ice sheets, stable operation of Atlantic's circulation system is not possible (Broecker, et al., 1990b). Rather, because the ice sheets constitute a tremendous source of fresh water, circulation in the Atlantic tends to flip back and forth between the conveyor-on and conveyor-off modes. When the conveyor is operative its heat output tends to melt back the ice, releasing large amounts of fresh water to the Atlantic. The conveyor also efficiently exports excess salt from the Atlantic. The combination of meltwater dilution and salt export drives down the density of waters in the Atlantic until the point is reached where the conveyor can no longer function. It goes off. With the conveyor inoperative, the export of salt and dilution with meltwater are reduced to the point where water vapor export once again begins to enrich salt in the Atlantic. The salt content and hence also the density of Atlantic waters steadily rise until the conveyor turns on again. This cycle repeats over and over again.

Although a rigorous demonstration that such an oscillator did operate is not possible, the facts we do have lend strong support to this scenario. First of all, we know that the 40 million cubic kilometers of excess ice present in the ice sheets of the northern hemisphere during peak glacial time began to melt about 13,000 years ago and was gone by about 8000 years ago. Thus the average flux of meltwater during this interval must have been about 0.25 Sverdrups. Assuming this to be the magnitude of the melting rate during the proposed conveyor-on episodes then the dilution of salt due to meltwater would have been comparable to today's rate of vapor export. If the assumption is made that then as now the conveyor was exporting salt at a rate comparable to the rate it was being enriched through vapor export, then during the conveyor-on episodes, salt would have been diluted at a rate corresponding to the input of meltwater. When the conveyor was off the dominant important term in the salt budget would be vapor export. As yet we do not have an adequate estimate of the rate of water vapor export in the presence of an ice sheet. However, since vapor export is dictated by the interaction of planetary winds with mountain ranges, the rate may not have been very different from today's.

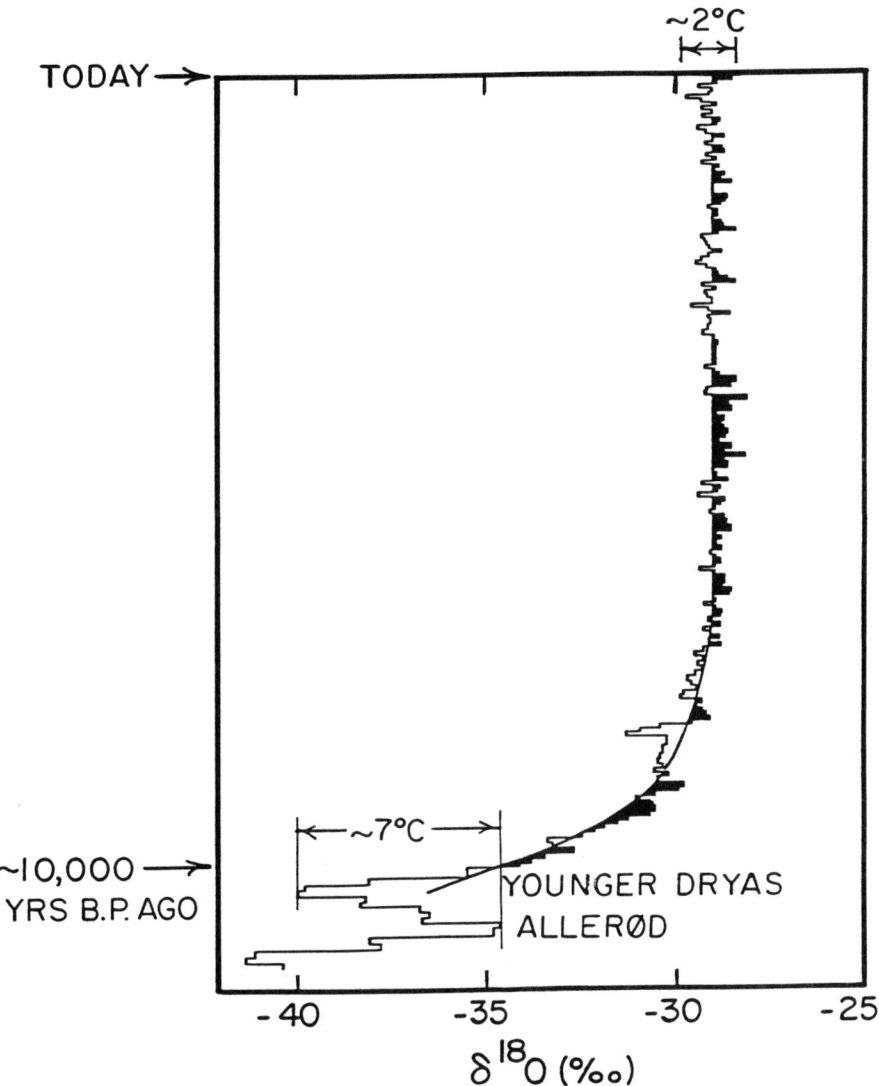

Figure 16. Oxygen isotope record for the Camp Century Greenland ice core (Dansgaard et al., 1971 covering the period from about 13,000 radiocarbon years B.P. to the present.

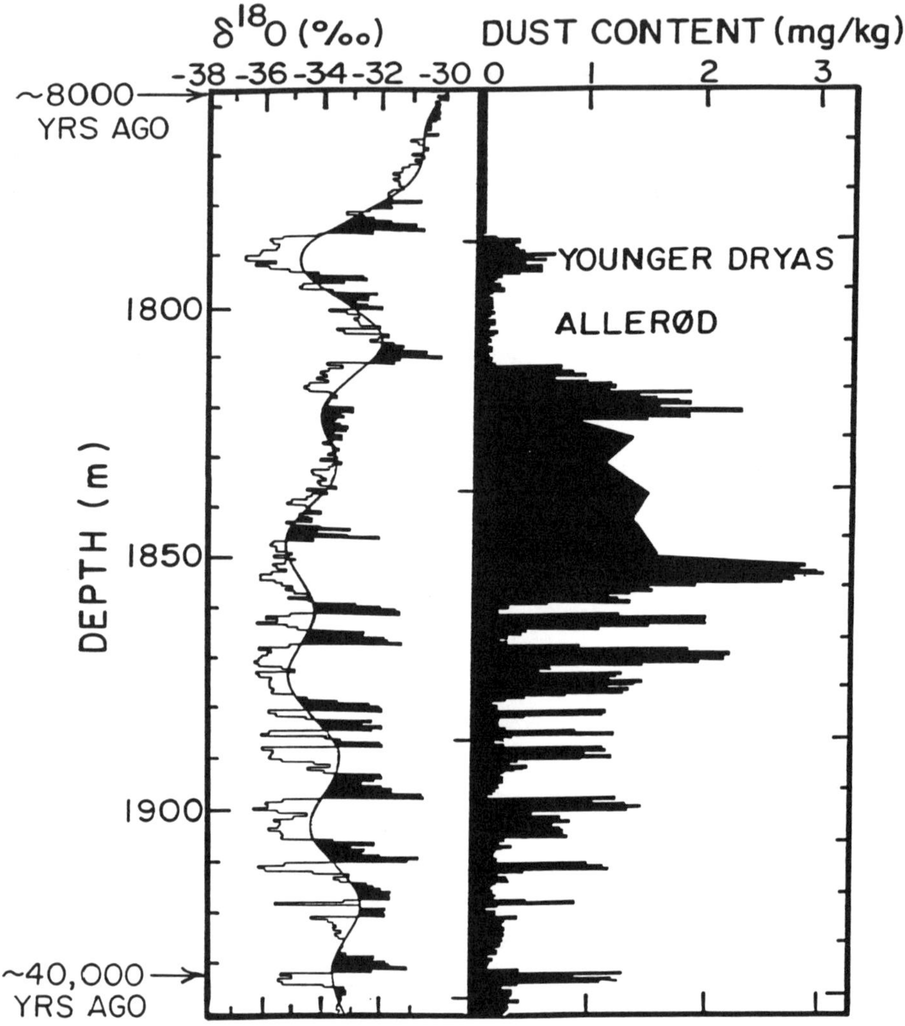

154 The Great Ocean Conveyor

Figure 17. Oxygen isotope and dust records for that part of the Dye 3 Greenland ice core covering the time period ~45,000 to ~8,000 years B.P. (Hammer et al., 1985). Note that many of the oxygen isotope events are characterized by rapid warmins followed by more gradual coolings.

Additional support comes from the observation that the average duration of individual warm and cold episodes recorded in Greenland ice ranges from about one half to two millennia. This spacing is consistent with that expected from the salt oscillation hypothesis. A vapor loss of 0.35 Sverdrups removes a layer 15 cm thick from the Atlantic each year. If uncompensated by salt export, the salt content of Atlantic waters will rise 1.4 gm/liter per millennium! As a salt buildup or reduction of 1 to 2 gm/liter for the Atlantic is about what is needed to tip the balance between conveyor-on and conveyor-off, a match exists between the observed timing and that predicted timing (Birchfield and Broecker, 1990). This is a strong point in favor of the oscillator hypothesis!

For an oscillator to function requires a combination of a long time constant drift (in this case the buildup or drawdown of salt) and a short time constant stabilization mechanism. In the previous section it was shown that a tendency exists for fresh water to pool at the surface of the northern Atlantic. When the conveyor comes on this pool is quickly destroyed raising the salinity in NADW source region. This stabilizes the conveyor in its on-position. Similarly when the conveyor stops the pool quickly reappears stabilizing the conveyor in its off position.

As can be seen in figure 17, many of the Greenland temperature cycles are characterized by abrupt warmings followed by more gradual coolings. The salt oscillator hypothesis provides a natural explanation for this shape (see Broecker, in press b). The abrupt warmings are caused by turn-ons of the conveyor. Immediately following such a reinitiation, the conveyor runs with extra vigor. The reason is that in order to overcome the fresh water pool present in the northern Atlantic when the conveyor is inoperative, the salinity of the Atlantic would have to rise above the level required for steady state operation. Thus when the conveyor comes on the buoyancy contrast between deep water formed in the northern Atlantic and deep water present in the remainder of the ocean will be unusually large. This excess density will drive the conveyor at an unusually high rate. As a consequence, a greater amount of heat will be released to the atmosphere over the north Atlantic. However, once operative, the conveyor's strength will steadily wane. The reason is that the combination of dilution with meltwater and export of excess salt will lessen the buoyancy contrast between deep waters inside and outside the Atlantic. As the strength of the conveyor wanes, the amount of heat given off to the atmosphere over the northern Atlantic will also decrease, causing air temperature to drop. Eventually the conveyor will shut down, abruptly cutting off the supply of ocean heat. The atmospheric temperature cycle generated in this way resembles that seen in the ice core record (see figure 18).

Only for the most recent of these cycles do we have sufficient auxiliary evidence to add muscle to this scenario. The abrupt warmings at about 12,700 10,000 radiocarbon years ago, provide smoking guns in this regard. Not only the oxygen isotope record in ice cores (Dansgaard, et al., 1989) but also that in lake sediments on the European continent (Lötter and Zbinden, 1989) demonstrate that both of these warmings were accomplished in only 50 years! Further, the geographic pattern of these temperature changes associated with Younger Dryas is as expected if they were caused by conveyor turn ons. Pronounced changes are confined to latitudes greater than 40°N and extend from the maritime provinces of Canada and the ice cap of Greenland on the west across the northern Atlantic, the British Isles and Scandinavia into Russia on the east (see Rind, et al., 1986 for summary). Finally, Boyle and Keigwin 1987, have shown, based on carbon isotope and cadmium concentration measurements on benthic foraminifera shells from a deep sea core from the vicinity of Bermuda, that water of Antarctic origin flooded the western basin of deep Atlantic during Younger Dryas time, confirming that a shutdown of deep water formation accompanied this cold event.

ITS FUTURE

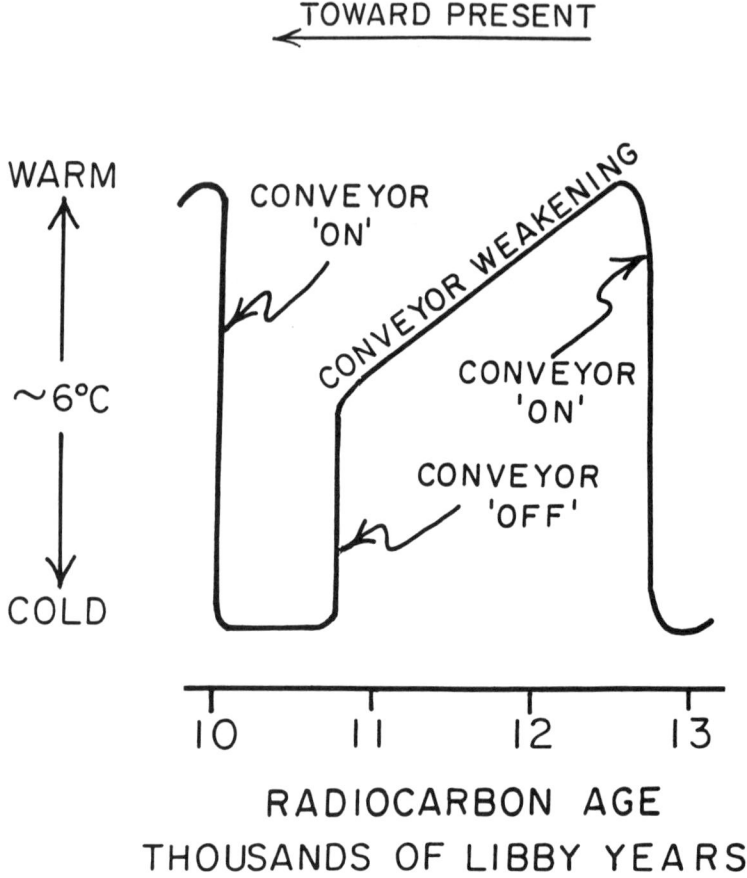

Figure 18. Diagrammatic representation of the temperature record for the northern Atlantic basin from just before 13,000 to just after 10,000 radiocarbon years before present. The features shown in this diagram appear in records of ^{18}O in ice (Dansgaard et al. 1971), of ^{18}O in lake sediment (Lötter et al. 1989), of ocean planktonic foraminifera (Ruddiman and McIntyre, 1981), of pollen and of beetles (Atkinson et al., 1987). Of particular interest is rapidity of the warmings at 12,700 years BP and at 10,000 years BP. Detailed measurements in the Dye 3 ice core and in varved lake sediments reveal that these warmings took place in about 50 years (Dansgaard et al. 1989)!

When the Natural History article containing the great global conveyor belt diagram appeared, the editor put a sales "stimulator" on the cover which state "Europe beware: The big chill may be coming." At the time I was much annoyed because no mention of the conveyor's future was made in the article. To make matters worse, even after reading the article itself, many people were left with the impression that I was warning of an imminent conveyor shutdown. The fact is that at that time I thought that the coming greenhouse warming would, if anything, strengthen the conveyor by increasing the rate of vapor loss from the Atlantic Basin. I had not given serious thought to the question as to whether any changes associated with man's activities might threaten the conveyor.

The first activity which comes to mind in this regard is the rerouting of water for agricultural use. Irrigation projects increase the recycling of water on the continents and thereby change the point at which a given water molecule reenters the ocean. Of particular interest in this regard is the Russian proposal to divert the great northward flowing Siberian Rivers to the south for agricultural use. The result of such a diversion would be to increase the vapor loss from the Atlantic Basin for instead of flowing out of river mouths into the Arctic, the water would move through the atmosphere across Asia into the Pacific Basin. The longterm result would be to strengthen the conveyor.

In addition to increasing vapor export from the Atlantic Basin, the greenhouse warming will increase the transport of fresh water to the northern Atlantic. On the short term (i.e. decades), the salinity decrease created in northern surface waters would be more important than the Atlantic-wide salinity increase caused by increased vapor loss from the Atlantic Basin. The reason is that the replacement time for waters in the northern Atlantic is shorter than the replacement time for waters in the upper limb of the conveyor. So if a threat to the conveyor is in the making, it is most likely to come in this way. To be on guard we should pay close attention to the climate and oceanography of the northern Atlantic Basin. The finding by Brewer et al., 1983 that the salinity of Atlantic deep waters to the north of 50°N declined between 1972 and 1981 and the finding by Schlosser et al., 1991 that deep ventilation of the Greenland Sea was shutdown during the 1980's are indications that changes do occur. Unfortunately we have no way to tell whether these changes signal natural fluctuations or anthropogenically driven trends.

CONCLUSIONS

The conveyor is only one of many elements which together constitute the Earth's climatic system. It stands out because of its dramatic impact on the climate of a single region on our planet. We must keep in mind however, that the abrupt global warmings which heralded the termination of the last major glaciation can certainly not be explained by the conveyor alone (Broecker and Denton, 1989, 1990). Rather, elements of the system, such as the Hadley cell, which influence cloudiness and atmospheric water vapor content must also have been involved. The challenge of the Global Change Research Initiative is to understand the complex web of interactions which tie together the operation of these diverse elements.

ACKNOWLEDGEMENTS

I would like to thank the Exxon Corporation and Lawrence Livermore National Laboratory for their generous support of my research. Instead of requiring me to write long proposals and reports, they encourage me instead to put this effort into articles such as this. My research on ocean chemistry is supported by DOE's CO_2 program grant no. LLNL B130547; that on the climate history of the Atlantic by NSF Climate Dynamics grant no. ATM 89-21306 and by NOAA grant no. NA90-AA-D-AC520; and that on atmospheric vapor transport by EPRI grant RP 2333-6.

REFERENCES

Atkinson, T.C., K.R. Briffa and G.R. Coope, 1987; Seasonal temperatures in Britain during the past 22,000 years, reconstructed using beetle remains, Nature, vol. 325.

Baumgartner, A. and Reichel, E., 1987 Die Weltwasserbilanz, Oldenbourg, München.

Birchfield, G.E. and W.S. Broecker, 1990; A salt oscillator in the glacial northern Atlantic? Part II: A 'Scale Analysis' Model, Paleoceanography, 5 (6), 835-843.

Boyle, E.A. and L. Keigwin, 1987; North Atlantic thermohaline circulation during the past 20,000 years linked to high-latitude surface temperature, Nature, 330, 35-40.

Brewer, P.G., W.S. Broecker, W.J. Jenkins, P.B. Rhines, C.G. Rooth, J.H. Swift, T. Takahashi and R.T. Williams, 1983; A Climatic freshening of the deep Atlantic north of 50°N over the past 20 years, Science, 222,1237-1239.

Broecker, W.S. and T.H. Peng, 1982; Tracers in the Sea, Eldigio Press, Palisades, NY, 690 pp.

Broecker, W.S., D. Peteet and D. Rind, 1985; Does the ocean-atmosphere have more than one stable mode of operation? Nature, 315, 21-25.

Broecker, W.S., 1987; The Biggest Chill, Natural History Magazine, 74-82.

Broecker, W.S., and G.H. Denton, 1989; The role of ocean-atmosphere reorganizations in glacial cycles, Geochimica et Cosmochimica Acta, 53, 2465-2501.

Broecker, W.S. and G.H. Denton, 1990; What Drives Glacial Cycles? Scientific American 49-56.

Broecker, W.S., T.-H. Peng, J. Jouzel, and G. Russell, 1990a; The magnitude of global fresh water transports of importance to ocean circulation, Climate Dynamics, 4, 73-79.

Broecker, W.S., G. Bond, M. Klas, G. Bonani, and W. Wolfli, 1990; A salt oscillator in the glacial North Atlantic? 1. The Concept, Paleoceanography, 5 (4), 469-477.

Broecker, W.S., S. Blanton, and T. Takahashi, W. Smethie and G. Ostlund, 1991a; Radiocarbon decay and oxygen utilization in the dep Atlantic Ocean, Global Biogeochemical Cycles, 5, 87-117.

Broecker, W.S., A. Virgilio, and T.-H. Peng, 1991b; Radiocarbon Age of Water in the Deep Atlantic Revisited, Geophys. Res. Lett., 18 (1), 1-3.

Broecker, W.S., The Strength of the Nordic Heat Pump, 13,500 to 9500 B.P., in press, Erice volume.

Chamberlin, T.C., 1906; On a possible reversal of deep-sea circulation and its influence on geologic climates, The Journal of Geology, 363-373.

Dansgaard, W., S.J. Johnsen, H.B. Clausen, and C.C. Langway, Jr., 1971; In The Late Cenozoic Glacial Ages, (ed. by Karl K. Turekian), New Haven and London, Yale Univ. Press, pp. 37-56.

Dansgaard, W., J.W.C. White, and S.J. Johnson, 1989; The abrupt termination of the Younger Dryas climate event, Nature, 339:532-533.

Gordon, A.L., and A.R. Piola, 1983; Atlantic Ocean upper layer salinity budget, Jour. Phys. Oceanogr., 13, 1293-1300.

Gordon, A.L., 1985; Indian-Atlantic Transfer of Thermocline water at the Agulhas retroflection, Science, 227, 1030-1033.

Gordon, A., 1986, Interocean Exchange of Thermocline Water, Jour. Geophys. Research, 91 (C4), 5037-5046.

Hammer, C.U., H.B. Clausen, W. Dansgaard, A. Neftel, P. Kristinsdottir, and E. Johnson, Continuous impurity analysis along the Dye 3 deep core. In Greenland Ice Core: Geophysics, Geochemistry, and the Environment (eds. C.C. Langway, H. Oeschger, and W. Dansgaard); Amer. Geophys. Union Mon. 33, pp. 90-94.

Levitus, S., 1982; Climatological Atlas of the World Ocean, NOAA Professional Paper No. 13, (U.S. GPO, Washington, DC).

Lotter, A.F. and H. Zbinden, 1989; Late-Glacial pollen analysis, oxygen-isotope record, and radiocarbon stratigraphy from Rotsee (Lucerne), Central Swiss Plateau, Eclogae geol. Helv. 82 (1), 191-202.

Maier-Reimer, E. and U. Mikolajewicz, 1989; Experiments with and OGCM on the cause of the Younger Dryas, in Oceanography, (A. Ayala-Castanares, W. Wooster, and A. Yanez-Arancibia, ed.), 87-100, UNAM Press, Mexico.

Manabe S., and R.J. Stauffer, 1988; Two stable equilibria of a coupled ocean-coupled ocean-atmosphere model, J. Climate, 1, 841-866.

Oort, A.H., 1983; Global atmospheric circulation statistics, 1958-1973, NOAA Professional Paper 14.

Rind, D. D. Peteet, W.S. Broecker, A. McIntyre and W. Ruddiman, 1986; The impact of cold North Atlantic sea surface temperatures on climate: Implications for the Younger Dryas cooling (11-10k), Climate Dynamics, 1, 3-33.

Ruddiman, W.F. and A. McIntyre, 1981; The North Atlantic Ocean during the last deglaciation, Paleogeogr., Palaeoclim., Palaeoecol, 35, 145-214.

Schlosser, P, G. Bönisch, M. Rhein and R. Bayer, 1991; Reduction of Deepwater Formation in the Greenland Sea during the 1980s: Evidence from Tracer Data, Science, 251, 1054-1056.

Wüst, G. 1935; Die Stratosphäre. Wissenschaftliche Ergebnisse der deutschen atlantischen Expedition "Meteor", 6, 109:288.

Wüst, G. and A. Defant, 1936; Atlas zur Schichtung und Zirkulation des Atlantischen Ozeans. Schnitte und Karten von Temperatur, Salzgehalt und Dichte. Wissenschaftliche Ergebnisse der deutschen atlantischen Expedition "Meteor", 6 (Atlas), 103pp.

Zaucker, F. and W.S. Broecker, Atmospheric water vapor transport from a general circulation model, in press, Jour. Geophysical Research.

TRACE GASES IN THE ATMOSPHERE: TEMPORAL AND SPATIAL TRENDS

Donald R. Blake
Department of Chemistry
University of California
Irvine, CA 92717

Our planet is continuously bathed in solar radiation. Although we who are confined to a fixed location on the globe experience day and night, the earth does not. It is always day in the sense that the sun is shining on half of the globe. Much of the incoming solar radiation, about 30 percent, is scattered back to space by clouds, atmospheric gases and particles, and objects on the earth. The remaining 70 percent is, therefore, absorbed mostly at the earth's surface. This absorbed radiation gives up its energy to whatever absorbed it, thereby causing its temperature to increase. Because solar radiation is absorbed continuously by the earth, it might be supposed that its temperature should continue to increase. It does not, of course, because the earth also emits radiation, the spectral distribution of which is quite different from that of the incoming solar radiation. The higher the earth's temperature, the more infrared radiation it emits. At a sufficiently high temperature, the total rate of emission of infrared radiation equals the rate of absorption of solar radiation. Radiative equilibrium has been achieved, although it is a dynamic equilibrium, and absorption and emission go on continuously at equal rates. The temperature at which this occurs is called the radiative equilibrium temperature of the earth. This is an average temperature, not the temperature at any one location or at any one time. It is merely the temperature that the earth, as a blackbody, must have in order to emit as much radiant energy as the earth absorbs solar energy.

All objects emit radiation continuously. If their temperatures are typical of those on the surface of the earth, most of the radiation will be at infrared wavelengths. Some of this radiation is absorbed by the atmosphere, and this radiation is reemitted to the Earth, making its atmosphere warmer than it would otherwise be. Downward radiation from the atmosphere in turn makes the ground warmer than it would otherwise be. The extent to which the atmosphere absorbs infrared radiation is, therefore, of great importance in determining the temperature of the earth's surface. (See the chapter by Thomas Ackerman.) If the atmosphere absorbed no infrared radiation, the average surface temperature, as well as the air in contact with it, would be well below freezing, about

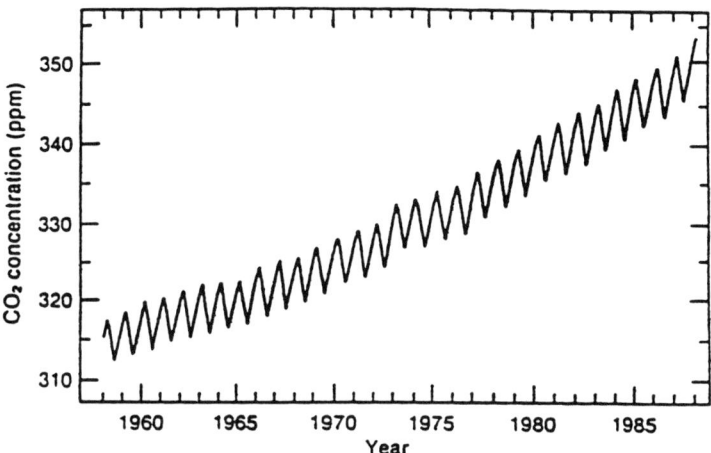

Fig. 1. Variation with time of the concentration of atmospheric carbon dioxide in parts per milion (ppm) of dry air, observed with a continuously recording, nondispersive infrared gas analyzer at Mauna Loa Observatory, Hawaii. (From ref. 1.)

255 K (-18 C). However, the atmosphere does absorb infrared radiation and, as a consequence, surface temperatures are higher than they would otherwise be. This warming of the atmosphere and earth's surface has come to be known as the greenhouse effect.

Not all atmospheric gases absorb infrared radiation to the same degree. Indeed, the most abundant ones by far, nitrogen and oxygen, are the least absorbing. Of far greater importance are water vapor and much less abundant gases such as carbon dioxide (CO_2) and ozone (O_3). The atmospheric concentration of CO_2 is about 355 parts per million by volume (ppmv) and is increasing at a yearly rate of about 2 ppmv. A plot of the CO_2 concentration with time is shown in figure 1. Fossil fuel combustion is the major source of man-made carbon dioxide. About 60% of this man-made carbon dioxide remains in the atmosphere while the rest is absorbed by the oceans or biosphere. (The chapter by James Kasting and James Walker and the chapter by Mark Trexler discuss in further detail the geochemical and biological cycles of carbon.) Of major concern is whether the oceans and biosphere can maintain pace with carbon dioxide production, thus keeping the fraction at 60%. If not, the yearly increase in carbon dioxide will continue to rise.

Table 1 lists the greenhouse gases that are on the rise. The table summarizes their preindustrial and current

Table 1. Concentrations of Important Greenhouse Gases

	CO_2	CH_4	CFC-12	CFC-11	N_2O
Atmospheric Concentration in 1990 (ppmv)	355	1.72	.0005	.00027	.31
Pre-Industrial Concentration (ppmv)	280	0.8	0	0	.288
Current Rate of Change (ppmv) (% per year)	1.8 0.5%	.013 1%	.00002 4%	.0001 4%	.001 0.3%
Atmospheric Lifetime (years)	50-200	10	130	65	150

concentrations, their current rates of increase and their atmospheric lifetimes. At first glance at this table, one might remark that the impact of a gas such as CCl_2F_2 (CFC-12), whose concentration levels are about 0.0005 ppmv and are increasing annually at 0.00002 ppmv, would be trivial compared to that of carbon dioxide, whose concentration levels are around 355 ppmv. However, the greenhouse effect occurs through the absorption of infrared radiation into specific vibrational frequencies of polyatomic molecules in the atmosphere. The 355 ppmv of CO_2 already in the atmosphere strongly absorbs the infrared radiation that is at the right wavelengths to stimulate its molecular vibrations. Increasing the concentration to 356 ppmv will increase the probability of absorption a little, but it is hard to increase the total absorption very much. On the other hand, the amounts of halocarbons (CFC-12), CFC-11 (CCl_3F), CFC-13 (CCl_2FCClF_2), methylchloroform (CH_3CCl_3), carbon tetrachloride (CCl_4), nitrous oxide (N_2O) and methane (CH_4) in the atmosphere are much smaller, and absorption at their vibrational frequencies is very far from saturation. Figure 2 shows the absorption spectra of some of these gases. Some of the absorptions happen to fall into infrared "windows", in which absorption by H_2O, CO_2 and O_3 are all very weak or nonexistent, and the individual molecules can be as much as 20,000 times more effective in retaining infrared radiation than additional CO_2 molecules when the mixing ratio of the latter is already 355 ppmv. When one considers the effects from

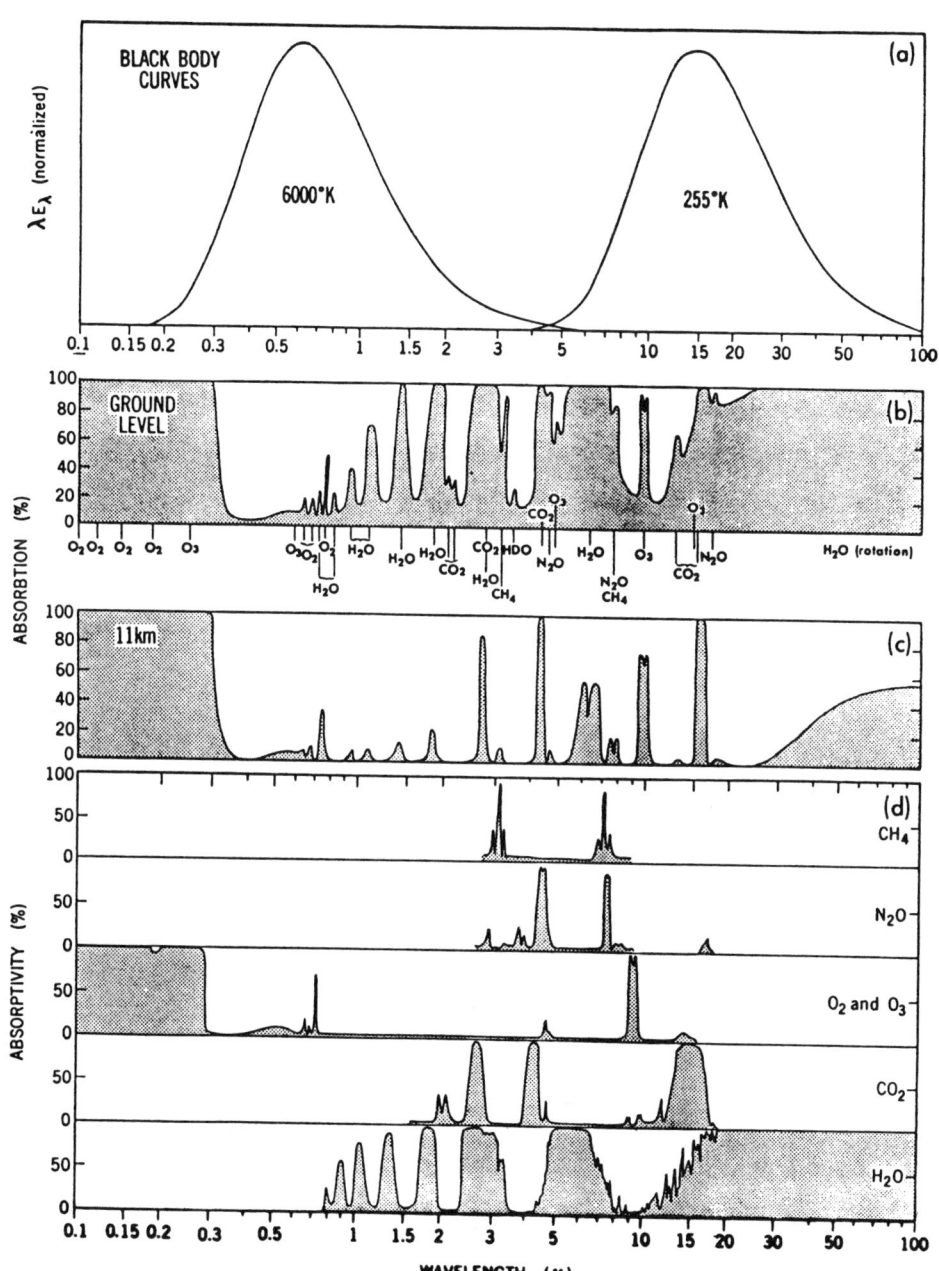

Fig. 2. Emission and absorption spectra. (a) Black-body curves for the sun (6000 K) and Earth (255 K). (b) and (c) Absorption by the atmosphere at two altitudes. (d) Absorptivity by atmospheric gases. (From ref. 2.)

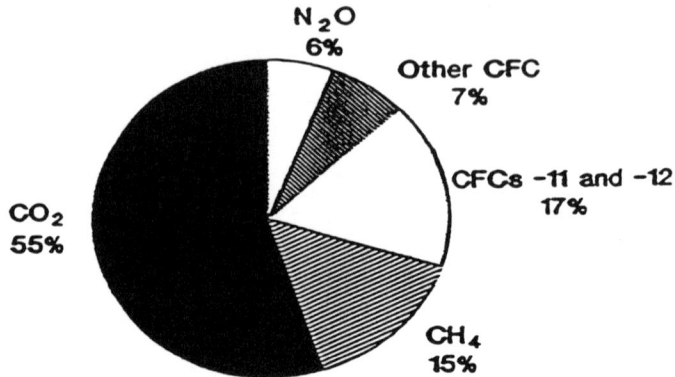

Fig. 3. Contribution to future global warming for selected greenhouse gases. These values reflect absorption properties, emission rates and atmospheric residence times for the various gases. (From ref. 3.)

increasing concentrations of these halocarbons, N_2O, CH_4 and tropospheric O_3, etc., the combined absorptions from increments in these gases over the past decade become about as important as the increment in CO_2 in contributions to the anticipated overall increase in the greenhouse effect. The relative contributions are shown in the pie chart in figure 3.

Trace gas concentrations in the atmosphere reflect, in part, the overall metabolism of the biosphere and the broad range of human activities such as agriculture, production of industrial chemicals, and combustion of fossil fuels and biomass. There is dramatic evidence that the composition of the atmosphere is now changing due to increased gaseous emissions associated with human activities.

Halocarbons
    Detailed measurements of various halocarbons have been conducted since 1978 by a number of research groups throughout the United States. Although the absolute concentrations of individual species may vary by as much as 10% between research groups, there is generally good agreement concerning the overall trends.
    Chlorofluorocarbons 12 and 11 are the most widely used CFCs at this time. They are globally used in many and varying applications, including aerosol propellants, refrigerants and foaming agents for plastics. Concentration versus time plots for these species are given

in Figures 4 and 5. Both CFC-12 and CFC-11 are increasing at about 4% per year and are now found in the atmosphere at concentrations of about 500 pptv and 270 pptv, respectively. (One pptv is one part per trillion by volume, or 10^{-12}.) These gases linger in the atmosphere for 130 and 65 years, respectively. But the production and consumption of these substances and a number of other halogenated CFCs and halons will soon be limited by the Montreal Protocol, which many governments signed in 1987. (See the chapter by Richard Benedick.)

Carbon tetrachloride is used predominantly as a chemical feedstock in the production of CFCs 11 and 12, leading to relatively little emission of carbon tetrachloride to the atmosphere. A small fraction is still used as a solvent in chemical and pharmaceutical production processes. Its use as a grain fumigant is declining. The observed rate of increase is about 2% and the current atmospheric concentration is approximately 120 pptv.

Stabilized methylcholoroform has been marketed since the early 1960s. Its principal use has always been the industrial degreasing of metallic or metaloplastic pieces. It is widely used for cold-cleaning processes in the engineering industry. It is also used as a solvent in adhesives, varnishes and paints where low flammability and low toxicity are important. Sales of methylchloroform grew rapidly in the 1960s and early 1970s, when it replaced tri- and perchloro-ethylene and carbon tetrachloride in many industrial applications. The atmospheric concentration of methylchloroform in 1991 is

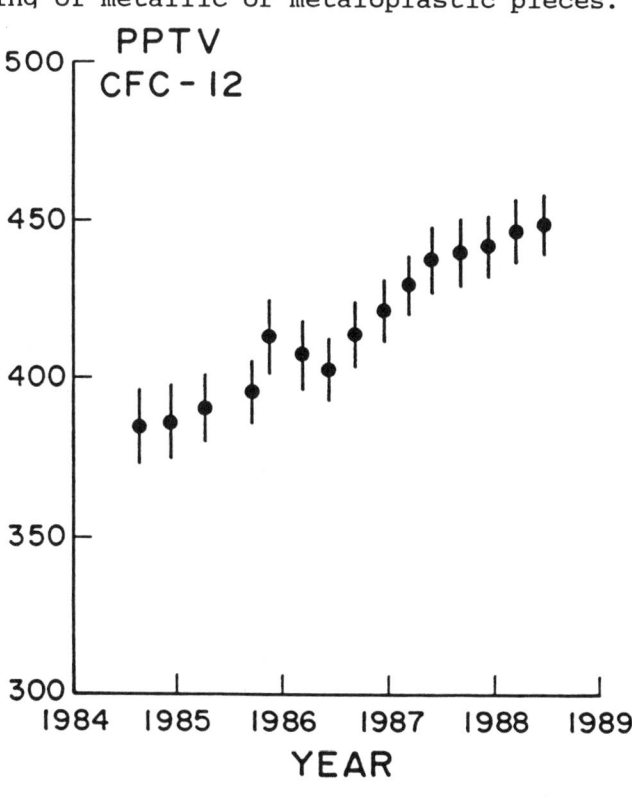

Fig. 4. CFC-12 mixing ratio vs. time.

about 150 pptv and is increasing at about 5 pptv per year. Global emissions for methylchloroform are actually higher than those of CFC-12 yet the overall rate of increase is less. This reflects the difference in atmospheric lifetime between these two species: 120 years for CFC-12 and about 7 years for methylchloroform. The major removal process is by hydroxyl radical attack in the troposphere as shown in reaction (1).

$$CH_3CCl_3 + HO \longrightarrow CH_2CCl_3 + H_2O \qquad (1)$$

The only known removal process for CFC-12 is photolysis in the mid to upper stratosphere. The relatively short lifetime of methylchloroform is attributable to its carbon-hydrogen bonds, which are relatively reactive with OH. Thus, if the currently used CFCs are replaced by compounds which contain a hydrogen bond, the atmospheric lifetimes will be greatly reduced.

The most recent addition to CFC monitoring among many research groups is that of CFC-113. This compound is used largely as a solvent to clean and deflux sophisticated electronic assemblies and components. Its future use is vulnerable to competing systems, changes in electronics technology and possible requirements to reclaim the solvent. One can see from the plot of concentration versus time given in Figure 6 that the large scale use of this particular species began in the late 1970s, about the time regulation of CFCs 11 and 12 began in the US. But

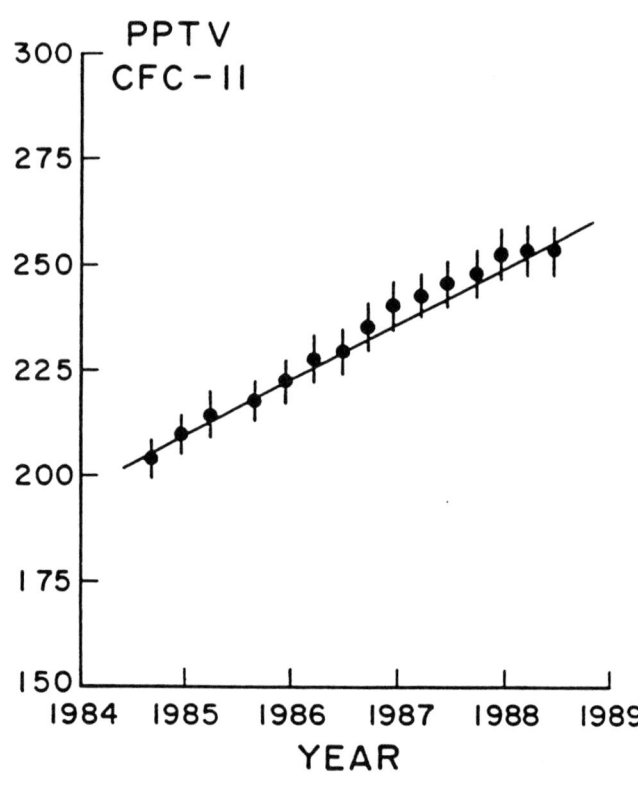

Fig. 5. CFC-11 mixing ratio vs. time.

CFC-113 is not likely to be a beneficial substitute for CFC-11 or CFC-12 because, like these other CFCs, its atmospheric lifetime is long--over 100 years. Although CFC-113 is a difficult species to measure accurately, the atmospheric concentration at the beginning of 1991 was about 80 pptv and it has had a yearly increase of about 6 pptv.

All of the CFCs are not only very efficient at absorbing terrestrial infrared radiation but also play a major role in transporting chlorine atoms into the stratosphere, where they can participate in the catalytic destruction of the ozone layer. All of these gases have very long atmospheric lifetimes, ranging from 50 to 120 years. Thus these species drift around until they reach the upper stratosphere. During this lengthy journey, these compounds absorb outgoing radiation emitted by the earth and redirect a large portion back towards the ground. Once these gases are high enough in the stratosphere, they continue to absorb radiation, but this time they absorb solar ultraviolet radiation. In this absorption process, a chlorine atom is ejected from the halocarbon molecule.

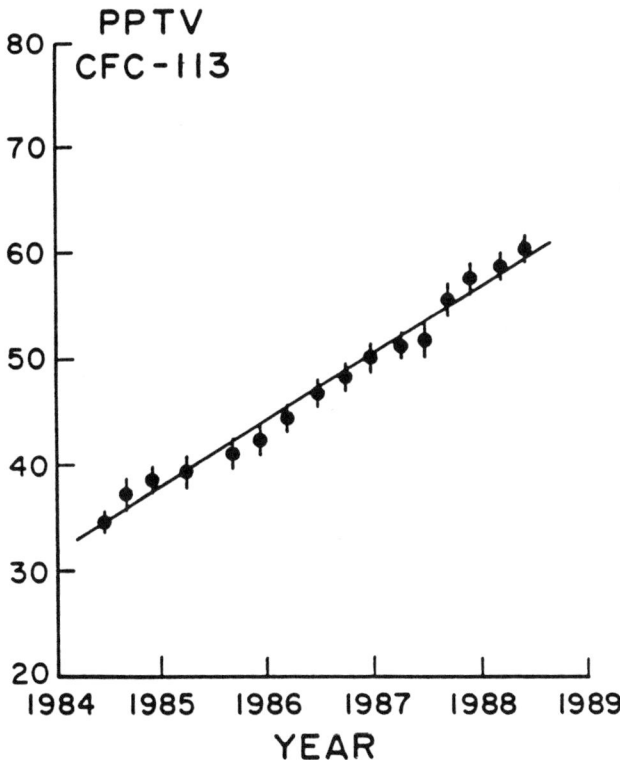

Fig. 6. CFC-113 mixing ratio vs. time.

At this point, we switch from a greenhouse problem to an ozone layer problem. However, because of the extremely long atmospheric lifetimes for these halocarbons, they will have contributed to the greenhouse warming for decades or even centuries before their chlorine begins the catalytic destruction of the ozone layer.

Nitrous Oxides and Methane
Like the CFCs, nitrous oxide has a long lifetime, about 150 years, and is destroyed in the stratosphere. There, nitrous oxide reacts with oxygen atoms to form nitric oxide, which can catalyze destruction of ozone. However, unlike the halocarbons, whose origins are unquestionably industrial, nitrous oxide has mostly natural sources. The main natural source of nitrous oxide is bacterial. Humans may be increasing nitrous oxide emissions from soils by using ammonia and urea fertilizers. The added nitrogen eventually is returned to the atmosphere, partly as nitrous oxide. Burning coal, which contains organic nitrogen compounds, is another anthropogenic source of nitrous oxide as is the manufacture of nylon. The rate of increase observed for nitrous oxide is about 0.3% or 1 ppbv.

The final greenhouse gas to be discussed here is methane. Bacteria produce methane by anaerobic fermentation in wet locations where oxygen is scarce: swamps, peat bogs, other natural wetlands, paddies, and the intestinal tracts of cattle, sheep and termites. Oil and natural gas exploitation appears to be a significant source. Leakage in the transmission of natural gas has recently been postulated as a possible significant source of atmospheric methane. Studies of the carbon-14 content of atmospheric methane indicate that a least 80% must have a biological origin. The sources and possible strengths are displayed in Figure 7.

Increases in atmospheric methane could be due to the greater area devoted to rice cultivation or the increased number of cattle. However, increases in atmospheric methane may also be caused by rising levels of carbon monoxide from combustion processes. Both methane and carbon monoxide are removed from the atmosphere largely by reaction with hydroxyl radicals, as shown by reactions 2 and 3, below:

$$HO + CO \longrightarrow H + CO_2 \qquad (2)$$

$$HO + CH_4 \longrightarrow CH_3 + H_2O \qquad (3)$$

With both gases competing for a limited amount of hydroxyl, the average lifetime of a methane molecule may be getting longer. A plot of the methane concentration versus time is given in Figure 8, while a latitudinal distribution is

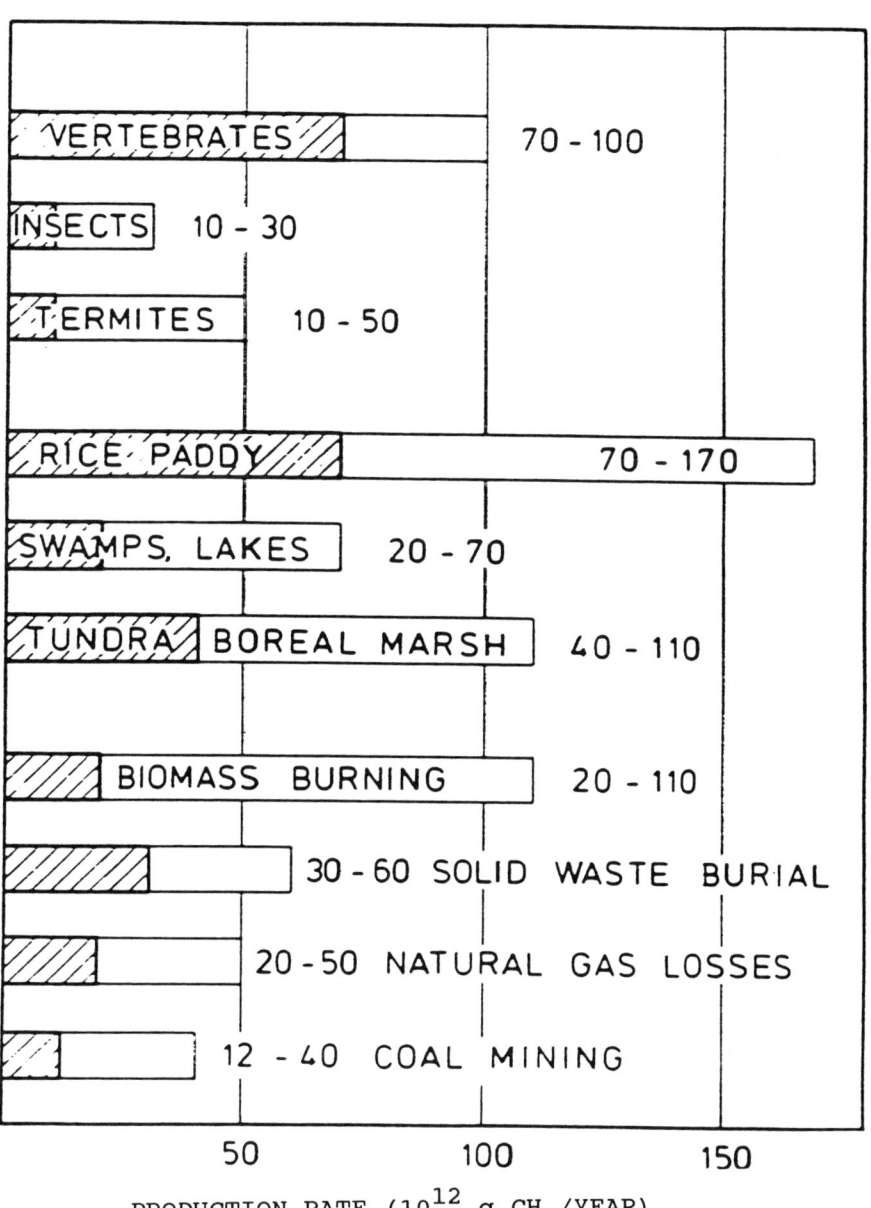

Fig. 7. Major sources of methane production.

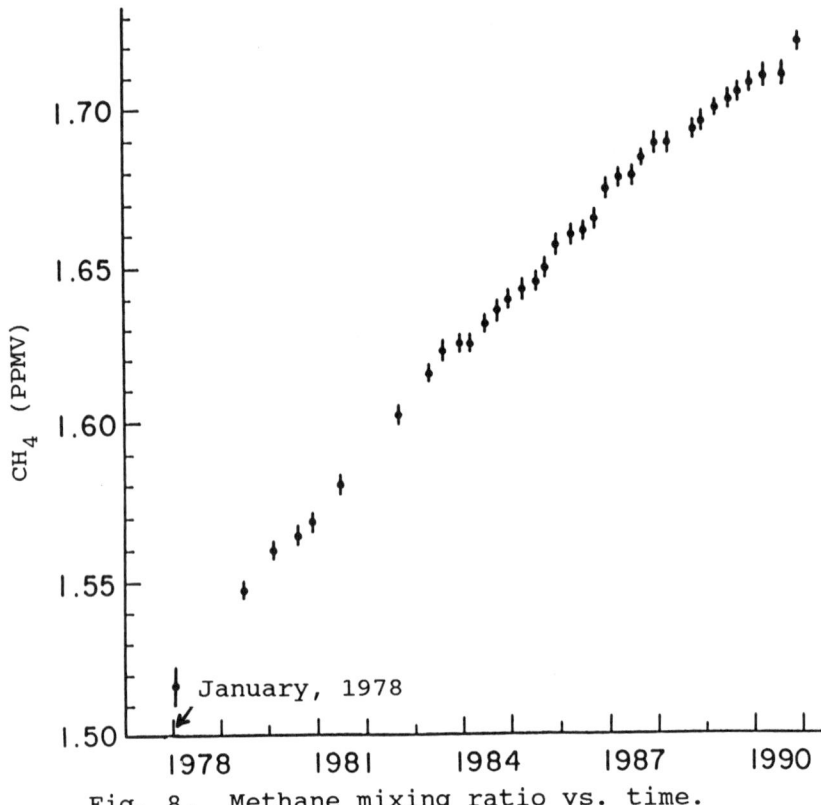

Fig. 8. Methane mixing ratio vs. time.

shown in Figure 9. The April 1991 world average concentration of atmospheric methane is 1.720 ppmv, while the recent yearly increase observed is about 0.013 ppmv, or about 1%.

The major removal of atmospheric methane occurs in the troposphere; however, about 10% drifts up into the stratosphere. Once there, the methane is oxidized, with the consequent formation of water molecules. Two important routes exist for depositing water vapor into the statosphere. The first is direct upward transport of water vapor in a parcel of air; the second is the carrying of hydrogen atoms upward in the chemical form of methane and the release of water molecules as the methane molecule is oxidized. In the 1980s atmosphere, these two processes delivered approximately equal amounts of hydrogen to the stratosphere. However, with atmospheric methane concentrations increasing at a rate of about 1% per year, the average amount of water vapor in the stratosphere is presumably steadily increasing with time. An increase in stratospheric water vapor has a binary effect: Water itself is a greenhouse gas and, thus, increasing stratospheric water vapor levels will add to the overall global warming. Secondly, the more water vapor in the stratosphere, the more water there is available to form polar stratospheric clouds, which have been linked to the dramatic losses in ozone over Antarctica during the austral spring.

Conclusion

Carbon dioxide, halocarbons, nitrous oxide, methane and many more trace gases not mentioned here are all greenhouse gases that can lead to global warming. This warming is not just a theory but reality waiting to happen. These gases also play a major role in determining stratospheric ozone levels. With the unprecedented decrease in ozone above Anarctica in austral spring and recent satellite data indicating a decrease in northern hemispheric ozone levels by as much as 7%, ozone depletion resulting from increasing levels of halocarbons is real. The gases that affect these two phenomena are rapidly increasing in atmospheric concentration due, in part, to the extremely long lifetimes of these species. At this point in time, there is no indication from the data being

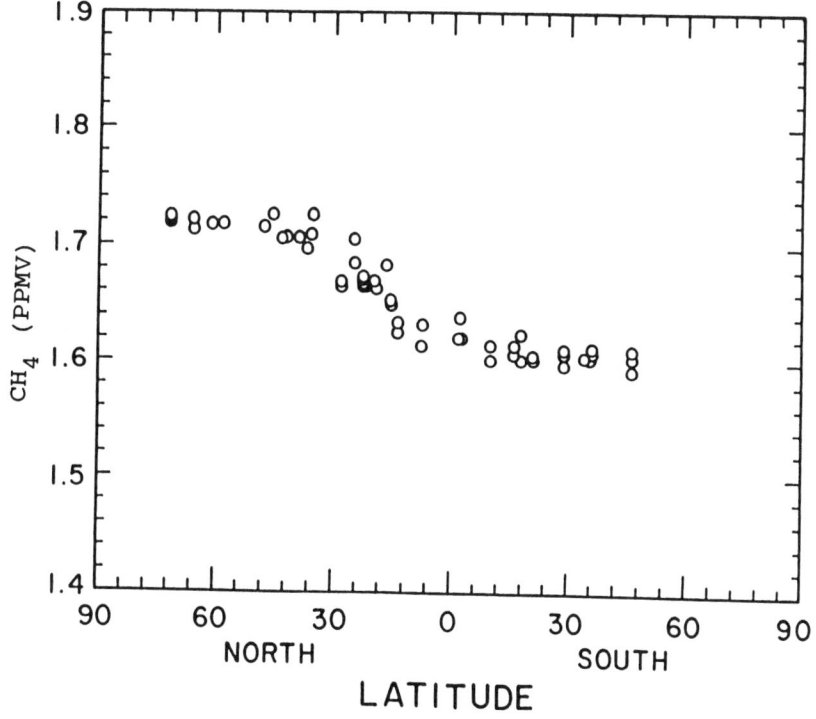

Fig. 9. Methane mixing ratio vs. latitude. Measurements were made during the summer, 1985.

collected that the rate of increase for these gases will decline significantly before the end of the century. The long lifetimes of some of these gases mean that, even if all emissions to the atmosphere were completely halted today, many of these species would still have an effect on ozone levels and global warming into the twenty-second century.

References.

1. C. D. Keeling et al, Geophysical Monographs 55, D.H. Peterson, ed., American Geophysical Union, Wash., D.C. (1989), pp. 165-236.

2. J. P. Peixoto and A. H. Oort, Physics of Climate, American Institute of Physics, New York, to be published, figure 6.2.

3. Intergovernmental Panel on Climate Change, Scientific Assessment of Climate Change, World Meteorological Organization/U.N. Environmental Program (Cambridge, MA: Cambridge University Press, 1990).

THE GEOCHEMICAL CARBON CYCLE AND THE UPTAKE OF FOSSIL FUEL CO_2

James F. Kasting
Department of Geosciences, Penn State University, University Park, PA 16802

James C. G. Walker
Space Physics Research Laboratory, University of Michigan, Ann Arbor, MI 48109

ABSTRACT

Atmospheric carbon dioxide levels are controlled over long time scales by the transfer of carbon between the atmosphere, oceans, and sedimentary rocks -- a process referred to as the CO_2 geochemical cycle. Carbon dioxide is injected into the atmosphere-ocean system by volcanism; it is removed by the weathering of silicate rocks on the continents followed by the deposition of carbonate minerals on the sea floor. Humans are currently perturbing the natural carbon cycle by burning fossil fuels and deforesting the tropics, both of which add CO_2 to the atmosphere. The effects of human activities on future atmospheric CO_2 levels can be estimated by including anthropogenic emissions in a model of the long-term carbon cycle. The model predicts that CO_2 concentrations could increase by a factor of six or more during the next few centuries if we consume all of the available fossil fuels. Preserving existing forests and/or reforesting parts of the planet could mitigate the CO_2 increase to some extent, but cannot be depended on to make a significant difference. Because the removal processes for atmospheric CO_2 are slow, the maximum CO_2 level reached is relatively insensitive to the fossil fuel burning rate unless the burning rate is many times smaller than its present value. The model also predicts that hundreds of thousands of years could pass before atmospheric CO_2 returns to its original, preindustrial level. Implications of these results for future energy and land use policies are discussed.

INTRODUCTION

Most scientists and other informed citizens are aware that atmospheric carbon dioxide levels are currently increasing as a consequence of human activities. The increase has been carefully monitored since 1958 by a series of measurements performed at Mauna Loa, Hawaii[1] (Fig. 1) and, more recently, at other locations around the globe. The Mauna Loa data show that atmospheric CO_2 has increased more or less linearly from 315 parts per million (ppm) by volume in 1958 to about

176 The Geochemical Carbon Cycle

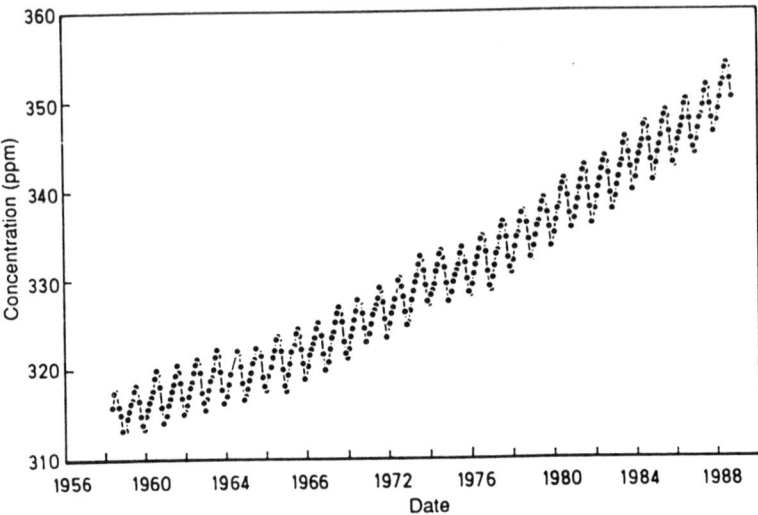

Fig. 1 Measurements of atmospheric CO_2 made at Mauna Loa, Hawaii. This graph is often referred to as the 'Keeling curve,' because the measurements were begun by Charles Keeling. (From the Carbon Dioxide Information Analysis Center, Oak Ridge, Tennessee).

350 ppm today. Superimposed on this secular trend is a 5-6 ppm seasonal cycle caused by terrestrial vegetation in the northern hemisphere. The seasonal cycle results from a relative excess of photosynthesis (which consumes CO_2) in northern hemisphere summer, which is largely balanced by an excess of respiration (which produces CO_2) during the winter. We shall not say too much about this biological carbon cycle here, since that is the subject of a separate chapter in this volume[2]. Some discussion of biology is unavoidable, however, as the increase in atmospheric CO_2 levels over the past century, and perhaps over the next few centuries as well, is tied to both the geochemical and biological carbon cycles. In what follows, we shall try to illustrate the interrelationship between these cycles and to identify their relative importance for the fossil fuel CO_2 problem.

Fewer people are aware that atmospheric CO_2 levels have probably varied in the past for reasons that have nothing to do with human influence. The best documented case is the last glacial/interglacial cycle. Measurements of CO_2 in air bubbles trapped in polar ice indicate that atmospheric CO_2 increased from ~200 ppm at the peak of the last glaciation, 18,000 years ago, to ~280 ppm at the beginning of the current interglacial period, ~10,000 years ago[3] (Fig. 2). The atmospheric CO_2 concentration remained at roughly 280 ppm until the middle of the last century, when

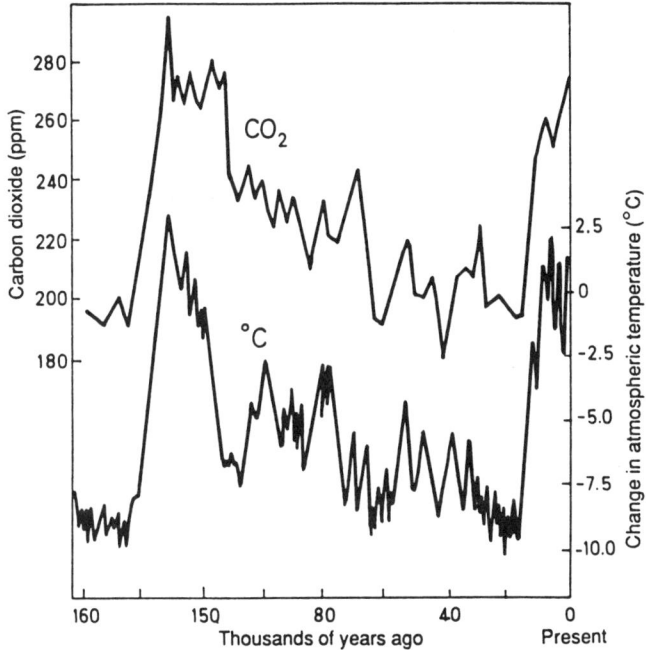

Fig. 2 Measurements of CO_2 trapped in air bubbles in the core from Vostok, Antarctica. The temperature curve is inferred from measurements of the D/H ratio of the ice. (From Barnola et al.[3], as modified by Gribbin[34].)

it began to rise towards its current higher level[4,5]. The cause of the glacial-to-interglacial CO_2 variation is poorly understood. Most likely, it involves a change in the partitioning of dissolved carbon between the surface and deep ocean, possibly induced by a change in deep ocean circulation[6,7] or by changes in the amount of calcium carbonate buried on continental shelves[8].

On a somewhat longer time scale, there is reason to believe that atmospheric CO_2 levels may have been significantly higher than today during the Cretaceous Period, which lasted from about 135 million years (m.y.) to 65 m.y. ago. Global surface temperatures appear to have been as much as 10°C warmer than at present based on the oxygen isotopic composition of carbonate sediments sampled by deep sea cores[9]. (Oxygen has two important isotopes, ^{16}O and ^{18}O. $CaCO_3$ becomes enriched in ^{18}O compared to dissolved bicarbonate as seawater becomes colder. The ocean itself becomes enriched in ^{18}O in glacial periods because the ice in the polar caps is relatively enriched in ^{16}O.) A likely cause for this warmth is increased atmospheric CO_2[10,11]. On still longer time scales, greatly increased atmospheric CO_2 levels (100 to 1000 times present) may have been required to compensate for reduced solar luminosity early in the Earth's history[12,13]. Direct evidence for these earlier fluctuations in atmospheric CO_2 levels is lacking; however, there are good theoretical reasons why CO_2 levels might have changed. The argument in each case hinges on the CO_2 geochemical cycle, which controls atmospheric CO_2 levels over long time scales. Since this same cycle is also involved in the uptake of fossil fuel CO_2, it is worthwhile to consider how it works.

THE CO_2 GEOCHEMICAL CYCLE

Carbon at the Earth's surface is stored in seven major reservoirs (Table 1). Of these, kerogen (dispersed, insoluble organic carbon in rocks) is probably the least important component from the standpoint of carbon cycling because it is relatively inert. (Much of the kerogen released by weathering is thought to be redeposited in sediments without undergoing oxidation.) Most of the fossil fuel reservoir consists of coal; oil and natural gas constitute less than 10 percent of the total[14]. The carbonate rock reservoir consists primarily of two different minerals: calcite ($CaCO_3$) and dolomite ($CaMg(CO_3)_2$). The ocean contains three inorganic carbon species: carbonic acid (H_2CO_3), bicarbonate ion (HCO_3^-), and carbonate ion ($CO_3^=$). (Chemical oceanographers prefer to denote carbonic acid as $(CO_2)_{aq}$; we use H_2CO_3 here to simplify the exposition.) Carbonic acid is produced when atmospheric CO_2 dissolves in rainwater or seawater; bicarbonate and carbonate ion are produced when this (weak) acid dissociates

Table 1. Carbon reservoirs

Reservoir	Size (10^{16} moles C)
Atmosphere	5.5
Living biosphere[a]	5
Soils[a]	13
Fossil fuels[b]	35-50
Ocean	330
Kerogen[c]	~10^5
Carbonate rocks[c]	~10^6

Notes:
[a]Ref. 15
[b]Higher estimate from ref. 16
[c]Ref. 17
Other estimates from ref. 14

$$CO_2 + H_2O \rightleftharpoons H_2CO_3 \quad (1)$$

$$H_2CO_3 \rightleftharpoons HCO_3^- + H^+ \quad (2)$$

$$HCO_3^- \rightleftharpoons CO_3^= + H^+ \quad (3)$$

The equilibrium constants for these reactions are well known and are tabulated in

Broecker and Peng[14] (p. 151).

The CO_2 geochemical cycle describes how the various inorganic carbon species move from one reservoir to another. Which transfer processes are most important depends on the time scale of interest. The time required to effectively transfer CO_2 between the atmosphere and the ocean is of the order of the deep ocean mixing time, ~1000 years. Over periods much longer than this, it is convenient to consider the atmosphere and ocean as a single combined reservoir. The CO_2 content of this reservoir is determined by the balance between the weathering of silicate minerals on the continents and the release of CO_2 by volcanism. If one represents all silicates by the mineral wollastonite, $CaSiO_3$, the process of silicate weathering can be written as

$$CaSiO_3 + 2\,CO_2 + H_2O \rightarrow Ca^{++} + 2\,HCO_3^- + SiO_2 \qquad (4)$$

The calcium and bicarbonate ions released by weathering are carried by streams and rivers to the oceans. There, organisms use them to make shells of calcium carbonate

$$Ca^{++} + 2\,HCO_3^- \rightarrow CaCO_3 + CO_2 + H_2O \qquad (5)$$

When these organisms die, their shells fall toward the ocean floor. Many of them redissolve on the way down, but some (primarily those that fall in shallower regions) survive and are buried in sediments. The net effect of these two processes can be determined by adding reactions (4) and (5) to get

$$CaSiO_3 + CO_2 \rightarrow CaCO_3 + SiO_2 \qquad (6)$$

The combination of these two processes removes CO_2 from the atmosphere-ocean system and sequesters it in carbonate sediments. The SiO_2 in reaction (6) is used by diatoms to make shells of opal; this opal is later transformed into the common mineral quartz.

If this were all that were happening, the story of Earth's climate evolution would be short and unhappy. The current rate of calcium carbonate deposition is sufficient to exhaust the carbon in the atmosphere-ocean system in about 400,000 years. Without CO_2, the atmosphere would become extremely cold, the oceans would freeze, and the Earth would become uninhabitable. Evidently, some process must return carbon to the system so that this does not happen. The return process is called carbonate metamorphism or, equivalently, silicate reconstitution. It works in the following manner: The sea floor is continually spreading outward from the midocean ridges as part of the process of plate tectonics. In some areas, old, carbonate-laden sea floor is subducted down into the mantle. When this occurs, the carbonate sediments are heated and metamorphosed; calcium carbonate reacts with quartz to reform calcium silicate, i.e. reaction (6) is reversed, and gaseous CO_2 is released. Some of this CO_2 makes its way back to the Earth's surface and is vented into the atmosphere through volcanos and hot springs. The total carbon content of the atmosphere-ocean system remains constant provided that the rate of CO_2 outgassing is equal to the rate of consumption of CO_2 by silicate weathering followed by carbonate deposition.

Observationally, one can estimate the weathering rate by measuring the concentrations of dissolved calcium, bicarbonate, and silica in river water and multiplying by the total discharge rate of the world's rivers. The measured values correspond to CO_2 consumption rates of about $(6-7) \times 10^{12}$ moles yr^{-1} [10,17]. The CO_2 outgassing rate is difficult to measure directly; recent attempts to do so[18,19] indicate that it is within a factor of two of the CO_2 consumption rate. The usual assumption made by modelers (including us) is that these two rates are equal, so that the system is in steady state on long time scales.

With these thoughts in mind, let us return briefly to the two, long-term, paleoclimate problems mentioned in the previous section. Could the geochemical carbon cycle explain the warmth of the Cretaceous? Berner and coworkers[10] have argued that it could. Seafloor spreading rates appear to have been faster at that time, as evidenced by the positions of magnetic lineations on the ocean floor. Faster seafloor spreading should lead to faster rates of carbonate metamorphism and increased volcanic outgassing of CO_2. It should also have caused sea level to increase, in agreement with evidence from the geologic record. Higher sea level implies less continental area exposed to weathering and, hence, a reduced loss rate for atmospheric CO_2. The combination of high CO_2 production and low CO_2 loss could have produced atmospheric CO_2 concentrations several times higher than the present value. This, in turn, might account for the observed climatic warmth.

What about the faint young sun problem? Could the geochemical carbon cycle have prevented the oceans from freezing early in Earth's history? Walker and coworkers[20] have shown that it might have. Consider the extreme case first: If the oceans had frozen completely, silicate weathering would have virtually ceased and volcanic CO_2 would have simply accumulated in the atmosphere. Eventually, the greenhouse effect produced by this CO_2 would have become large enough to melt the ice. (Recall from Table 1 that the amount of CO_2 tied up in carbonate rocks is very large.) In a more realistic model, global glaciation would probably never occur; atmospheric CO_2 concentrations would increase in such a way as to ensure the continued presence of liquid water. Indeed, increased volcanism and reduced continental area in the Earth's early history could have resulted in extremely high atmospheric CO_2 levels and a climate that was warmer than today, despite the weaker solar input[21,22].

To sum up: The general outlines, if not the details, of the geochemical carbon cycle are reasonably well understood. The operation of this cycle over long time scales can explain patterns in Earth's paleoclimatic history that might otherwise be difficult to understand. In the remaining parts of this chapter, we will show that the geochemical carbon cycle is also relevant to the present problem of global warming caused by the release of fossil fuel and biospheric CO_2.

MODERN CO_2 SOURCES AND SINKS

The rise in CO_2 concentrations illustrated in Figure 1 is a consequence of human perturbations to the natural carbon cycle. The primary anthropogenic CO_2

sources are the burning of fossil fuels and tropical deforestation (Table 2). The rate of fossil fuel consumption is well known; the rate of deforestation is highly uncertain. The highest estimates of the deforestation rate place it on a par with the burning of coal and oil, but the actual rate could be lower by a factor of as much as seven. The number to which both of these anthropogenic CO_2 sources should be compared is the CO_2 outgassing rate from volcanos; this represents the net input rate of CO_2 to the atmosphere-ocean system

Table 2. Sources of CO_2

Source	Magnitude (10^{14} moles CO_2 yr^{-1})
Fossil fuel burning[a]	
Coal	2
Oil	2
Other	1
TOTAL	5
Tropical deforestation[b]	0.3-2
Volcanic outgassing[c]	~0.05-0.11

[a]Ref. 23
[b]Ref. 15
[c]Refs. 10, 17, 18, and 24

prior to human influence. That rate is 50 to 100 times smaller than the current rate of fossil fuel usage. This is the first, and possibly the most important, lesson to be learned from considering the current global warming problem in the context of the long-term geochemical carbon cycle. To reduce the rate of atmospheric CO_2 increase to near zero would require roughly a hundred-fold reduction in the rate of fossil fuel consumption, all other things being equal. This constraint is much more severe than other estimates that have appeared in both the scientific and popular literature. It has obvious implications for future energy policy, to which we will return at the end.

The CO_2 added to the atmosphere by fossil fuel burning and deforestation will be removed by a variety of processes. We have already discussed the ultimate sink for this CO_2: silicate weathering followed by carbonate deposition (reactions 4-6 above). The previous discussion, however, treated the atmosphere and ocean as a single reservoir and neglected the biological carbon cycle entirely. Since we are concerned at present with both the long- and short-term response of the system to human activities, it is necessary to consider as well the processes that transfer CO_2 from the atmosphere to the ocean and to the terrestrial biosphere.

One process that can potentially remove CO_2 from the atmosphere is photosynthesis which, from a physicist's point of view, can be written as

$$CO_2 + H_2O \rightarrow CH_2O + O_2 \tag{7}$$

Here, 'CH_2O' is shorthand for more complex forms of organic matter. Photosynthesis removes approximately 1/10 of all the atmosphere's CO_2 every year[17]; however, most of this is returned almost immediately by respiration, which is just the reverse of reaction (7). A growing tree photosynthesizes slightly more than it respires, resulting in a net uptake of CO_2. This CO_2 is returned to the atmosphere when the tree dies and decays. Thus, a steady-state forest is neither a source nor a sink for CO_2. In order for photosynthesis to act as a net CO_2 sink, it is necessary to upset this steady state. If the amount of forested land increases, or if the trees in extant forests grow larger, or if soil carbon storage increases, CO_2 can be effectively sequestered into the biospheric and soil carbon reservoirs. Table 1 shows that these two reservoirs currently contain about 1/2 to 1/3 times as much carbon as is stored in fossil fuels. Thus, if one were able to double the size of these reservoirs through extensive reforestation, one could conceivably soak up half of the CO_2 that may eventually be released from the burning of fossil fuels. If, on the other hand, the present deforestation trend continues, the transfer of carbon will be in the opposite direction and the biosphere will act as a net source for atmospheric CO_2. Obviously, current land use practices will have to change dramatically if photosynthesis is to become a net CO_2 sink.

Another way that fossil fuel CO_2 can be removed is by dissolution in the oceans. Reactions (1)-(3) above show what happens chemically when CO_2 dissolves in water. But there is an added complication when one considers the atmosphere-ocean system: the ocean already contains a large amount of dissolved inorganic carbon (60 times the amount in the atmosphere) in the form of bicarbonate and carbonate ions (Table 1). If the extra fossil-fuel-generated CO_2 simply dissolved according to reactions (1)-(3), the ocean would rapidly become more acidic, i.e. its pH would decrease. This would shift the equilibria in reactions (2) and (3) to the left, resulting in a conversion of dissolved bicarbonate and carbonate ion into carbonic acid. Since atmospheric CO_2 is in approximate equilibrium with the carbonic acid in surface water, this feedback would make it difficult for the ocean to absorb much additional CO_2. What actually happens, therefore, is somewhat more complicated. Most of the anthropogenic CO_2 that dissolves in the ocean reacts with carbonate ions to yield bicarbonate

$$CO_2 + CO_3^= + H_2O \rightleftharpoons 2 HCO_3^- \qquad (8)$$

This reaction buffers the ocean's pH and allows it to absorb substantial quantities of anthropogenic CO_2. The amount that can be taken up in this manner is approximately equal to the total amount of carbonate ion in the oceans, about 12×10^{16} moles[14]. This is roughly 1/3 to 1/4 of the amount of carbon available in fossil fuels.

Additional CO_2 can be transferred from the atmosphere to the ocean by taking advantage of the buffering capacity of carbonate sediments and rocks. CO_2-rich water can dissolve calcium carbonate according to the reaction

$$CO_2 + CaCO_3 + H_2O \rightarrow Ca^{++} + 2 HCO_3^- \qquad (9)$$

This reaction is similar to reaction (8), except for the addition of calcium to the reactants and to the products. The calcium carbonate that participates in reaction (9) can come from either seafloor sediments or from exposed carbonate rocks on the continents. (Dolomite, $CaMg(CO_3)_2$, can be dissolved by a reaction similar to reaction

Table 3. Sinks for CO_2

Sink	Uptake capacity (10^{16} moles CO_2)	Timescale (yrs)
Photosynthesis	−18 to +18[a]	50 − 100
Dissolution in the ocean	12[b]	~10^3
Dissolution of carbonate sediments	40[b]	$10^3 - 10^4$
Carbonate weathering on land	5×10^5 [c]	$10^4 - 10^5$
Silicate weathering on land	~10^6 [c]	$10^5 - 10^6$

[a] See text
[b] Ref. 14
[c] Ref. 10

(9)). It is generally thought that dissolution of seafloor carbonate sediments will provide the greatest amount of CO_2 buffering over the next few centuries. The amount of carbonate that might eventually be exposed to corrosive seawater is difficult to estimate because it depends on such factors as the depth to which worms burrow into the sediments. Broecker and Peng[14] place the figure at about 40×10^{16} moles. This is enough buffering capacity to neutralize most or all of the carbon in fossil fuels, were it all to be effectively utilized. Even more buffering capacity is available from carbonate rocks stored on the continents, but utilization of that capacity is expected to be slower because it relies on the slow process of rock weathering. Some models nevertheless predict that this could be an important factor on the 1,000 to 10,000-year time scale (see next section).

It is instructive, at this point, to say a few more words about time scales. The time scale for anthropogenic CO_2 release can be estimated by dividing the fossil fuel inventory (Table 1) by the current consumption rate (Table 2). This gives a release time of 700 to 1000 years, depending on the estimated size of the recoverable fossil fuel reservoir. This release time is admittedly uncertain because it assumes that the present rate of burning will continue into the future even though the human population is expanding, economies are developing, and the oil reserves will probably be exhausted within the next century or two. Other, possibly more realistic, burning rates are considered in the modeling discussed below. The time scales for CO_2 uptake differ from one process to another. For photosynthesis, the relevant uptake time is the lifetime of a growing tree, roughly 50 to 100 years. For the ocean, there are two separate time scales: the mixing time of the surface ocean (a few months) and the mixing time of the deep ocean (~1000 years). Since most of the volume of the ocean and, hence, most of its CO_2 uptake capacity, is in the deep ocean, the longer time scale is more relevant to the fossil fuel problem. The time scale for seafloor carbonate dissolution is several thousand years, because CO_2-enriched seawater will probably have to circulate over these sediments several times to induce them to dissolve[14]. The characteristic time scale for weathering of carbonate and silicate minerals on the continents is tens to hundreds of thousands of years[10].

The point of this discussion is the following: The probable CO_2 release time is of the order of centuries; the CO_2 uptake time is thousands to hundreds of thousands of years. In view of this mismatch, it should not be surprising that that the atmospheric CO_2 concentration is increasing; moreover, it should be expected that CO_2 levels will continue to increase over the next few centuries if we continue to burn fossil fuels. The next section attempts to quantify this prediction by describing a recent computer simulation of the long-term effects of fossil fuel burning.

MODELING THE UPTAKE OF FOSSIL FUEL CO_2

The CO_2 release and uptake processes described in the previous section can be incorporated into computer models to simulate the rise in atmospheric CO_2 over the past two centuries and to try to predict future CO_2 levels as a function of fossil fuel consumption and patterns of land use. Here, we describe the results from a recent modeling study by Walker and Kasting[24]. This study is unique in that it is the only modeling study to date that incorporates all aspects of the geochemical carbon cycle; hence, it is the only such study that is capable of tracking the fossil fuel CO_2 pulse from beginning to end. It should be emphasized from the start that the detailed predictions of this, or any other, existing model are uncertain because the CO_2 uptake processes are difficult to parameterize and because the computer models are, at best, highly simplified approximations to the real atmosphere-ocean system. Nonetheless, the basic nature of the predictions should be reasonably robust because the results depend to a large extent on relative reservoir sizes and approximate transfer times, both of which are fairly well understood.

The Walker and Kasting model consists of a set of 8 boxes representing the atmosphere, terrestrial biosphere (forests plus soils), warm and cold surface ocean, thermocline, and three deep ocean reservoirs (Fig. 3). The independent variables tracked by the model include atmospheric CO_2 concentration, biospheric carbon content, and five quantities in each ocean box: total dissolved inorganic carbon ($\Sigma \equiv [H_2CO_3] + [HCO_3^-] + [CO_3^=]$), alkalinity ($A \equiv [HCO_3^-] + 2[CO_3^=]$), dissolved phosphorus, and two isotopes of carbon (^{13}C and ^{14}C). The variables Σ and A are the

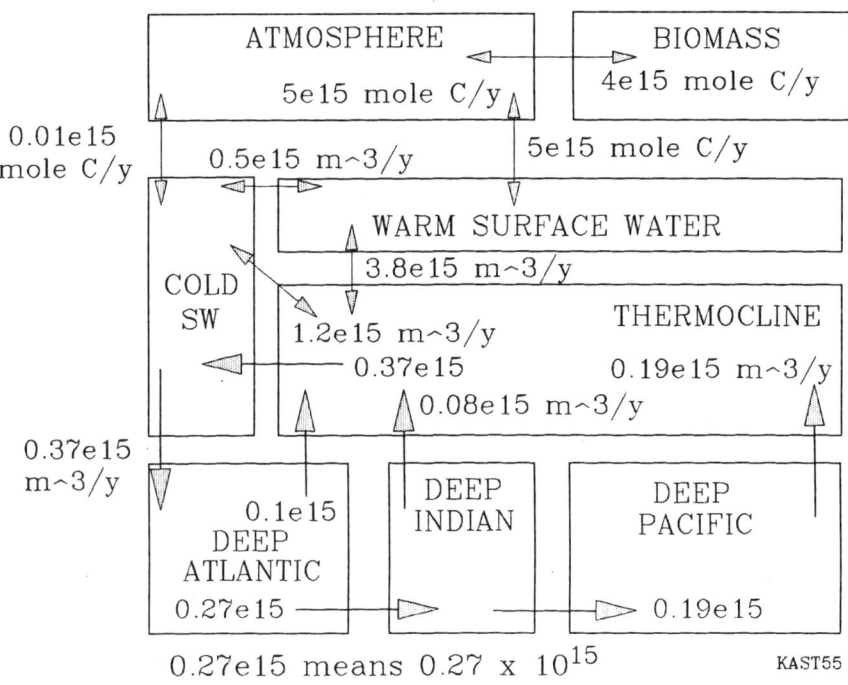

Fig. 3 Reservoirs and exchange fluxes in the Walker and Kasting box model. Exchange fluxes between the various ocean boxes are given in m³/yr; fluxes between the atmosphere and other reservoirs are in moles C/yr. 1e10 means 1×10^{10}.

chemical oceanographer's way of keeping track of dissolved carbonate species. Since the concentration of carbonic acid is negligible compared to that of bicarbonate and carbonate ion at typical ocean pH values (7.5 to 8), one can easily demonstrate that

$$[HCO_3^-] \approx 2\Sigma - A \qquad (10)$$

$$[CO_3^=] \approx A - \Sigma \qquad (11)$$

Two parameters are sufficient to completely characterize the dissolved inorganic carbon species; Σ and A happen to be the two easily measurable quantities. Phosphorus is tracked in the model because it, along with fixed nitrogen, is a limiting nutrient in much of the surface ocean. Thus, phosphorus availability controls the rate of organic productivity. The carbon isotopes are included as tracers of organic productivity and oceanic circulation.

In addition to the circulation fluxes shown in Figure 3, the computer model includes a flux of particulate matter from the warm surface ocean to the thermocline and deep ocean basins. The particles, which represent the remains of marine organisms, consist of a mixture of organic carbon and calcium carbonate. Most (>90 percent) of these particles dissolve (or are oxidized) as they descend through the water column. As such, they constitute a 'biological pump' that tends to remove dissolved CO_2 from surface water and to concentrate it in the deep ocean. This CO_2 is eventually brought back up to the surface by upwelling (Fig. 3); nonetheless, an appreciable vertical gradient in H_2CO_3 concentrations is maintained by this process. Some fraction of the calcium carbonate in the descending particles survives the passage through the ocean and is buried in sediments. The amount buried depends on the carbonate saturation depth which, in this model, represents the boundary above which calcite ($CaCO_3$) is stable and below which it dissolves. The area of ocean floor above the carbonate saturation depth is estimated from the calculated carbonate ion content of deep water, the solubility product of calcite, and the average hypsommetry (depth-area relationship) of the ocean basins[25]. In the real ocean, a transition zone is observed over which the extent of dissolution of $CaCO_3$ increases with depth and the percentage of $CaCO_3$ in sediments decreases. That complication is ignored in the model.

Finally, the model also includes a simplified representation of the rock cycle. The magnitude of the present silicate and carbonate weathering rates (reactions 4 and 9) is estimated from the data of Meybeck[26]. The rate of carbonate weathering is assumed to vary linearly with pCO_2; the silicate weathering is assumed to vary as $pCO_2^{0.3}$. The CO_2 dependence of the silicate weathering rate is intermediate between that of Berner et al.[10] and Walker et al.[20]. The assumption of a linear CO_2 dependence for carbonate weathering is made so as to maximize the rate of uptake of fossil fuel CO_2. These weathering rate laws incorporate the idea that both surface temperature and possibly global runoff rates increase with increasing atmospheric CO_2, and that the combination of these factors tends to accelerate rates of chemical weathering.

After adjusting the oceanic circulation rates and particle fluxes to try to match the present system as closely as possible, the model was used to simulate the observed rise in atmospheric CO_2 during the past two centuries (Fig. 4). The boxes in Figure 4 show (annually averaged) CO_2 measurements from Mauna Loa; the crosses and triangles represent measured values derived from air bubbles trapped in ice cores[4,5]. The dashed curve is the fossil fuel burning rate[1]; the solid curve shows atmospheric CO_2 concentrations projected by the model. Although the model correctly predicts the overall increase in atmospheric CO_2, the shape of the curve does not agree with the measurements: the predicted rate of CO_2 increase between 1850 and 1950 is too slow, whereas the rate of increase since that time is too fast.

Part of the reason for the discrepancy is well understood[27,28]. Much of the observed atmospheric CO_2 increase between 1850 and 1950 is thought to have been caused by the clearing of forests in the northern hemisphere and in the tropics, rather

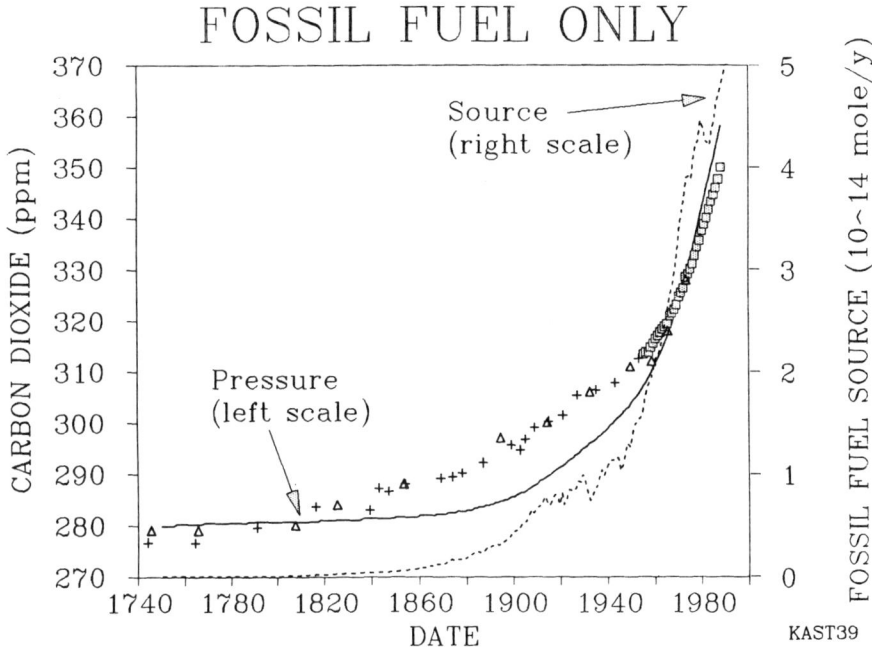

Fig. 4 History of atmospheric CO_2 during the last two centuries. Boxes denote annually-averaged measurements from Mauna Loa; triangles and crosses represent measurements of air bubbles trapped in polar ice. The dashed curve is the fossil fuel source, given by the right-hand scale. The solid curve shows the predictions of the box model when only the fossil fuel CO_2 source is included.

than by the burning of fossil fuels. This process is called the 'pioneer effect' in reference to the activities of colonial pioneers. Estimates of the rates of carbon release by deforestation during this time have been compiled by different authors[29,30]. They are shown here in Figure 5, along with a quantity termed 'biomass potential'[24]. Biomass potential is defined as normalized forest area, weighted by carbon storage; it is given a value of unity for the preindustrial Earth.

Including this additional anthropogenic CO_2 source in the model produces the results shown in Figure 6. Making this change brings the model into agreement with the ice core data up until about the year 1860 A.D. After this time, however, the model predicts much higher levels of atmospheric CO_2 than were actually observed. This is a problem that is common to all carbon cycle models (e.g. refs. 27 and 28):

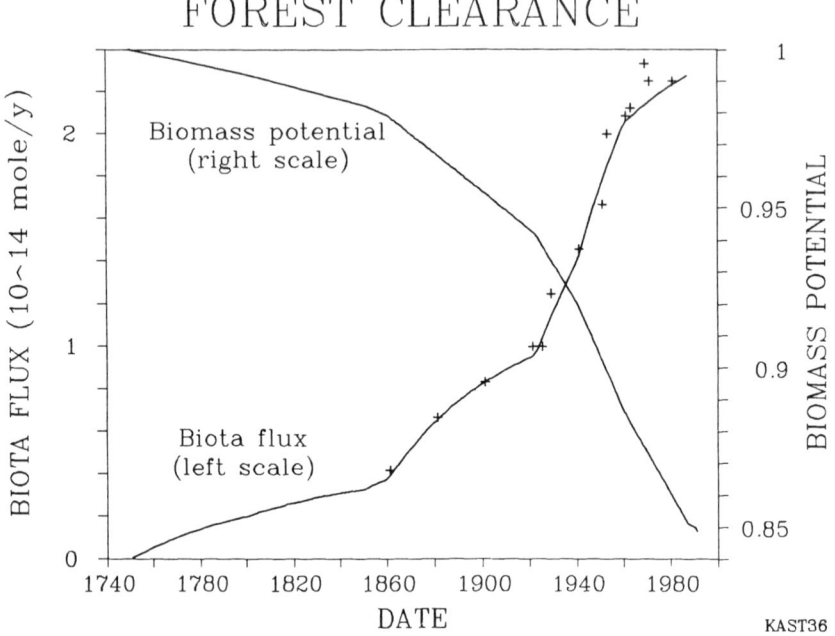

Fig. 5 The historical flux of carbon from the biota to the atmosphere caused by changes in land use. The crosses represent values deduced from land use studies. The quantity called 'biomass potential' (defined in the text) represents the amount of carbon stored in forests and soils, normalized to the preindustrial value. The preindustrial value is estimated from the biota flux.

there is evidently a missing sink for CO_2. The same problem is encountered when one tries to make an accurate accounting of the present global budget of atmospheric CO_2[31]. The most likely solution to the missing carbon problem involves a recent increase in the amount of carbon stored in temperate and boreal forests.

One reason that carbon storage in forests might be increasing is because C3 plants (which include most trees) grow faster under enhanced CO_2 levels. C4 plants, such as corn and most grasses, are much less sensitive to CO_2. (The terms 'C3' and 'C4' refer to the number of carbon atoms fixed in the primary photosynthetic step.) For C3 plants grown under controlled, greenhouse conditions, the relative growth rate can be approximated by the function[32]

$$F(CO_2) = 2.22(1 - \exp[-0.003(pCO_2 - 80)]) \qquad (12)$$

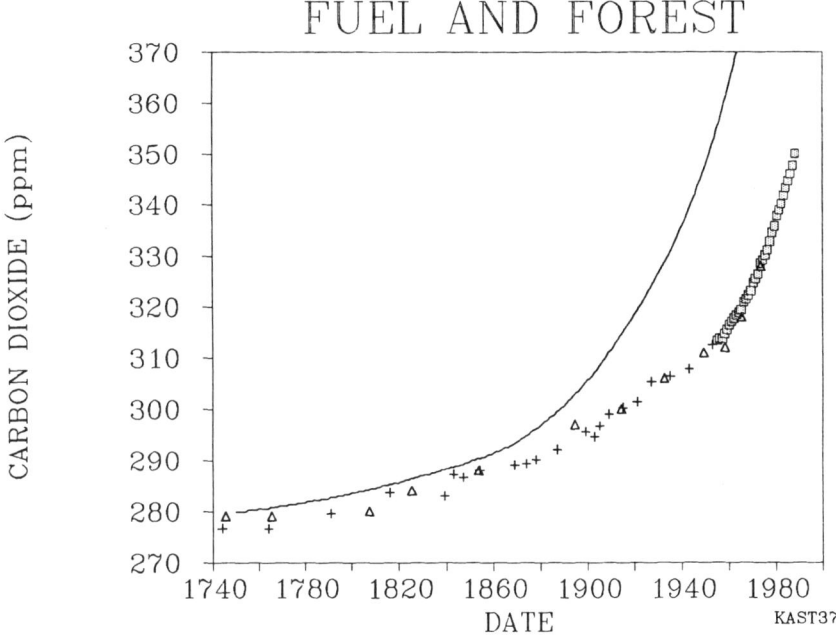

Fig. 6 Atmospheric CO_2 concentrations calculated when the estimated biota flux from land use changes (Fig. 5) is included in the box model.

This function has been adjusted to yield a relative growth rate of unity at the preindustrial CO_2 level of 280 ppm. Note that the growth rate drops to zero at a CO_2 level of 80 ppm. This is the compensation point, below which many plants respire faster than they photosynthesize. Whether or not this same fertilization function applies to natural ecosystems, where factors other than CO_2 availability may limit the growth rate, is a matter of current debate.

Adding CO_2 fertilization to the model allows it to reproduce the historical trend in atmospheric CO_2 levels reasonably well (Fig. 7). This does not, of course, prove that CO_2 fertilization is the correct explanation for the missing CO_2 sink or that this effect will continue in the future. Let us assume for the moment, however, that CO_2 fertilization is real and operative. One can then use the model to predict future atmospheric CO_2 levels for different assumptions concerning rates of fossil fuel consumption and patterns of deforestation/reforestation. Some possible scenarios are shown in Figure 8. In the 'save fuel' scenario (bottom panel) the present rate of fossil fuel burning, 5×10^{14} moles C yr^{-1}, is assumed to continue for the next 700 years until

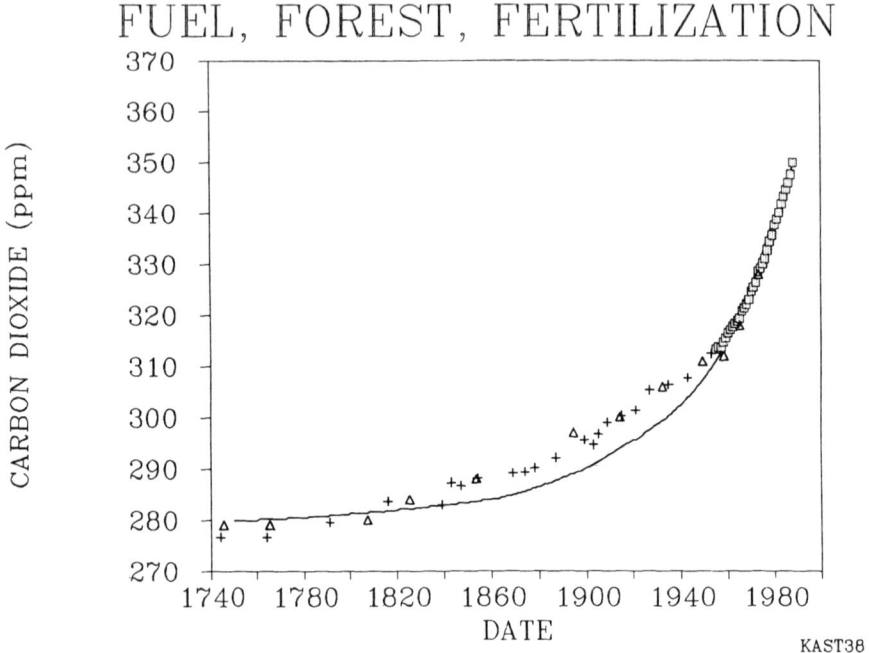

Fig. 7 Atmospheric CO_2 concentrations calculated when both the biota flux and the CO_2 fertilization effect are included in the box model.

the entire 35×10^{16} moles of carbon in the 1982 recoverable reserves are exhausted. The 'burn fuel' scenario assumes that the same total amount of fossil fuel is consumed in less than 400 years. The 'save forests' scenario (upper panel) assumes that deforestation ceases immediately and that the areal extent of forests remains constant into the indefinite future. The 'burn forests' scenario assumes that the present trend of deforestation continues until only 30 percent of the original, preindustrial forest area remains.

The various combinations of the above assumptions yield four different cases that were examined with the model (Fig. 9). The most pessimistic (but, arguably, the most realistic) of these is represented by the scenario labelled 'burn both,' in which the fossil fuels are burned rapidly and the forests are assumed to disappear. Atmospheric CO_2 concentrations are predicted to reach ~2100 ppm in this case; the peak occurs in about the year 2350 A.D. when the fossil fuel reserves are nearly exhausted. Somewhat surprisingly, the 'burn forests, save fuel' results are not tremendously different: atmospheric CO_2 peaks at ~2000 ppm in the year 2700.

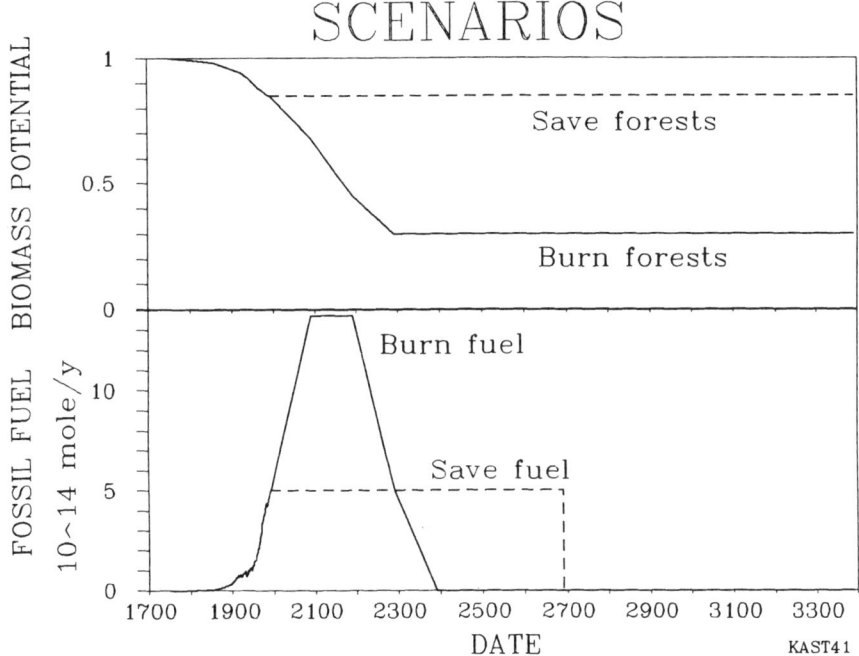

Fig. 8 Hypothetical scenarios for future rates of fossil fuel burning and patterns of land use. The 'save fuel' curve corresponds to burning the existing fossil fuel reserves at the present rate. The total amount of fossil fuel consumed is the same in both cases.

Burning the fossil fuels at a slower rate delays the CO_2 increase by 300-400 years but does little to diminish its eventual magnitude. Given a little additional thought, this result is not so surprising after all. In both cases, the time scale for CO_2 input is significantly shorter than the time scale for CO_2 removal. A significant decrease in the present fossil fuel consumption rate would be needed to reduce the magnitude of the atmospheric CO_2 peak.

What is remarkable in this model is the large reduction in future atmospheric CO_2 levels that results from preserving the forests. The two lower curves in Figure 9 indicate that the peak CO_2 level is reduced to ~1100 ppm in both the 'burn fuel' and 'save fuel' scenarios. Roughly 1000 ppm of CO_2, or three times the amount now present in the atmosphere, has been absorbed by the biomass reservoir as a consequence of CO_2 fertilization. Examination of Table 1 shows that this implies a doubling of the total amount of carbon stored in forests and soils. CO_2 fertilization

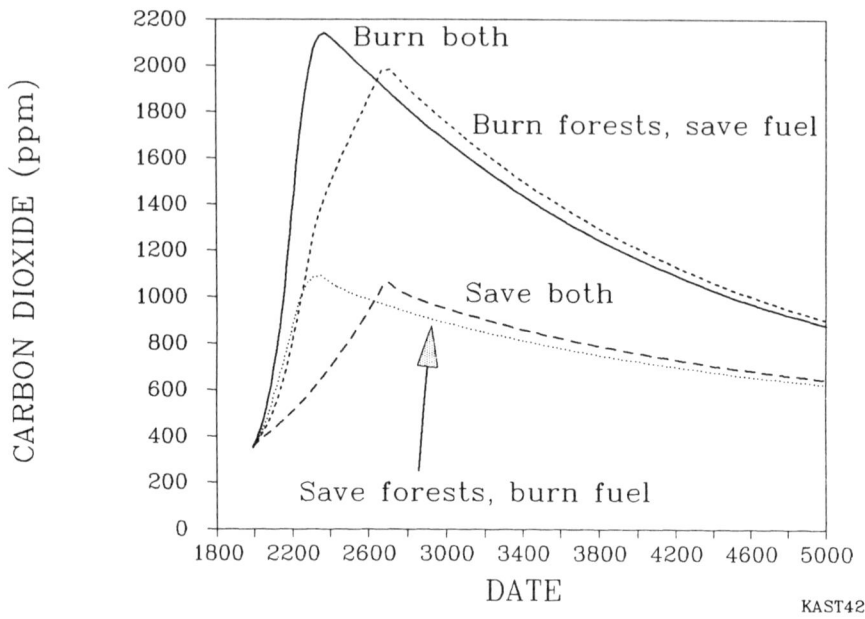

Fig. 9 Atmospheric CO_2 concentrations predicted by the box model for the different scenarios shown in Fig. 8. All of these calculations assume that CO_2 fertilization is real and will continue to affect growth rates in the future.

is also assumed to apply in the 'burn forests' scenarios, but its effect on atmospheric CO_2 is much smaller because of the smaller area of forested land.

Taken at face value, these results would seem to indicate that preserving the forests is more important in the long term than is conserving fossil fuel. This conclusion may be premature, however, for two reasons: First, as mentioned earlier, CO_2 fertilization may be less effective for natural ecosystems than it is in a controlled, greenhouse environment. Second, even if trees do grow faster in a high-CO_2 environment, it is not obvious that soil carbon storage will increase proportionately. Indeed, Houghton and Woodwell[15] have argued (convincingly, we believe) that soil carbon storage will likely decrease as the climate warms. Their prediction is based partly on the observation that tropical soils today contain much less carbon than do soils in temperate and boreal ecosystems. The reason is that organic material decays much faster in a warm climate. Thus, if global warming converts temperate ecosystems into subtropical ones, much of the organic carbon stored in these soils could be returned to the atmosphere.

This phenomenon is crudely simulated in the computer model by increasing the decay rate of biomass by a factor of two for a 10°C rise in surface temperature. The model predicts that global surface temperatures will rise about 2°C for each doubling of atmospheric CO_2; hence, the maximum temperature increase for the scenarios considered in Figure 9 is about 6°C. The corresponding increase in the decay rate is $2^{0.6}$, or a factor of roughly 1.5. Including this 'enhanced decay' rate assumption in the model yields the results shown in Figure 10. The curves shown are for the 'save forests, burn fuel' scenario; the terms 'enhanced growth' and 'constant growth' indicate whether or not CO_2 fertilization has been included in the calculation. It can be seen from Figure 10 that the beneficial effects of preserving the forests are greatly diminished when either the fertilization effect is eliminated or the decay rate is enhanced. Only an optimist would conclude that forests will save us from the most pronounced effects of fossil fuel burning.

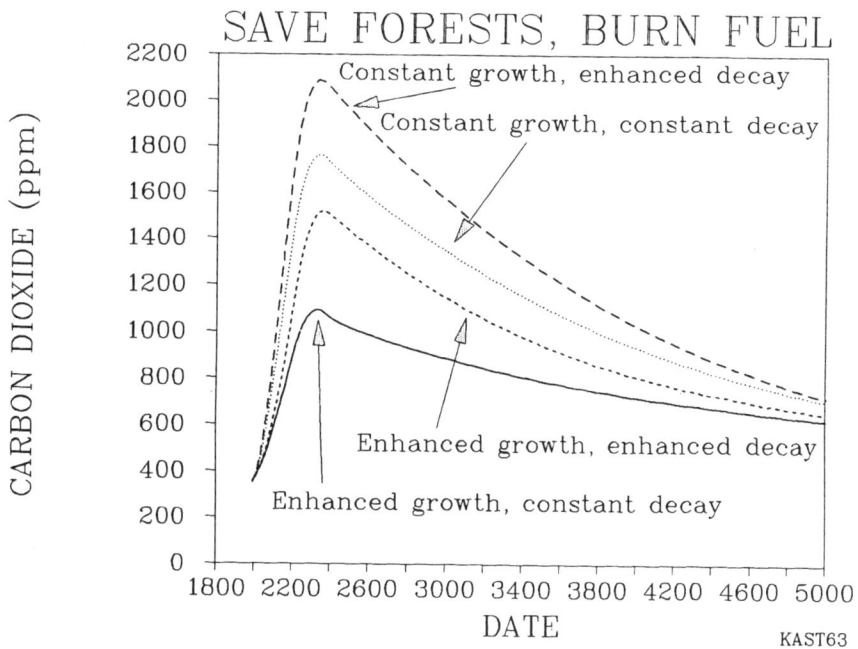

Fig. 10   Predicted atmospheric CO_2 concentrations for the 'save forests, burn fuel' scenario of Fig. 9, but with various assumptions concerning CO_2 fertilization and soil carbon decay rates. The solid curve is the same as in Fig. 9; the other curves show the effects of suppressing CO_2 fertilization and/or increasing the decay rate.

The discussion at this stage has evolved away from the geochemical carbon cycle and towards the biological carbon cycle. But the geochemical cycle underlies all of these future CO_2 projections in a fundamental way. This point is brought out by the last three figures. Figure 11 shows the near-term effect of assuming different fossil fuel consumption rates in the Walker and Kasting[24] model. The solid curve is for the 'burn fuel' scenario of Figure 8; the dashed curves are for a constant burning rate of 5×10^{14} moles C yr^{-1} (the current value) and for immediate decreases of 20 percent and 50 percent from that value. All curves are for the optimistic, 'enhanced growth, constant decay' scenario of Figure 10.

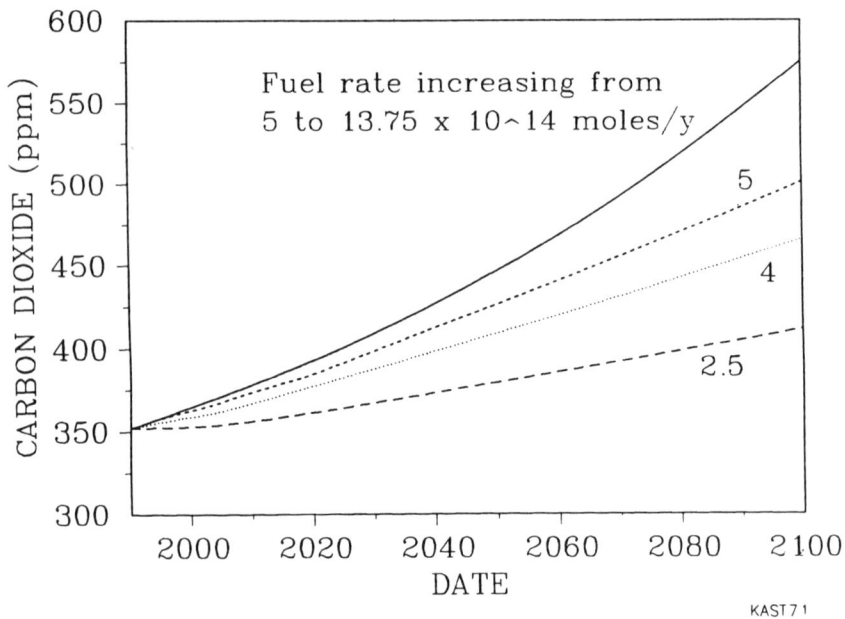

Fig. 11 Predicted atmospheric CO_2 concentrations during the next century for the 'save forests' scenario with various assumed rates of fossil fuel burning: 1) solid curve -- 'burn fuel' scenario from Fig. 8; 2) short dashes -- present burning rate; 3) dots -- 80 percent of the present burning rate; 4) long dashes -- 50 percent of the present rate. CO_2 fertilization is included in all of these calculations.

The results indicate that modest decreases in fossil fuel consumption cannot stabilize atmospheric CO_2, contrary to statements that have occasionally appeared in the literature. Such decreases can, however, slow the rate of global warming. According to this model, a 20 percent reduction in fossil fuel emissions produces a 23 percent reduction in the rate of CO_2 increase; a 50 percent emissions reduction produces a 60 percent reduction in the rate of CO_2 increase. The effect is thus slightly nonlinear; one gets proportionately more benefit from large emission rate reductions than from small ones. But an anthropogenic emission rate of half the present value is still more than enough to overwhelm the natural carbon cycle over the next century.

Figure 12 shows the longer-term effect of even more drastic decreases in the fossil fuel burning rate. The assumption, as before, is that the total amount of fossil fuel consumed is 35×10^{16} moles; the assumptions concerning forests are the optimistic ones described above. The slowest burning rate considered (solid curve) represents an immediate, 25-fold decrease in the rate of fossil fuel burning.

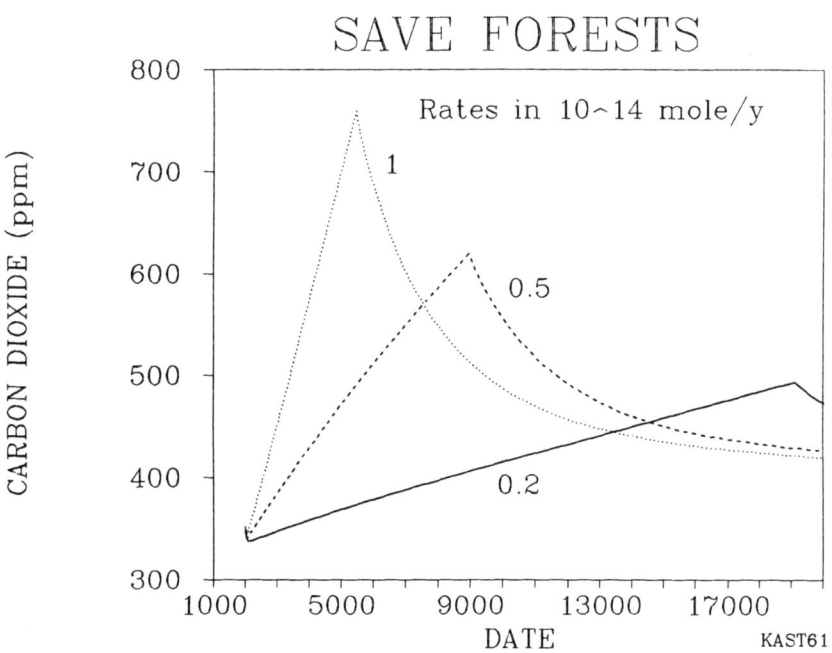

Fig. 12 As in Fig. 11, but with substantially lower rates of fossil fuel burning. The total amount of fossil fuel consumed is fixed at 35×10^{16} moles; the burning rate ranges from 1/5 to 1/25 of the present value.

Even in this case, atmospheric CO_2 concentrations increase to ~500 ppm before beginning to decline. The peak CO_2 level occurs 17,000 years in the future, which is sufficiently far off that we might not worry about it very much. However, the calculation illustrates an important point: atmospheric CO_2 concentrations should continue to rise as long as the rate of anthropogenic CO_2 input exceeds, or is comparable to, the natural input of CO_2 to the atmosphere-ocean system. The natural CO_2 source, as discussed earlier, is volcanic outgassing; its magnitude is roughly 100 times less than the present rate of fossil fuel burning. Burning fossil fuels at even a few percent of the present rate should therefore be enough to keep atmospheric CO_2 levels on the rise.

A second point that was made earlier is the long time scale required for the geochemical CO_2 removal processes to be effective. This point is illustrated in Figure 13, which shows the recovery of the atmosphere-ocean system to its assumed preindustrial steady state. (The long time-scale simulation is made possible by using

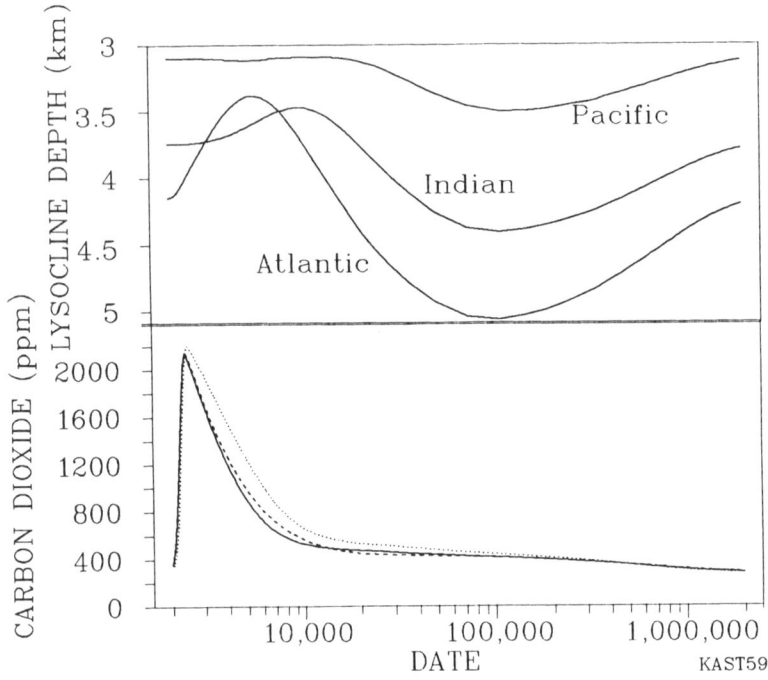

Fig. 13   The long-term recovery of atmospheric CO_2 after the injection of fossil fuel and forest carbon in the 'burn both' scenario of Fig. 9. The solid curve is the standard model; the dashed curved shows the effect of suppressing seafloor carbonate dissolution; the dotted curve assumes a carbonate weathering rate that is less sensitive to increases in atmospheric CO_2. The upper panel shows the calculated depth of the lysocline in the three deep ocean basins for the standard case.

an implicit integration method.[33]) The solid curve in the bottom panel shows atmospheric CO_2 levels predicted for the pessimistic, 'burn both' scenario of Figure 9. The other curves show the consequences of suppressing seafloor carbonate dissolution (dashed curve) or assuming a weaker, square-root, dependence of the carbonate weathering rate on atmospheric CO_2 (dotted curve). In each case, the bulk of the anthropogenic CO_2 is removed within a few thousand years. The important processes on this time scale are neutralization of dissolved carbonate ion (reaction 8), dissolution of seafloor carbonate sediments, and weathering of carbonate rocks on land (both represented by reaction 9). During this time, the carbonate saturation (or 'lysocline') depth moves up, especially in the Atlantic where the deep water is forming. But the system cannot return all the way to steady state as a consequence of these processes alone. The ocean at this stage contains an excess of bicarbonate ion produced from reactions (8) and (9). The only way it can lose this bicarbonate is by an increase in silicate weathering on the continents (reaction 4) caused by higher atmospheric CO_2, followed by excess deposition of carbonate sediments on the seafloor (reaction 5). The results of the weathering rate increase are manifested by an increase in deep ocean alkalinity and a corresponding deepening of the carbonate saturation depth. The deepening saturation depth allows carbonate sediments to be deposited over a greater area of the ocean floor; this increased deposition, in turn, finally removes the excess anthropogenic CO_2. The entire process requires more than a million years because it is limited by the slow rate of weathering of silicate rocks.

Once again, it should be emphasized that the details of the CO_2 uptake processes are poorly understood. Thus, the precise shapes of the curves shown in Figure 13 are subject to considerable uncertainty. If global temperatures increase more than assumed in the model, or if the rate of silicate weathering increases more rapidly with higher atmospheric CO_2, the long-term recovery of the system may be faster than shown here. On the other hand, if deep ocean circulation rates decrease, or if carbonate sediments dissolve more slowly than assumed, the decline of the fossil fuel CO_2 pulse may be significantly delayed. In spite of these uncertainties, it is safe to conclude that it will take thousands to hundreds of thousands of years for atmospheric CO_2 to return to its preindustrial level if we consume most of the available fossil fuels. Whether the actual recovery time lies towards the short or long end of this window is a question of mainly academic interest.

IMPLICATIONS FOR FUTURE ENERGY AND LAND USE POLICY

The implications of our model results for future global energy and land use policy are clear. Energy conservation can buy us time, but it cannot by itself prevent global warming. The required reductions in CO_2 emissions are simply too large. If we wish to avoid the excessive atmospheric CO_2 increases seen in Figures 9 and 13, it will be necessary to reduce the rate of fossil fuel consumption to at most a few percent of the present value. Preserving existing forests, or reforesting areas that were previously cleared, could help to take up some of the fossil fuel CO_2; however, one should not count on this to alleviate global warming because the net transfer of carbon between the atmosphere and the biosphere could very easily go in the wrong direction.

Exactly how one might implement these suggestions is not at all clear. If we are to stop consuming fossil fuels, we will need to develop alternative energy sources and we may have to change our lifestyles. Ideally, CO_2 emissions could be reduced or eliminated simply by encouraging energy conservation and by exploiting renewable energy sources such as solar, wind, and hydro power. More realistically, we may need to rely heavily on nuclear power for the next few decades and on new technologies (satellite solar power, solar-hydrogen systems, fusion?) for the period beyond that. Satellite solar power may be the best bet. In addition to being inexhaustible and non-polluting, satellite solar energy could be beamed down anywhere on Earth, including those developing countries that are projected to account for a major fraction of future CO_2 emissions. Developing this resource would require a substantial international commitment of time, money, and creative thinking.

CONCLUSION

Over long time scales the CO_2 content of the atmosphere is controlled by the geochemical carbon cycle, in which carbon dioxide is exchanged between the atmosphere, the ocean, and carbonate rocks. The basic outlines of this cycle are reasonably well understood and have been used to explain why atmospheric CO_2 levels may have varied in response to past changes in solar luminosity and to variations in seafloor spreading rates. Some progress has been made toward understanding why atmospheric CO_2 levels varied during the past glacial/interglacial cycle.

The same processes that control the atmospheric CO_2 content on long time scales will be involved in the removal of CO_2 produced by human activities. Comparing the present fossil fuel consumption rate with the volcanic outgassing rate of CO_2 shows that humans are currently overwhelming the natural CO_2 source by a factor of ~100. Atmospheric CO_2 levels should continue to rise as long as this imbalance continues. Because the amount of carbon in recoverable fossil fuels is much larger than the amount in the atmosphere, and because the removal processes for CO_2 are much slower than the rate at which CO_2 is being produced, atmospheric CO_2 could conceivably increase by a factor of six or more during the next few centuries. The eventual magnitude of the increase is only weakly dependent on the fossil fuel burning rate, unless the burning rate can be reduced to only a few percent of the present value. Forests could play an important role in taking up fossil fuel CO_2, but only if forested lands are preserved, CO_2 fertilization is important, and soil carbon storage increases as the forests grow. Most of the CO_2 produced by human activities will be removed from the atmosphere within the next few thousand years, but the combined atmosphere-ocean system may require hundreds of thousands of years to return to its preindustrial steady state. Prudence would suggest that we halt the experiment before we find out if this prediction is correct.

ACKNOWLEDGEMENTS

This work was supported in part by NASA Grant NAGW-176 to the University of Michigan.

REFERENCES

1. C. D. Keeling, R. B. Bacastrow, A. F. Carter, S. C. Piper, T. P. Whorf, M. Heimann, W. G. Mook and H. Roeloffzen, in Aspects of Climate Variability in the Pacific and the Western Americas, ed. by D. H. Peterson (Amer. Geophys. Union, Washington, D.C., 1989), pp. 165-236.
2. M. Trexler, The biological carbon cycle, (American Physical Society), this volume.
3. J. M. Barnola, D. Raynaud, Y. S. Korotkevich and C. Lorius, Nature 329, 408-414 (1987).
4. H. Friedli, H. Lotscher, H. Oeschger, U. Siegenthaler and B. Stauffer, Nature 324, 237-238 (1986).
5. A. Neftel, E. Moor, H. Oeschger and B. Stauffer, Nature 315, 45-47 (1985).
6. W. S. Broecker and T.-H. Peng, Global Biogeochemical Cycles 3, 215-239 (1989).
7. W. S. Broecker, Glacial-to-interglacial CO2 variations (American Physical Society), this volume.
8. B. N. Opdyke and J. C. G. Walker, The return of the coral reef hypothesis: Glacial to interglacial partitioning of basin to shelf carbonate and its effect on Holocene atmospheric pCO_2, Geology, submitted, 1991.
9. S. Savin, Ann. Rev. Earth Planet. Sci. 5, 319-355 (1977).
10. R. A. Berner, A. C. Lasaga and R. M. Garrels, Amer. J. Sci. 283, 641-683 (1983).
11. E. J. Barron and W. M. Washington, in The Carbon Cycle and Atmospheric CO2: Natural Variations Archean to Present, ed. by E. T. Sundquist and W. S. Broecker (Amer. Geophys. Union, Washington D.C., 1985), pp. 546-553.
12. T. Owen, R. D. Cess, and V. Ramanathan, Nature 277, 640-642 (1979).
13. J. F. Kasting, Palaeogeogr., Palaeoclimat., Palaeoecol. 75, 83-95 (1989).
14. W. S. Broecker and T.-H. Peng, Tracers in the Sea (Lamont-Doherty Geological Observatory, Palisades, New York, 1982), 690 pp.
15. R. A. Houghton and G. M. Woodwell, Scientific American 260, 36-44 (April, 1989).
16. R. B. Bacastrow and A. Bjorkstrom, in Carbon Cycle Modelling, SCOPE 16, ed. by B. Bolin (Wiley, New York, 1981), pp. 29-79.
17. H. D. Holland, The Chemistry of the Atmosphere and Oceans (Wiley, New York, 1978), 351 pp.
18. R. A. Berner, Geochim. Cosmochim. Acta 54, 2889-2890 (1990).
19. T. M. Gerlach, EOS 72, 249-255 (1991).
20. J. C. G. Walker, P. B. Hays and J. F. Kasting, J. Geophys. Res. 86, 9776-9782 (1981).
21. J. C. G. Walker, Origins of Life 16, 117-127 (1985).
22. J. F. Kasting and T. P. Ackerman, Science 234, 1383-1385 (1986).

23. R. M. Rotty and C. D. Masters, in Atmospheric carbon dioxide and the global carbon cycle, ed. by J. R. Trabalka (U.S. Dept. of Energy, Washington, D.C., 1985), DOE/ER-0239.
24. J. C. G. Walker and J. F. Kasting, Effect of forest and fuel conservation on future levels of atmospheric carbon dioxide, Global and Planet. Change, in press.
25. H. W. Menard and S. M. Smith, J. Geophys. Res. 71, 4305-4325 (1966).
26. M. Meybeck, Amer. J. Sci. 287, 401-423 (1987).
27. I. G. Enting and G. I. Pearman, In The Changing Carbon Cycle: A Global Analysis, ed. by J. R. Trabalka and D. E. Reichle (Springer-Verlag, New York, 1986), pp. 425-458.
28. W. M. Post, T.-H. Peng, W. R. Emanuel, A. W. King, V. H. Dale and D. L. DeAngelis, Amer. Sci. 78, 310-326 (1990).
29. R. A. Houghton, J. E. Hobbie, J. M. Melillo, B. Moore, B. J. Peterson, Shaver G.R. and G. M. Woodwell, Ecol. Monogr. 53, 235-262 (1983).
30. R. P. Detwiler and C. A. S. Hall, Science 239, 42-47 (1988).
31. P. P. Tans, I. Y. Fung and T. Takahashi, Science 247, 1431-1438 (1990).
32. G. Esser, Tellus 39B, 245-260 (1987).
33. J. C. G. Walker, Numerical Adventures in Earth History, Oxford University press, New York, in press.
34. J. Gribbin, Hothouse Earth (Grove Weidenfeld, New York, 1990), 273 pp.

FORESTRY AND GLOBAL WARMING:
THE PHYSICAL AND POLICY LINKAGES

M. C. Trexler[*]
World Resources Institute, Washington, D.C. 20006

ABSTRACT

The potential for biotically mitigating global warming is receiving a great deal of policy and technical attention around the world. Elements of the political community are drawn to the notion that land-use patterns can be modified more easily than energy consumption patterns, and some modelers suggest that the potential for storing carbon in terrestrial ecosystems is very large. Most work to date, however, uses only physical criteria in estimating how much land might be available for reforestation. Accounting for social and economic constraints is much more difficult, resulting in daunting uncertainty about what could actually be accomplished. Furthermore, our relative ignorance of the functioning of the global carbon cycle makes attempting to manipulate it for human purposes questionable at best. Nevertheless, there are many reasons besides global warming to pursue a radical restructuring of land-use patterns around the world. Such a restructuring should be undertaken in conjunction with many other measures to slow global warming, most immediately in the energy sector.

INTRODUCTION

Global warming and its mitigation is assuming a higher and higher international profile. One way to slow and moderate global warming might be to manipulate the planet's biotic carbon cycle. Deforestation could be slowed (thus slowing the release of carbon to the atmosphere), carbon could be drawn out of the atmosphere by increasing the storage capacity of terrestrial vegetation and soils, and biomass could be substituted for fossil fuels in producing needed energy. The management of pasture, agricultural, and other lands around the world can also be modified to slow the accumulation of carbon dioxide (CO_2) in the atmosphere. Figure 1 illustrates the many carbon flows associated with land-use change and management; some of these flows could constructively be manipulated.

Policy options for mitigating global warming through biotic means are being explored intensively, partially because some analysts suggest that a global reforestation program could affect dramatically the rate at which CO_2 is accumulating in the atmosphere.[1] Because of perceptions regarding land availability and relative tree planting costs, it is often suggested that forestry as a mitigation strategy be pursued most intensively in the tropics, where it is commonly asserted that as much as one billion hectares (1 ha = 10^4 m^2 or 2.47 acres) of previously

---

[*]Dr. Mark C. Trexler directed the World Resources Institute's work on global warming mitigation through forestry until May, 1991. Now living in Portland, Oregon, he continues to work with the Institute on forestry, energy policy, and global warming issues.

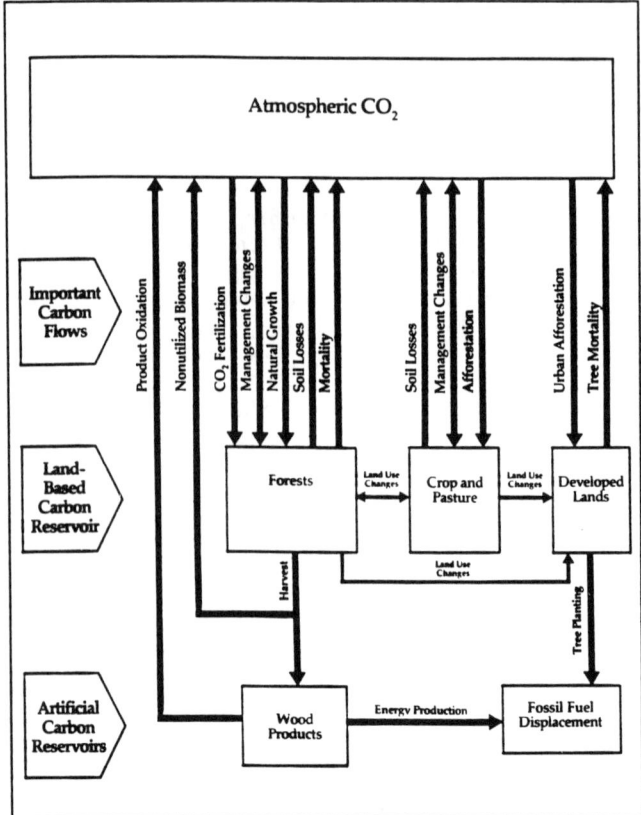

Fig. 1: Important Biotic Carbon Flows

forested but now degraded or abandoned land ought to be available for reforestation.[2] Reforestation of this land could in principle remove billions of metric tons of carbon from the atmosphere every year. As a result, power companies are talking about planting or protecting trees to offset their CO_2 emissions,[3] legislation has been introduced in Congress to formalize trees-for-CO_2 trading, and one of the few global warming initiatives embraced domestically by the United States government is the planting of an additional one billion trees per year as part of the "America the Beautiful" Program.[4] Internationally, negotiations are underway on a global forestry convention to protect and enhance the world's forests, and hence their carbon-storing capacity.

The overall potential and cost-effectiveness of biotic policy options remain poorly understood; what is appropriate in one country might not work well in another. Slowing deforestation would be valuable in several tropical countries, but less so in temperate countries. Countries where large areas of marginal or degraded lands are at least theoretically available for reforestation often face intense competition for any land that might be returned to economic usefulness through such efforts. Many tropical countries also lack the governmental and private infrastructures needed to make large-scale reforestation and forest protection work over the long run. In contrast, temperate industrialized countries such as the United States often lack large areas of severely degraded or abandoned lands that could support trees.[5] Temperate countries, however, are often better positioned to employ market and management mechanisms to increase carbon storage on those lands that are available.

The complexity of assessing the carbon implications of particular forestry efforts can be illustrated with an example from the Pacific Northwest region of the United States. The debate over old-growth harvesting, particularly in the Tongass National Forest, has recently been cast in terms of its global warming implications. Senator Murkowski of Alaska, for example, has argued that 48 trees grow back for every old-growth tree removed, concluding that replacing some of the old forest with young trees increases the amount of CO_2 removed from the atmosphere.[6] The basic fallacy of his argument is made evident by understanding that where there was one tree to begin with, there will eventually be room for only one tree in the future. The other 47 seedlings will no longer exist. In fact, scientists calculate that harvesting one hectare of old-growth forest reduces net terrestrial carbon storage by almost 300 metric tons, even after new trees have had a chance to grow for 60 years, and even if 42 percent of the originally harvested timber is assumed to be incorporated into long-term (carbon-storing) products.[7]

THE BIOTIC CARBON CYCLE

Other chapters in this volume introduce and discuss the global carbon cycle in more detail than will be undertaken here. The basic statistics are well known. In just two centuries, CO_2 concentrations have risen by over 25 percent from approximately 280 ppm in 1750 to more than 350 ppm in 1989. Today the atmosphere contains roughly 755 gigatons (1 x 10^9 metric tons) of carbon, some 180 gigatons more than it did prior to the industrial revolution. This unprecedented rate of change continues, with CO_2 concentrations rising by about 1.4 ppm or 0.5 percent per year.

Human activities are presumed to be the primary force behind the carbon cycle's current instability. Even centuries before the onset of the industrial revolution, for example, humans were releasing large quantities of CO_2 into the atmosphere as they converted temperate forests and grasslands to agricultural uses. Cumulatively, emissions related to land-use change probably exceeded cumulative fossil fuel emissions until just a couple of decades ago. This is not because biotic emissions have decreased (during the 1980s they were probably larger than ever before), but because fossil fuel emissions have accelerated so dramatically over the last four decades. Today, between seven and eight billion metric tons of carbon are being released annually through human activities, almost six billion of them through fossil fuel consumption.[8]

The role of the terrestrial carbon cycle in contributing to rising CO_2 concentrations is not fully understood. In fact, many questions remain with regard to the size of biotic carbon reservoirs in the first place; estimates of the quantity of carbon present in global vegetation range from under 500 to more than 800 gigatons.[9] The quantity of carbon stored in the world's soils is also subject to uncertainty, but is commonly asserted to be about 1,500 gigatons. Scientists have challenged estimates of carbon in vegetation at the higher end of the range in recent years, arguing that prevailing sampling methods have significantly overstated average biomass concentrations, particularly for forests.[10]

Given uncertainty in the basic carbon content of the world's biomass, it is not surprising that estimates of carbon release through land use change would vary as well. Houghton estimates that between 90 and 120 gigatons of carbon were released to the atmosphere between 1840 and 1980 through land-use change, but other estimates range widely.[11] Nobody knows, for instance, exactly how many hectares of forest have been or are currently being converted to alternative land-uses, or how much carbon is released per hectare from vegetation and soils during the conversion process. The deforested land might have been primary or secondary forest, dry or moist. Table I illustrates how widely carbon content can vary by ecosystem type. The entire January, 1991 issue of the <u>Canadian Journal of Forest Research</u> is dedicated to assessing the magnitude of the carbon emissions resulting from land-use change.

As a result of the many uncertainties, global deforestation carbon-release estimates for 1980 range from 0.4 - 2.5 billion metric tons in several different studies,[12] and relatively wide ranges persist in estimates of emissions into the 1980s.[13] Table II estimates emissions for 1980 from key tropical countries according to Houghton. Adding to the uncertainty over biotic emissions are processes other than deforestation that no doubt also contribute to biotic carbon emissions, but that are much harder to quantify. Among the many processes missed by formal deforestation statistics are:

Forest Type	Carbon (Tons/ha)
Tropical Premontane wet	690
Tropical montane wet	375-416
Tropical wet	180-500
Tropical moist	320-480
Subtropical wet	272
Subtropical moist	160-290
Subtropical dry	80-90

Estimates of carbon content by ecosystem vary widely due to natural variability, modelling assumptions, and quite limited data. These figures should be seen as broadly representative.

Table I. Carbon Loading By Forest Type

Source: S. Brown and A. Lugo, Biomass of Tropical Forests: A New Estimate Based on Forest Volumes, Science 223, 1290 (1984).

- forest degradation without deforestation (fuelwood collection, fire, logging, climate change);
- soil carbon losses (oxidation, erosion, desertification).

From a climate change perspective, however, gross CO_2 emissions estimates for land-use change are not the end of the story. There are, for example, a number of processes under way that would tend on net to draw carbon out of the atmosphere. These include:

- natural forest regeneration and regrowth (logged areas, abandoned lands);
- soil carbon recovery (abandoned lands, sustainable agriculture);
- intentional reforestation (farm forestry, industrial plantations);
- CO_2 fertilization as a result of rising CO_2 concentrations.[14]

Brown, for example, suggests that far larger areas of forest were disturbed through human activity in past centuries than has been commonly assumed.[15] She suggests that rather than being in the basic carbon balance assumed for mature forests, large areas of forest in the tropics are still recovering from these past intrusions and may be able to accumulate considerable quantities of additional carbon in the future. The recent modeling of global carbon flows has also raised many questions about the net impact of terrestrial ecosystems on atmospheric CO_2 concentrations. Questions are even being raised as to whether the carbon emissions from deforestation around the world are more than offset by net carbon absorption in the remaining terrestrial ecosystems.[16] An affirmative answer to these questions does not decrease the damage from CO_2 emissions from deforestation and other land-use changes, for in their absence atmospheric CO_2 concentrations no doubt would be

COUNTRY	NET CARBON RELEASE	SHARE OF TOTAL
	(10^6 tons)	(percent)
Brazil	336	20
Indonesia	192	12
Colombia	123	7
Ivory Coast	101	6
Thailand	95	6
SUBTOTAL	847	51
Others	812	49
TOTAL	1,659	100

* Figures presented are midpoints of the estimated ranges.

Table II: Leading Sources of Biotic Carbon

Source: Houghton, et al., 1987.

increasing more slowly. But it does illustrate the difficulty of making sweeping conclusions about the behavior of the biotic carbon cycle based on what we perceive to be a good understanding of just individual elements of the cycle, e.g. plantation growth rates.

Questions regarding the net implications of terrestrial biotic carbon flows on global carbon fluxes arise because only some 3 of the 7 to 8 gigatons of carbon being emitted through human activities remain permanently resident in the atmosphere. Obviously, a large fraction of the carbon being emitted finds its way into sinks other than the atmosphere. It was historically assumed that the oceans constituted the primary sink for this carbon. In 1990, however, Tans, Fung, and Takahashi looked at the empirical north-south atmospheric concentration gradients for CO_2 and concluded that the bigger sinks for CO_2 had to be in the northern hemisphere.[17] In the northern hemisphere, however, they found that the differential partial pressures between atmospheric and water-borne CO_2 were insufficient to suggest the absorption of the very large amounts of CO_2 that were unaccounted for, and that as little as one gigaton was actually disappearing into the oceans.

Tans and colleagues postulated that a lot of the "missing" CO_2 must be going into terrestrial ecosystems and that temperate vegetation was already absorbing 2-3.4 gigatons of carbon per year during the 1980s. Keeling, too, has postulated that the terrestrial biota as a whole has generally been a net sink for carbon, although his models suggest this situation has changed in recent years.[18] Both hypotheses are hard to verify since actual increases in growth rates across the large temperate forests have not yet been observable in empirical stand data.

Our ignorance with respect to the basic functioning of the carbon cycle, combined with uncertainty over the magnitude of future emissions, makes it difficult to project how the concentration of CO_2 in the atmosphere could change in coming decades. The quantity of carbon that ultimately could be released through fossil fuel combustion is huge, measuring in the trillions (1×10^{12}) of metric tons. If nothing is done, says the Intergovernmental Panel on Climate Change, carbon emissions probably will increase by roughly 2 to 3 percent annually, going from some 6 gigatons today to more than 8 gigatons by 2000 and to more than 13 gigatons by 2025.[19] Biotically, hundreds of gigatons of carbon in forest vegetation, and hundreds more in the world's soils, theoretically could be released to the atmosphere through future land-use change. If deforestation were to track population growth, for example, biotic emissions could climb to five gigatons per year in the next century. Fortunately, the most rapid population growth is not co-located with the large remaining tracts of tropical forest, so biotic emissions are not expected to increase in parallel.

A key question in projecting CO_2 concentrations is what fraction of emitted carbon will remain resident in the atmosphere in the future. If CO_2 fertilization of temperate forests is accounting for as big of a carbon sink as is postulated by Tans and colleagues, there is the real danger that another nutrient or climatic constraint will become limiting, potentially shutting off the sink within a short period. Even CO_2 uptake by the oceans could prove markedly non-linear.[20]

MANIPULATING THE BIOTIC CARBON CYCLE TO SLOW GLOBAL WARMING

Carbon dioxide emissions from human activities account for just a small fraction of the more than 200×10^{12} metric tons of carbon that cycle annually between the atmosphere and terrestrial biota as a result of photosynthesis, plant respiration, and the decay of plant matter. But because these activities are responsible for the unwanted buildup of CO_2 in the atmosphere, modifying them should help moderate the rate and ultimate magnitude of global warming. Reducing fossil fuel CO_2 emissions in the first place is the most obvious place to start, but changing land-use practices and patterns ought to be able to play a role as well. The size of the naturally occurring annual flux between the living biota and the atmosphere, however, also suggests that it might be possible to modify the carbon cycle to encourage the increased uptake of carbon in forests, soils, and other biotic sinks. It is the only point in any of the major global cycles at which large quantities of a greenhouse gas can be practically removed from the atmosphere by human means.

Some plants do in fact have extraordinary potential for fixing CO_2 in vegetative matter; a hectare of water hyacinths can extract 60 metric tons of carbon from the air per year, while algae can extract even more. Nevertheless, it is not only the immediate carbon fixation rate that is important to the goal of mitigating global warming. Instead, the fixed carbon must be kept out of the atmosphere for a period of time long enough to further the policy objective, or must be used to displace a comparable quantity of other carbon from being emitted to the atmosphere. Plants such as water hyacinths or annual shrubs present the problem of annually disposing of huge quantities of harvested biomass in such a way as to prevent the release of the fixed carbon to the atmosphere through normal decay processes. Conversely, a long-lived tree species such as Douglas fir can serve as a natural carbon repository for more than a hundred years, albeit at a much slower rate of carbon accumulation. Another option is to extract liquid alcohol from quick rotation trees or even water hyacinths, and to burn it in place of fossil fuels. While biofuels release carbon dioxide in similar proportion to fossil fuels on a gallon per gallon basis, a sustainably managed biofuels program would add no carbon to the atmosphere on a net basis.

While looking at biotic options for global warming mitigation, a set of caveats should be borne in mind. First, we are currently moving in quite a different direction than global warming mitigation would require. Deforestation has become much worse rather than better over the last decade, and population pressures, the need for agricultural land, and the degradation of existing land will all contribute to more and more pressure on remaining land and forest resources. Another practical problem with using global forestry to mitigate global warming is the potentially serious impact that the global warming we are already committed to may have on existing forests. By affecting not only temperature but also moisture regimes, even moderate global warming will significantly affect the growth and regeneration capabilities of some standing forests.[21] Global warming also will increase respiration and decay rates in many regions, accelerating the release of currently sequestered carbon to the atmosphere.

These major caveats aside, biotic options do exist that ought to be able to slow the accumulation of CO_2 in the atmosphere from what otherwise would occur. The caveats discussed do not change the physics of the five land-use-based approaches that can be used to slow the buildup of CO_2 in the atmosphere:

- slowing or stopping the loss of existing forests, thus preserving current carbon reservoirs;
- adding to the planet's vegetative cover through reforestation or other means, thus enlarging living terrestrial carbon reservoirs;
- increasing the carbon stored in nonliving carbon reservoirs such as agricultural soils;
- increasing the carbon stored in artificial reservoirs, including timber products; and
- substituting sustainable biomass energy sources for fossil fuel consumption, thus reducing energy-related carbon emissions.

These approaches are all based on the same basic premise: adding to the planet's net carbon stores in vegetative cover or soil, or preventing net losses elsewhere in the biota, will help moderate global warming by keeping atmospheric CO_2 levels lower than they would be otherwise. Manipulating the type and extent of global vegetation cover is far from a homogeneous problem: different climates support very different types of vegetation, growth rates, and soil capabilities, while different countries are characterized by different population pressures, land availability, rates of vegetation loss, policy and administrative infrastructures, and financial resources.

A simple comparison of the carbon sequestration implications of alternative regions illustrates this issue. A Douglas fir forest in the United States could sequester carbon for more than 100 years. Barring events such as catastrophic forest fires or insect infestations, the amount of carbon likely to be fixed in the trees and soils of a new forest would be relatively predictable, with little chance of its being lost back to the atmosphere prematurely as a result of human action. Indeed, the utilization of much of the produced lumber in long-term applications such as construction materials would lengthen considerably the time period during which the absorbed carbon is kept sequestered from the atmosphere.

Carbon sequestration forestry projects in the tropics, on the other hand, present a more complex assessment problem.[22] Driving the design of tropical forestry projects should be the requirement that benefits from a project must accrue or at least be perceived as accruing to local people within a short time frame in order to elicit local participation and cooperation. Existing human population densities, projected population growth, and widespread rural poverty in tropical countries will often make planting large stands of long-rotation new trees impractical. This suggests the use of fast-growing short-rotation tree species. Simply growing and harvesting new trees every 5-10 years, however, would do little more than marginally delay the buildup of atmospheric CO_2. Indirectly simulating the long-term carbon sequestration benefits of long-term forestry, even as individual trees are repeatedly cut and regrown, is the challenge for such projects. Two ways to achieve this are to take pressure off standing forest by using the new trees or to substitute the newly-grown biomass for fossil fuels. Reliably projecting the actual carbon sequestration benefits of a tropical sustainable development project is not a simple matter. In protecting standing forest, for example, carbon offset findings have to be predicated on assumptions regarding the carbon content of the forest, how much of the standing forest actually would have been lost in the absence of protective measures, and how much the spread of agroforestry or woodlots really would prevent incursions into standing forest. Such estimates are dependent on assumptions about local sources of deforestation, the magnitude of fuelwood or other wood-product deficits, and the impacts of social forestry on land use as well as population migration patterns. Estimating how much biofuels production realistically would offset fossil fuels combustion, rather than just increasing energy consumption, depends on assumptions about the likely future availability of fossil fuels, existing and future energy infrastructures, and fuel price elasticities. Ultimately, the robustness of any of these estimates may be suspect when dealing with a developing country facing tremendous population growth and serious economic problems. The only carbon sequestration project yet undertaken in such a context is described in Box 1.

The plausible assessment of the costs of biotic options to mitigate global warming is not yet possible. Apart from the difficulties of assessing the true societal costs of changing land uses, other variables complicate the process further. The initial costs of forestry options can vary widely, as can the recurring costs of managing trees. The carbon benefits of tree planting are themselves subject to uncertainties that policy implementors cannot control -- among them, long-term biomass growth rates and natural disasters. And further, the conventional economic payoff of planting trees is so far away, often 50 to 75 years, that the potential revenues associated with biotic options are often ignored in estimating their current cost-effectiveness. For these reasons, the costs of sequestering or displacing carbon through biotic options can vary so much that comparing the apparent cost-effectiveness of such options to each other and to other mitigation options (such as appliance efficiency) can be frustratingly difficult and even misleading.

Costs aside, two primary approaches exist for estimating what it might be possible to accomplish biotically in terms of carbon cycle manipulation. First is a system equilibrium approach, attempting to use historical data and carbon cycle models to project physical potentials. Only a few such estimates have been made. The second is a land-availability approach, which at first blush appears more empirical but in reality is still at roughly the same stage of abstraction.

Carbon Cycle Equilibrium Estimates of Carbon Storage

As previously noted, carbon modelers differ considerably in their estimates of what is now happening with the biotic side of the carbon cycle, and what might happen in the future. Keeling suggests that the biota has recently gone from being a net sink of several hundred million metric tons annually to a net source.[23] Tans and colleagues, however, suggest that terrestrial biota is today taking up 2-3.4 gigatons of carbon on net. Schlesinger estimates that the maximum potential sink in soils is not likely to exceed 0.4 gigatons per year.[24] Prentice and Fung suggest that land vegetation could act as large carbon sink in the potentially warmer climate of future, storing more than 2×10^9 additional metric tons per year.[25] They see a doubled CO_2 climate as leading to increased carbon storage of 245 to 338 gigatons (the equivalent of between 122 and 169 ppm of atmospheric CO_2). Prentice and

---

An Example of Carbon Storage Forestry

Although planting trees to offset carbon emissions from fossil fuel consumption was first proposed in 1977, the idea was not acted upon until 1988, when Applied Energy Services (AES) began building a new 183 MWe coal-fired powerplant in Connecticut.

AES, an independent power producer with a contract to sell power to Connecticut Power and Light, was concerned about how releasing approximately 15×10^6 metric tons of carbon into the atmosphere would contribute to global warming. Approached for advice on how to neutralize the global warming impacts of the plant's CO_2 emissions, the World Resources Institute ultimately recommended that AES fund a sustainable agriculture and agroforestry project proposed for Guatemala by the relief organization CARE. Under the final terms of the agreement, AES contributed $2 million to the project, making it possible for CARE to attract additional financial and in-kind support for the project valued at nearly $15 million.

WRI analysis concluded that the project will more than offset the 15×10^6 metric tons of carbon emitted over the plant's lifetime. In addition, significant benefits will accrue to local residents. Over 10 years, the project will bring private and community woodlots, agroforestry, alley-cropping, live-fencing, and soil-conservation practices to about 40,000 Guatemalan farm families. If the processes put into action by the project last 40 years, the life span of the power plant, well over 100×10^6 trees will be planted. Nevertheless, the project's most important carbon-storage benefit lies in protecting the region's existing standing forest. This protection is to be achieved by increasing the productivity of agricultural lands, thus reducing the need to destroy neighboring forest to secure fuelwood and make way for agriculture.

Source: M.C. Trexler, P.E. Faeth, and J.M. Kramer, Forestry as a Response to Global Warming: An Analysis of the Guatemala Agroforestry and Carbon Sequestration Project. World Resources Institute (1989).

Box 1: The Guatemala Carbon Forestry Project

Fung's estimates, of course, have to be interpreted cautiously in estimating the potential of using the biosphere to mitigate the warming in the first place, since their estimates are based on a doubling of CO_2 concentrations from pre-industrial times. Prentice and Fung's estimates are particularly striking if taken in the context of the conclusion of Adams and colleagues that the amount of carbon present in terrestrial ecosystems today is already considerably more than double that present during the Last Glacial Maximum some 18,000 years ago.[26] Can the quantity of carbon in terrestrial ecosystems really almost be doubled again in the face of the impacts of the climatic and other perturbations that would accompany a doubling of atmospheric CO_2?

PLANTATION TYPE:*	PLANTED AREA 10^6 ha	ASSUMED YIELD Tons/ha/yr
High yield sycamore:	800	7.5
Best tropical:	400	15.4
Good tropical:	600	10.1
Marginal land:	1,900	3.2

*Plantation growth rates in metric tons of carbon per unit area per year assume a total biomass accumulation equal to 1.6 times stemwood accumulation.

Table III: Added Plantations Required to Offset 5.9×10^9 Tons of Carbon

Land Availability Estimates of Carbon Storage

Freeman Dyson first quantified in 1977 the area of new forest or tree planation that would be required to absorb the quantity of fossil-fuel carbon being emitted through combustion each year.[27] He concluded that 730×10^6 hectares of trees, an area roughly equal to 80 percent of the land area of the entire United States, would be required to offset some 5 gigatons of emitted carbon. More recently, but based on a more in-depth analysis of achievable tree growth rates, Marland arrived at similar numbers.[28] Sedjo, in turn, concludes that global warming could be mitigated by planting less than 500×10^6 hectares of plantations, albeit at a cost of $200 to $400 billion in establishment costs alone.[29] It is important to note that some of the estimates assume the need to only remove 50 percent of carbon emissions from the atmosphere through forestry, assuming that other sinks will continue to remove the roughly 50 percent they do today. Our ignorance over how the system functions, however, makes this an optimistic assessment of the situation.

A large part of any land area of the magnitude being discussed by these authors will be marginal land with less than ideal growth rates. In light of this, more pessimistic but still perfectly plausible estimates of land requirements could exceed 1×10^9 hectares. Table III frames a variety of forestry scenarios given such variables, and Table IV summarizes growth rates for a range of species and forest types. Citing both land and input requirements such as fertilizers, Edmonds and colleagues largely dismiss the viability of a forestry-based response to global warming.[30]

Beyond the problem of assessing how much land might ultimately be available based on physical criteria, it remains difficult to assess the practical significance of such numbers. Billions of tons juxtaposed against billions of hectares have little relevance to policy decisions. Tables V - VI attempt to provide some practical significance to the numbers discussed above. Table V estimates the land area required under various forestry scenarios to mitigate the carbon emissions of a 100 MWe power plant. To bring these figures down to an even more practical

SPECIES OR FOREST TYPE	CARBON FIXATION RATE Metric Tons C/ha
Forest Productivities	
Average U.S. commercial forest:	1.3
U.S. non-commercial forest:	0.3
Natural Douglas fir stands:	3.2
Natural loblolly pine stands:	2.1
Tropical moist forest:	4.2
Plantation Productivities	
American sycamore:	7.5
Douglas fir:	5.1
Loblolly pine:	6.0
SE Asia Dipterocarp:	4.5
Tropical pines:	12.2
Tropical eucalyptus:	15.7
Silvicultural Target Productivities	
Loblolly pine:	17.3
Douglas fir:	14.4

*Carbon fixation rates depend on many site specific variables, and listed rates should only be seen as roughly indicative of what can be achieved on good quality sites. Estimates account for total biomass accumulation assumed to be equal to 1.6 times stemwood accumulation. Figures represent mid-points in published productivity figures.

Table IV: Representative Carbon Fixation Rates by Tree Species and Forest Type

Source: Marland, 1988; World Resources Institute, 1989.

scale, Table VI provides a variety of ways in which the global warming impact of U.S. per capita CO_2 emissions could be offset.

The figures in Tables V and VI help illustrate the scale of the issue, but provide little insight into whether and by how much global warming could actually be slowed through global afforestation and reforestation efforts. They are even less suggestive of how one actually would design and implement a global reforestation effort. The number of variables relevant to bounding the actual potential of large-scale forestry-based strategies for combatting global warming is quite large. Not only do strictly physical variables such as land availability, tree growth rates, and planetary albedo need to be estimated, but social and economic variables such as demographic trends, land tenure, and governmental infrastructural capabilities must be carefully considered in future research as well.

The failure of many past reforestation and afforestation projects can be explained by lack of attention to factors such as lend tenure problems, corrupt agencies, and a basic lack of expertise. There are many variables and issues about

FORESTRY MITIGATION MEASURE*	APPROXIMATE LAND REQUIREMENT
Doubling U.S. commercial tree productivity:	170,000 ha
Rescuing doomed tropical forest**:	65,000 ha
Planting temperate Douglas fir plantation:	45,000 ha
Planting temperate sycamore plantation:	30,000 ha
Planting tropical eucalyptus plantation:	14,000 ha
Planting next-generation loblolly pine:	12,000 ha

Assumes 100 MWe, baseload operation, 90% capacity factor.

*Assumes total biomass accumulation equal to 1.6 times stemwood accumulation. Land areas specified would offset 100 MWe of capacity for a 40-year period, assuming growth rates in Table IV could be maintained.

**Assumes that the hectares would have otherwise been deforested with a release of 150 metric tons of carbon from biomass and soil per hectare. Offsetting 100 MWe of emissions requires the protection of some 1,790 hectares of such (otherwise destroyed) forest per year.

Table V: Forest Measures to Offset the Carbon Production of 100 MWe (Coal)

which much more will need to be known before the global potential of forestry for increased carbon sequestration can reliably be assessed. Estimates of carbon storage potentials that realistically incorporate these various issues do not yet exist. Even on a purely physical basis it is easy to misinterpret the carbon benefits of proposed forestry projects. Box 2 lays out the necessary steps to perform a realistic carbon accounting.

What does exist already are preliminary looks at land availability based on just some of the important variables. U.S. Forest Service scientists, for example, estimate conventional tree planting and forest management by themselves to be capable of storing an additional 700×10^6 metric tons of carbon per year in the United States.[31] Yet technical potential is often quite different from practical reality. Indeed, the average commercial productivity of U.S. timberlands is just 1.4 metric tons of carbon per hectare per year, compared to a technical potential of more than 7 metric tons of carbon per hectare per year on good temperate soils. Table VII summarizes research that attempts to differentiate between the theoretically possible and practically achievable for a series of biotic policy options that could be pursued in the United States. While the United States may well possess the land resources and technology needed to pursue any individual policy option to the absolute limit of its theoretical potential, competing policy objectives and rising marginal costs would make such a course inappropriate. Arriving at what may be practically achievable by any given date, 2030 in the case of Table VII, therefore involves a significant element of subjectivity. The economics of various biotic policy options are not well understood and can change rapidly. Marginal cost estimates will often differ substantially depending on whether the marginal costs reflected are private or societal, short-term or long-term. The long-term economic and environmental feedbacks of implementing the identified policy options have also not yet been analyzed, and the global commodity status of timber will

MITIGATION STRATEGY*	MITIGATION REQUIREMENT
Planting temperate sycamore plantation:	0.6 hectares
Planting tropical eucalyptus plantation:	0.3 hectares
Driving your car less (at 25 mpg):	50,000 fewer miles
Recycling rather than venting automobile air conditioning charges:	3 charge recyclings**
Replacing 75 W incandescent bulbs with 18 W fluorescent bulbs:**	approx. 120 bulbs

*Mitigation assumes CO_2 reduction (or the warming equivalent in other greenhouse gases). Areas of planted trees would mitigate annual emissions as long as growth rates are maintained. Energy efficient bulbs would mitigate annual emissions over their lifetimes. The other alternatives require the mitigating measure to occur every year.

**Assumes a 2 lb CFC charge, and a molecule for molecule CFC:CO_2 warming effectiveness ratio of 20,000:1.

Table VI: Options to Offset U.S. Per Capita Carbon Emissions of 5 Tons/Year

complicate such analysis. Market feedbacks from the large-scale implementation of tree planting anywhere in the world, for example, pose a potentially serious threat to the long-term effectiveness of carbon sequestration policies implemented unilaterally in the United States or any other country. Although energy policies pose some of the same problems -- that is, any significant policy-inspired fall in global energy demand probably would be partially offset by falling prices and resulting upward pressure on consumption -- these undesired market feedback effects are much easier to control in the energy sector; imposing an energy tax that is neither too high nor too low can prevent the feedback altogether. In the forestry sector, however, imposing a guaranteed price floor for timber (the functional equivalent of the energy tax in this case) poses far more vexing problems.

For the tropics, some analysts have begun to address the first set of issues, namely simple land availability. Houghton, for example, suggests that reforestation of tropical lands formerly forested but currently not used for cropland or settlements could stabilize, at least temporarily, the concentration of CO_2 in the atmosphere.[32] He estimates that 500×10^6 hectares would be available for reforestation: 100×10^6 hectares of mostly degraded grazing land in Latin America; 100×10^6 hectares of largely degraded grasslands in Asia; and in Africa, 300×10^6 hectares is estimated available, half of this in the savannas of Western Africa. If shifting agriculture were replaced with low-input, permanent cropping which was sustainable and produced sufficient food, then an estimated 85 percent, or an additional 365×10^6 hectares, of this area could return to full forest cover. Under Houghton's analysis, then, a total of 865×10^6 hectares in the tropics could be available indefinitely for reforestation.

Grainger suggests that 758×10^6 hectares of degraded tropical lands have potential for forest replenishment.[33] Specifically, this includes 87×10^6 hectares of deforested watershed, 331×10^6 hectares of degraded drylands, 203×10^6 hectares of forest fallows in humid areas, and 137×10^6 hectares of logged rainforests. Grainger recommends that the drylands, montane areas, and humid forest fallows, totaling

> For a rough idea of how much carbon a particular forestry or other biotic policy will sequester, several steps must be taken:
>
> - Project realistic biomass growth rates, accounting for the species to be planted, likely soil conditions, and water availability. Also account for the possible impacts of climate change itself on water availability and growth rates at project sites.
>
> - Project the proportion of biomass accumulation likely to be lost to natural forces, including fires, droughts, and insect or disease outbreaks.
>
> - Account for carbon released during project implementation, including that in vegetation cleared from project sites, soil carbon released when lands are cleared or planted, and fossil fuel energy consumed during the planting, maintenance, and possible eventual harvesting on project sites. The energy consumed in producing fertilizers and pesticides for the project as well as that consumed in converting the biomass to liquid fuels or other energy forms, must also be accounted for.
>
> - Account for any carbon that would have accumulated on the project site in the absence of the project.
>
> - Account for the ultimate disposition of accumulated carbon. Biomass growth estimates are often premised on whole-tree biomass rather than just merchantable timber. If harvesting or thinning takes place, a significant proportion of the total carbon can be released back to the atmosphere -- as roots, parts of trees left behind, or lumber wastes decompose.
>
> - Avoid any double counting of carbon benefits. Delaying harvest of a forest containing 1×10^6 tons of carbon, for example, can "reduce" emissions over time by only 1×10^6 tons, regardless of how many years the delay is in force.
>
> Box 2: Performing A Realistic Carbon Accounting

621×10^6 hectares, would be most suitable for plantation establishment, while the remaining 137×10^6 hectares of cutover tropical forest would instead be managed for natural regeneration. The 621×10^6 hectares recommended for plantations is divided roughly equally between Asia (36 percent), Latin America (36 percent), and Africa (28 percent). Grainger points out, however, that this area is 50 times the total area of tropical forest plantations in 1980 of 11.5×10^6 hectares, 60 percent of which were for industrial wood production and the remainder for fuelwood production, forest protection, and other non-industrial purposes.

There is little doubt that tropical countries potentially have more land to reforest than do temperate countries, that deforestation is a bigger problem in the tropics, and that many tropical countries desperately need to reestablish tree or forest cover for watershed protection, energy supplies, and food and fodder production. Forestry ought also to be in principle cheaper in the tropics, partially

Option	Theoretical Carbon Benefit[a] (millions of tons/yr)	Practical Benefit to 2030 (millions of tons/yr)	($/ton)
Urban Forestry	15	3-5	$0-25
Private Land Conversions	400-900[b]	50-150	$0-50
Public Land Conversions	(not known)	(not known)	(not known)
Forestry Management	100-400	35-75	$0-100
Forgoing Old-Growth Harvests	10-20	5-10	$0-100
Biomass Energy Production	400-1,000[b]	20-150	$20-75
Increase Wood Use	(not known)	(not known)	(not known)
Soil Carbon Buildup	50-150	10-25	$0-10

a. The carbon benefit figure estimates carbon storage in biomass and in soils as well as fossil fuel carbon displaced through implementation of the biotic options. Figures represent an average annual carbon benefit and thus overstate the carbon benefits in the early years of a policy option's implementation.
b. These entries cannot be summed because they involve use of overlapping land bases.

Table VII: U.S. Biotic Policy Opportunities and Estimated Costs

because much higher biomass growth rates can be achieved. But tropical countries are often characterized by factors not conducive to long-term carbon storage. Land tenure policies that eliminate the incentive for careful land stewardship, widespread natural and intentional wildfires that maintain grass climax systems, public policies that encourage the conversion of "worthless" forest into "productive" land, forestry departments that have few staff, little money, or both, and forestry departments that are more forest police that forest managers and extensionists. Add to this political instability as well as climatic instability in many countries, and the rosy picture than can be painted for the tropics gets considerably grayer.

Researchers at the World Resources Institute are attempting to account for additional variables beyond physical land availability in the tropics. Beginning in the summer of 1990, WRI, with support from the U.S. Environmental Protection Agency, took the first step in compiling country-specific assessments of the role tropical forestry could play in slowing the accumulation of carbon dioxide in the atmosphere. The methodology of the research project is as follows:

- More than 50 countries were selected for study. Particularly small or arid countries were omitted from the sample. It was understood that the large number of countries selected would keep the analysis relatively crude. Nevertheless, it was thought important to perform at least a first-order assessment for the tropics as a whole.

- A data collection protocol was established, identifying which types of information out of an almost infinite universe of possibilities should be targeted for collection. Standardized country "profiles" were established for each of the selected countries using Grandview, a computer software package.

- Country-by-country information was collected through extensive literature reviews, through more than 150 in-depth interviews with governmental and non-governmental experts on the selected countries, and through the circulation of a questionnaire to several hundred foresters around the world.

- The large quantities of information collected were entered into the computer profiles. Researchers than analyzed the collected data and derived "best-estimates" of land availability and implementation rates. Specifically, they are attempting to derive 10 numbers relating to the total land availability and rate at which forestry efforts could be implemented in each country:

    - current and future deforestation rates, and potential for slowing them;
    - area available for forest regeneration, and annual rates;
    - area available for agroforestry, and annual rates;
    - area available for plantations, and annual rates.

- The numerical estimates of land availability are to be coupled in a spreadsheet with biomass density estimates for the different forestry approaches. The result will be estimates of 1) how much plausible changes in deforestation rates could slow the emission of carbon to the atmosphere in the first several decades of the next century, and 2) how much carbon could plausibly be removed from the atmosphere through increased reforestation, forest regeneration, and agroforestry early in the next century.

This study reflects the first attempt to incorporate social and economic variables into estimating the overall potential for tropical forestry to help slow global warming.[34] In addition, it is the first time the perspective of individuals in tropical countries and experts on those countries has been sought out to this end. Preliminary estimates of land availability are summarized in Table VIII.

Tree Cover*	Africa	Asia	L. America	Total
Forest Regeneration	59×10^6	36×10^6	8×10^6	150×10^6
Farm Forestry	30×10^6	20×10^6	7×10^6	56×10^6
Plantation Forestry	5×10^6	38×10^6	2×10^6	45×10^6

* Regional projections reflect hectares which it is estimated could be converted to the various tree cover types by the year 2050. Projections are preliminary and are subject to change. Projections of hectares likely to be deforested, and projections of the ability to avert this deforestation, are not included.

Table VIII: Estimated Tropical Land Availability By Type of Tree Cover

CONCLUSIONS

The linking of global warming mitigation to the tropical forestry sector is of interest to two quite separate policy communities. The first group consists of people primarily concerned about global warming who see forestry as one option in the portfolio of measures that need to be instituted to mitigate the problem. The second group is composed of people who see the concern over global warming as yet another but perhaps more urgent and better funded opportunity to mount a new comprehensive attempt to slow and even reverse detrimental tropical forestry trends.

There can indeed be a confluence of interests between efforts to slow global warming on one hand and efforts to address forestry sector crises in many tropical countries on the other. The variables contributing to this confluence include:

- Tropical deforestation accounts for a significant proportion of current anthropogenic CO_2 emissions to the atmosphere, and is the dominant source of anthropogenic CO_2 in many tropical countries.

- Huge areas of tropical forest, currently reservoirs for hundreds of billions of metric tons of carbon, remain threatened by future deforestation and forest degradation. The oxidation of the carbon now stored there could significantly increase atmospheric CO_2 concentrations.

- Hundreds of hectares of acres of previously forested tropical lands are theoretically available for reforestation or regeneration. In many cases these lands are proving incapable of sustaining agricultural or pastoral land-uses over the long-term.

- Forest loss and other forms of land degradation are imposing tremendous economic, social and environmental costs on the people and resource bases of many tropical countries. Reforestation efforts could advance sustainable development, energy production, and environmental goals in tropical countries, all while adding to terrestrial carbon stores.

Clearly there are many reasons to aggressively pursue forestry and land-use policy change around the planet. Forests and tree cover more generally are vital to watershed protection, biodiversity conservation, energy supplies, slowing desertification and soil erosion, food production, and sustainable development.

WRI research suggests that deforestation can be slowed significantly in many countries, but that it will take decades to achieve the full benefit. The research also suggests that considerably less land is plausibly available for reforestation than has been previously estimated based purely on physical variables. Other conclusions of the research include:

- Land availability per se is not the primary constraint to global warming forestry in the near to mid-term. Social, political, and infrastructural barriers will keep plausible reforestation rates very modest for at least a decade or so. Even then, the rate at which reforestation realistically can be accomplished is modest compared to the plausible availability of land in many countries of the tropics.

- The success stories that do exist in the tropical forestry sector tend to be site-specific. This suggests that achieving similar successes in other areas will depend on thoroughly understanding the local situation. This will dramatically slow our ability to move directly to the large-scale implementation of virtually any forestry management option.

- Any effort to simultaneously launch major efforts in all of the possible forestry arenas, for example, would in most cases overwhelm institutional and technical capacities. Money is far from the only constraint facing the mitigation of global warming through forestry.

- Global warming mitigation through forestry cannot successfully be tackled by the forestry community alone; foresters and forestry departments simply do not control many of the variables that will prove important to long-term success in this area. Yet many important sectors are not yet involved in the debate. The successful implementation of large-scale forestry management options will depend on large-scale policy reform in the land tenure and tax arena, on large-scale increase in agricultural productivity, and on long-term efforts to stabilize population and other pressures on the natural resource base. These linkages raise questions about the ability of an international forestry instrument to cope successfully with the overall problem. Drafters intended the Tropical Forestry Action Plan to account for these linkages, but the plan has not succeeded.

- The large-scale reforestation of degraded lands with industrial plantations is the most commonly discussed option for using tropical forestry to mitigate global warming. Implementation considerations, however, may in many cases suggest approaches to tree planting that have little to do with large-scale plantations. As already noted, slowing deforestation offers much more potential in the short term than any reforestation effort, and large-scale plantations usually will do little to help slow deforestation. In addition, the long-term sustainability of high-yield plantations on degraded tropical soils is open to considerable question. Even where reforestation is the appropriate measure, promoting the reliability of carbon sequestration benefits will require that projects be tailored to the specific situations and, in particular, to the interests of those individuals involved. In many cases this will suggest smaller-scale sustainable agriculture, agroforestry, and woodlot projects. In many cases, it will also require addressing underlying social issues such as land tenure.

For the foreseeable future many uncertainties will exist regarding what can be accomplished in the area of modifying land-use change patterns and improving land management to slow global warming. Even under the best circumstances deforestation rates are currently high and will have to be slowed incrementally, and reforestation programs are still relatively small and will have to grow over time. There is so much momentum built into the deforestation process in the form of population growth, logging concessions, and rural resettlement efforts, that actually reversing current trends poses mammoth difficulties. In light of past experience, it would be a remarkable achievement if a large enough effort could be mounted to bring the terrestrial biota into a CO_2 balance by ending or offsetting through reforestation those resulting from deforestation. It is therefore likely to be several decades before tropical land use change could successfully turn the tropics into a net sink for carbon. It is even harder to conceive of an effort large enough to offset a significant proportion of rapidly rising fossil fuel emissions of CO_2. In fact, these were among the conclusions of one of the first conferences held to address precisely this question.[35] This is far from a justification for abandoning the effort, since tropical land use change may otherwise contribute very large quantities of carbon to the atmosphere, and there are many other benefits associated with efforts to slow deforestation and increase reforestation.

Should even optimistic estimates of the potential of biotic options prove viable, their implementation will not reduce the need to change energy consumption patterns and trends. Even if the potential of forestry for absorbing CO_2 is significant, it pales next to the potential emissions of CO_2 implied by a continuing (and even increasing) global reliance on fossil fuels. Reforestation may be able to help during a transition period away from fossil fuels, and during that period may be used to offset significant CO_2 emissions from specific projects as well as displacing the need for additional fossil-fueled projects, but it should not be perceived as making a continued long-term reliance on fossil fuels environmentally acceptable.

REFERENCES

1. R.A. Houghton, "The Future Role of Tropical Forests in Affecting the Carbon Dioxide Concentration of the Atmosphere," Ambio 19(4), 204 (1990).
2. R.A. Houghton and A. Grainger, Modelling the Impact of Alternative Afforestation Strategies to Reduce Carbon Dioxide Emissions, Proceedings from Tropical Forestry Response Options to Global Climate Change Conference (Sao Paulo, Brazil, 1990).
3. M.C. Trexler, P.E. Faeth, and J.M. Kramer, Forestry as a Response to Global Warming: An Analysis of the Guatemala Agroforestry and Carbon Sequestration Project (World Resources Institute, Washington, D.C. 1989).
4. U.S. Department of Agriculture, U.S. Forest Service, America the Beautiful: National Tree Planting Initiative (U.S. Government Printing Office, Washington, D.C. 1990).
5. M.C. Trexler, Minding the Carbon Store: Weighing U.S. Forestry Strategies to Slow Global Warming (World Resources Institute, Washington, D.C. 1991); R. Moulton and K. Richards, The Costs of Carbon Sequestration in U.S. Forests (U.S. Forest Service, Washington, D.C. 1991).
6. F.H. Murkowski, "The Tongass," Christian Science Monitor (Mar. 14, 1989); Natural Resources Defense Council (1989).
7. M.E. Harmon, W.K. Ferrell, and J.F. Franklin, "Effects in Carbon Storage of Conversion of Old-Growth Forests to Young Forests," Science 247, 699-702 (1990).
8. W.M. Post, et al., "The Global Carbon Cycle" Amer. Scientist 78, 310 (1990); Emmanuel, W.R., et al., "Modeling the Global Carbon Cycle and Changes in the Atmospheric Carbon Dioxide Levels," in J.R. Trabalka, ed., Atmospheric Carbon Dioxide and the Global Carbon Cycle (U.S. Department of Energy, DOE/ER-0239, 1985), p. 141-173.
9. W.M. Post, et al. 1990.
10. For a review of the literature see S. Brown and A.E. Lugo, "The Storage and Production of Organic Matter in Tropical Forests and Their Role in the Global Carbon Cycle," Biotropica 13, 161-87 (1982); S. Brown and A.E. Lugo, "Biomass of Tropical Forests: A New Estimate Based on Forest Volumes," Science 223, 1290-93 (1984); R.P. Detwiler and C.A.S. Hall, "Tropical Forests and the Global Carbon Cycle," Science 239, 42-47 (1988); S. Brown, A.J. Gillespie, and A.E. Lugo, "Biomass Estimation Methods for Tropical Forests with Applications to Forest Inventory Data," Forest Science 35(4), 881; S. Brown, A.J. Gillespie, and A.E. Lugo, "Biomass of Tropical Forests of South and Southeast Asia," Can. J. of Forest Res. 21(1), 111 (1991).
11. R.A. Houghton, et al., "Changes in the Carbon Content of Terrestrial Biota and Soils Between 1860 and 1980: A Net Release of CO_2 to the Atmosphere," Ecol. Monographs 53, 235-62 (1983).
12. R.P. Detwiler, C.A.S. Hall, and P. Bodgonoff, "Land Use Change and Carbon Exchange in the Tropics: Estimates for the Entire Region," Envtl. Mgmt. 9, 335-44 (1985); R.A. Houghton, et al., "Changes in the Carbon Content of Terrestrial Biota and Soils Between 1860 and 1980: A Net Release of CO_2 to the Atmosphere," Ecol. Monographs 53, 235-62 (1983); R.A. Houghton, et al., "Net Flux of CO_2 From Tropical Forests in 1980," Nature 316, 617-20 (1985); R.A. Houghton, et al., "The Flux of Carbon from Terrestrial Ecosystems to the Atmosphere in 1980 Due to Changes in Land Use: Geographic Distribution of the Global Flux," Tellus 39B, 122-39 (1987).

13. R.A. Houghton, "Emissions of Greenhouse Gases," in N. Myers, ed., Deforestation Rates in Tropical Forests and Their Climatic Implications (Friends of the Earth 1990); V.H. Hale, R.A. Houghton, and C.A.S. Hall, "Estimating the Effects of Land-Use Change on Global Atmospheric CO_2 Concentrations," Can. J. of Forest Res. 21(1), 87 (1991); C.A.S. Hall and J. Uhlig, "Refining Estimates of Carbon Released from Tropical Land-Use Change," Can. J. of Forest Res. 21(1), 118 (1991).
14. P.J. Kramer and N. Sionit, "Effects of Increasing Carbon Dioxide Concentration on the Physiology and Growth of Forest Trees," in W.E. Shands and J.S. Hoffman, eds., The Greenhouse Effect, Climate Change, and U.S. Forests (Conservation Foundation, Washington, D.C. 1987).
15. S. Brown, personal communication.
16. W.M. Post, et al. 1990.
17. P.P. Tans, I.Y. Fung, and T. Takahashi, "Observational Constraints on the Global Atmospheric CO_2 Budget," Science 247, 1431-38 (1990).
18. C.D. Keeling, et al., "A Three-Dimensional Model of Atmospheric CO_2 Transport Based on Observed Winds: 1. Analysis of Observational Data," in D.H. Peterson, ed., Geophysical Monographs 55 (American Geophysical Union, Washington, D.C. (1989).
19. Intergovernmental Panel on Climate Change, Policy Makers Summary: The Formulation of Response Strategies (World Meteorological Organization and United Nations Environment Programme, Geneva, Switzerland, 1990).
20. W.M. Post, et al., 1990.
21. R. Sandenburgh, C. Taylor, and J.S. Hoffman, "Rising Carbon Dioxide, Climate Change, and Forest Management: an Overview," in W.E. Shands and J.S. Hoffman, eds., The Greenhouse Effect, Climate Change, and U.S. Forests (Conservation Foundation, Washington, D.C., 1987).
22. M.C. Trexler, et al., 1989.
23. D. Keeling, pers. comm. (1990).
24. W.H. Schlesinger, "Vegetation an Unlikely Answer," Nature 348, 679 (1990).
25. K.C. Prentice and I.Z. Fung, Nature 346, 48 (1990).
26. J.M. Adams, et al., Nature 348, 711 (1990).
27. F.J. Dyson, "Can We Control the Carbon Dioxide in the Atmosphere?," Energy 2, 287-91 (1977).
28. G. Marland, The Prospect of Solving the CO_2 Problem Through Global Reforestation (Office of Energy Research, Office of Basic Energy Sciences, U.S. Department of Energy, DOE/NBB-0082, 1988).
29. R.A. Sedjo, "Forests: A Tool to Moderate Global Warming?" Environment 31(1), 14-20 (1989).
30. J.A. Edmonds, et al., Future Atmospheric Carbon Dioxide Scenarios and Limitation Strategies (Noyes Publications 1986).
31. R.J. Moulton and K.R. Richards, 1991.
32. R.A. Houghton, 1990.
33. A. Grainger, 1990.
34. The study is scheduled for publication by the World Resources Institute in late 1991 or early 1992 under the authorship of M.C. Trexler and C. Haugen.
35. U.S. Environmental Protection Agency, Office of Research and Development, Proceedings of International Workshop on Large-Scale Reforestation, Corvallis, Oregon (May 8-10, 1990).

POLICY IMPLICATIONS OF GREENHOUSE WARMING

Rob Coppock*

Panel on Policy Implications of Greenhouse Warming
Committee on Science, Engineering, and Public Policy
National Academy of Sciences/National Academy of Engineering/
Institute of Medicine

ABSTRACT: A study panel of the National Academy of Sciences, National Academy of Engineering, and Institute of Medicine recently issued the report *Policy Implications of Greenhouse Warming*. That report examined relevant scientific knowledge and evidence about the potential of greenhouse warming, and assayed actions that could slow the onset of warming (mitigation policies) or help human and natural systems of plants and animals adapt to climatic changes (adaptation policies). The panel found that, even given the considerable uncertainties knowledge of the relevant phenomena, greenhouse warming poses a threat sufficient to merit prompt action. People in this country could probably adapt to the changes likely to accompany greenhouse warming. The costs, however, could be substantial. Investment in mitigation acts as insurance protection against the great uncertainties and the possibility of dramatic surprises. The panel found mitigation options that could reduce U.S. emissions by an estimated 10 to 40 percent at modest cost.

Introduction

Greenhouse warming may be the most important environmental issue for the next several decades. It currently is high on the national and international agendas. The National Academy of Sciences, National Academy of Engineering, and Institute of Medicine recently released an important study on the subject called "Policy Implications of Greenhouse Warming."

The report has been unusually well received. Resolutions have been introduced in both the U.S. Senate and the House of

---

* The opinions expressed in this paper are those of the author, and do not necessarily reflect the position of the National Academy of Sciences, the National Academy of Engineering, the Institute of Medicine, or any parts of those organizations.

Representatives endorsing the implementation of all its recommendations. The Science Advisor the President and the Secretary of Energy both issued statements commending the report and its recommendations. The Administrator of the U.S. Environmental Protection Agency stated that the administration would carefully study it. Environmental groups hailed it for recognizing that action is required now and that we need not wait for better data and analyses.

These are, of course, short-term reactions. Its real contribution may not be revealed for some time. The report may reflect a watershed in the national debate: a shift from a focus on whether or not greenhouse warming is occurring to what can be done to slow its onset and cope with its consequences. If the report helps stimulate such a shift, and provides a useful framework for guiding policy decisions, it will be of even greater value.

Study Purpose and Process

The House Report for the HUD-Independent Agencies Appropriations Act of 1988 called the study, requesting a broad examination with direct policy relevance. It asked for an assessment of the rate and magnitude of climate change and the projected impacts, and an evaluation of policy options of mitigating and responding to such changes.

This congressional mandate presented one of the stiffest challenges that has been undertaken by the Academy in recent years. Congress requested a comprehensive assessment, a task requiring an extraordinary breadth of knowledge. The study was conducted by 4 panels involving nearly 50 experts. They were asked to examine the full range of the ways to mitigate greenhouse warming, and to catalog the possible impacts and adaptive measures. This broad scope meant two things. First, the panels had to examine some areas where available knowledge is much more limited that for most Academy studies. Second, it required the panels to develop an approach that placed an exceptionally wide range of topics in perspective.

The four panels worked in parallel, but with considerable exchange of information and some overlap in membership. The Effects Panel examined what is knows about changing climatic conditions and related effects. The Mitigation Panel explored options for slowing or reversing the onset of greenhouse warming. The Adaptation Panel assessed the impacts of possible climatic changes on human and ecologic systems and policies that could help them adapt to those changes. The Synthesis Panel, reflecting both scientific expertise and decision making experience (see Table 1), developed overall findings and recommendations. They did so on the basis of the analyses of the other panels, other scientific and technical papers, and testimony from

Table 1. PANEL ON POLICY IMPLICATIONS OF GREENHOUSE WARMING--
Synthesis Panel

The Honorable Daniel J. Evans, Chair, Daniel J. Evans & Associates former U.S. Senator and governor from the State of Washington, and former president of The Evergreen State College

Robert McCormmick Adams, Secretary, Smithsonian Institution

George F. Carrier, T. Jefferson Coolidge Professor of Applied Mathematics, Harvard University and chairman of the 1985 NAS Committee on Atmospheric Effects of Nuclear Explosions

Richard N. Cooper, Professor of Economics, Harvard University and former Under-Secretary of State

Robert A. Frosch, Vice President, General Motors Research Labs and former Assistant Executive Director of the United Nations Environment Program

Thomas H. Lee, Professor Emeritus, Department of Electrical Engineering and Computer Sciences, Massachusetts Institute of Technolgy and former director of the M.I.T. Laboratory for Electromagnetic and Electronic Systems

Jessica Tuchman Maathews, Vice President, World Resources Institute and former director of the National Security Council's Office of Global Issues

William D. Nordhaus, Professor of Economics, Yale University and former member of the President's Council of Economic Advisors

Gordon H. Orians, Professor of Zoology, University of Washington and former director of the University's Institute for Environmental Studies

Stephen H. Schneider, Head, Interdisciplinary Climate Studies, National Center for Atmospheric Research

Maurice F. Strong, Chair, Strovest Holdings Inc. (resigned to serve as Secretary General of the United Nations Conference on Environment and Development) and former Under-Secretary-General of the United Nations

Sir Crispin Tickell, Warden, Green College, Oxford University and former Ambassador to the United Nations from the United Kingdom

Victoria J. Tschinkeld, Senior Consultant, Landers and Parsons and former Secretary of the Florida State Department of Environmental Regulations

Paul E. Waggoner, Distinguished Scientist, The Connecticut Agricultural Experiment Station invited experts.

The Greenhouse Effect and Greenhouse Warming

There is no doubt that the so-called greenhouse gases absorb energy—this can be demonstrated in the laboratory. Scientists also agree that these gases can trap energy when present in the atmosphere. Without the naturally-occurring greenhouse gases, mostly water vapor, carbon dioxide, and methane, the planet would be about 33°C colder than it is. The trapping of energy by trace gases—the greenhouse effect—thus is not necessarily bad. It makes earth habitable.

There also is consensus among scientists that atmospheric concentrations of greenhouse gases are increasing. This raises the possibility that the earth's temperature could be driven up to levels that would be detrimental for people and natural ecosystems. This could be bad, and the higher the temperature the worse it would be.

There is widespread, but not unanimous, agreement that global mean temperature has risen 0.3-0.6°C over the last 100 years. Measurements from before 1900 are not very reliable, since the adjustments necessary to make them comparable to current measurement are larger than the observed differences in temperature.

There is contentious debate, however, about whether this global average temperature can be attributed to atmospheric concentrations of greenhouse gases. There are two main issues. First, carbon dioxide (CO_2) is the dominant atmospheric greenhouse gas, but the location of more than 40 percent of the CO_2 emitted into the atmosphere is unknown. The capacity of this unidentified "sink" is not known, no is whether it will continue to absorb a constant portion of the CO_2 emissions. The inability to account for nearly half of the predominant greenhouse gas undermines confidence in understanding of the relevant phenomena.

Second, the historical temperature record does not correspond to our best estimates of atmospheric concentrations of greenhouse gases. The Northern Hemisphere temperature record, for example, shows a decrease from about 1900 to 1940, and an increase since then. Atmospheric emissions of greenhouse gases, however, have been increasing since the industrial revolution.

Thus, while much is known about greenhouse warming phenomena, there remain great uncertainties.

Table 2. Key Greenhouse Gases Influenced by Human Activity

	CO_2	CH_4	CFC-11	CFC-12	N_2O
Preindustrial atmospheric concentration	280 ppmv	0.8 ppmv	0	0	288 ppbv
Current atmospheric concentration (1990)[a]	353 ppmv	1.72 ppmv	280 pptv	484 pptv	310 ppbv
Current rate of annual atmospheric accumulation[b]	1.8 ppmv (0.5%)	0.015 ppmv (0.9%)	9.5 pptv (4%)	17 pptv (4%)	0.8 ppbv (0.25%)
Atmospheric lifetime (years)[c]	(50-200)	10	65	130	150

[a] The 1990 concentrations have been estimated on the basis of an extrapolation of measurements reported for earlier years, assuming that the recent trends remained approximately constant.

[b] Net annual emissions of CO_2 from the biosphere not affected by human activity, such as volcanic emissions, are assumed to be small. Estimates of human-induced emissions from the biosphere are controversial.

[c] For each gas in the table, except CO_2, the "lifetime" is defined as the ratio of the atmospheric concentration to the total rate of removal. This time scale also characterizes the rate of adjustment of the atmospheric concentrations if the emission rates are changed abruptly. CO_2 is a special case because it is merely circulated among various reservoirs (atmosphere, ocean, biota). The "lifetime" of CO_2 given in the table is a rough indication of the time it would take for the CO_2 concentration to adjust to changes in the emissions.

NOTES: Ozone has not been included in the table because of lack of precise data. Here ppmv = parts per million by volume, ppbv = parts per billion by volume, and pptv = parts per trillion by volume.

SOURCE: Committee on Science, Engineering, and Public Policy. <u>Policy Implications of Greenhouse Warming</u>. Washington, D.C.: National Academy Press. Table 3.1.

Current Emissions and Their Effects on the Earth's Radiative Balance

Expressed as the equivalent of carbon dioxide--the common standard for comparison--current concentrations of greenhouse gases are about 50 percent greater than before the industrial revolution. But the gases also remain effective in the atmosphere for different amounts of time. Table 2 summarizes current concentrations, emissions accumulation rates, and atmospheric lifetimes of key greenhouse gases.

Atmospheric concentrations of greenhouse gases are affected by most human activities. Table 3 shows estimated global greenhouse gas emissions from a variety of human activities for 1985. Emissions are shown both in total for each gas, and as CO_2-equivalent amounts. It should be noted that not all CFCs are included in the table, so that the percentage totals would probably change slightly with more complete

Table 3. Estimated 1985 Global Greenhouse Gas Emissions from Human Activities

	Greenhouse Gas Emissions (Mt/yr)	CO_2-equivalent Emissions[a] (Mt/yr)	
CO_2 Emissions			
Commercial energy	18,800	18,800	(57)
Tropical deforestation	2,600	2,600	(8)
Other	400	400	(1)
TOTAL	21,800	21,800	(66)
CH_4 Emissions			
Fuel production	60	1,300	(4)
Enteric fermentation	70	1,500	(5)
Rice cultivation	110	2,300	(7)
Landfills	30	600	(2)
Tropical deforestation	20	400	(1)
Other	30	600	(2)
TOTAL	320	6,700	(20)[b]
CFC-11 and CFC-12 Emissions			
TOTAL	0.6	3,200	(10)
N_2O Emissions			
Coal combustion	1	290	(>1)
Fertilizer use	1.5	440	(1)
Gain of cultivated land	0.4	120	(>1)
Tropical deforestation	0.5	150	(>1)
Fuel wood and industrial biomass	0.2	60	(>1)
Agricultural wastes	0.4	120	(>1)
TOTAL	4	1,180	(4)
TOTAL		32,880	(100)

[a]CO_2-equivalent emissions are calculated from the Greenhouse Gas Emissions column by using the following multipliers:

CO_2 1
CH_4 21
CFC-11 and -12   5,400
N_2O 290

Numbers in parentheses are percentages of total.
[b]Total does not sum due to rounding errors.

NOTE: Mt/yr = million (10^6) metric tons (t) per year. All entries are rounded because the exact values are controversial.

SOURCE: Committee on Science, Engineering, and Public Policy. Policy Implications of Greenhouse Warming. Washington, D.C.: National Academy Press. Table 2.1.

data.

Table 4 shows CO_2 emission per unit of economic activity for 1988/89. Note that the largest contributors in terms of total

Table 4. Estimated 1985 Global Greenhouse Gas Emissions from Human Activities

	Emissions (Mt CO_2/yr)	GNP (billions of $/yr)	Emissions/GNP Ratio (Mt CO_2/$1000 GNP)
China	2236.3	372.3[a]	6.01[b]
South Africa	284.2	79.0	3.60
Romania	220.7	79.8[a]	2.77[b]
Poland	459.4	172.4[a]	2.66[b]
India	600.6	237.9	2.52
East Germany	327.4	159.5[a]	2.05[b]
Czechoslovakia	233.6	123.2[a]	1.90[b]
Mexico	306.9	176.7	1.74
USSR	3982.0	2659.5[a]	1.50[b]
South Korea	204.6	171.3	1.19
Canada	437.8	435.9	1.00
United States	4804.1	4880.1	0.98
Australia	241.3	246.0	0.98
United Kingdom	559.2	702.4	0.80
Brazil	202.4	323.6	0.63
West Germany	669.9	1201.8	0.56
Spain	187.7	340.3	0.55
Italy	359.7	828.9	0.43
Japan	989.3	2843.7	0.35
France	320.1	949.4	0.34

[a]Estimates of GNP for centrally planned economies are subject to large margins of error. These estimates are as much as 100 times larger than those from other sources that correct for availability of goods or use free-market exchange rates.

[b]The emissions/GNP is also likely to be underestimated for centrally planned economies.

SOURCE: Committee on Science, Engineering, and Public Policy. Policy Implications of Greenhouse Warming. Washington, D.C.: National Academy Press. Table 2.3.

emissions are the United States (4804 million tons), the Soviet Union (3682 million tons), and China (2236 million tons). But in terms of emissions per unit of economic activity, the highest ranking countries are all developing countries or eastern block nations. This indicates that there are possibilities for slowing greenhouse warming in virtually all countries.

Since greenhouse gases have different atmospheric lifetimes, it is necessary to consider their effect over time. Figure 1 presents a simplified extrapolation of current atmospheric transformation rates. The curves show emissions in 2030 as percentages of 1990 levels (assuming linear change from 1990 to 2030). The "stars" indicate the so-called "business as usual" projections developed by the Intergovernmental Panel on Climate Change (IPCC). The left-hand

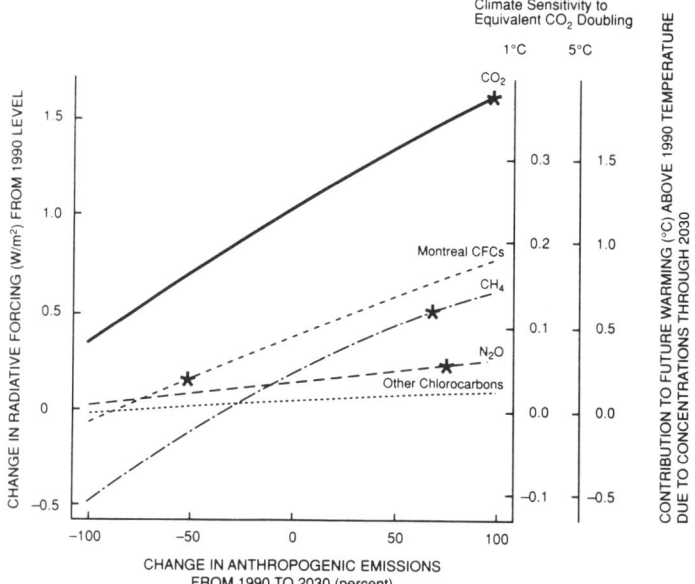

FIGURE 3.4 Commitment to future warming. An incremental change in radiative forcing between 1990 and 2030 due to emissions of greenhouse gases implies a change in global average equilibrium temperature (see text). The scales on the right-hand side show two ranges of global average temperature responses. The first corresponds to a climate whose temperature response to an equivalent of doubling of the preindustrial level of CO_2 is 1°C; the second corresponds to a rise of 5°C for an equivalent doubling of CO_2. These scales indicate the equilibrium commitment to future warming caused by emissions from 1990 through 2030. Assumptions are as in Figure 3.2.

To determine equilibrium warming in 2030 due to continued emissions of CO_2 at the 1990 level, find the point on the curve labeled "CO_2" that is vertically above 0 percent change on the bottom scale. The equilibrium warming on the right-hand scales is about 0.23°C (0.4°F) for a climate system with 1° sensitivity and about 1.2°C (2.2°F) for a system with 5° sensitivity. For CH_4 emissions continuing at 1990 levels through 2030, the equilibrium warming would be about 0.04°C (0.07°F) at 1° sensitivity and about 0.25°C (0.5°F) at 5° sensitivity. These steps must be repeated for each gas. Total warming associated with 1990-level emissions of the gases shown until 2030 would be about 0.41°C (0.7°F) at 1° sensitivity and about 2.2°C (4°F) at 5° sensitivity.

Scenarios of changes in committed future warming accompanying different greenhouse gas emission rates can be constructed by repeating this process for given emission rates and adding up the results.

SOURCE: Committee on Science, Engineering, and Public Policy. <u>Policy Implications of Greenhouse Warming</u>. Washington, D.C.: National Academy Press. Figure 3.4.

vertical scale shows the incremental energy absorption rates that would accompany various emission levels for each gas. The energy absorption is given in Watts per square meter. A change in absorption is called "radiative forcing." The right-hand vertical scales indicate committed equilibrium warming, with one scale showing the commitment to future warming for a "low" and the other a "high" degree of climate response to greenhouse gas concentrations. The total committed warming can be determined by adding the temperatures derived for the projected concentration of each gas.

Analytic techniques of this type need further development. The simplified approach in Figure 1, for example, does not account for chemical interactions among these trace gases in the atmosphere. Nevertheless, it can be used to produce first approximation estimates for various emissions scenarios.

Projecting Climate Change

The best tools for projecting climate change are called general circulation models (GCMs). They are huge computer models with hundreds of thousands of equations and dozens of variables. They attempt to account not only for atmospheric concentrations of greenhouse gases but for feedback mechanisms, such as those involving clouds, for ocean lags, and for other relevant phenomena. There are about a half-dozen different GCMs. Although GCMs project global mean equilibrium temperature increases of between 1.8° and 5.2°C, there is no scientific way to be sure that this range includes the actual mean temperature increases that would accompany a doubling of preindustrial CO_2 concentrations. Some scientists are unwilling to go further than to say that we have the direction of change right. However, GCMs do replicate many attributes of the global weather system (e.g., diurnal and seasonal variation, atmospheric phenomena like jet streams and tropical cell intrusions).

The global equilibrium temperature projected by GCMs would not be observed for many years after reaching the equivalent of double the preindustrial level of CO_2. The oceans, covering roughly 70 percent of the earth's surface, absorb energy from the sun and redistribute it to the deep oceans slowly. It will be decades, perhaps centuries, before the oceans and the atmosphere fully redistribute the absorbed energy and the equilibrium temperature is realized. This lag depends in part on the sensitivity of the climate system, and is believed to be between 10 and 100 years.

In the study, the panel concluded that there is clear evidence about several basic facts:

1. The atmospheric concentration of CO_2 has increased about 25 percent during the last century and is currently increasing at about 0.5 percent per year.

2. The atmospheric concentration of CH_4 has doubled during that period and is increasing at about 0.9 percent per year.
3. CFCs, which are man-made and have been released into the atmosphere in quantity only since World War II, are currently increasing at about 4 percent per year.
4. Current interpretations of temperature records reveal that the global mean temperature has increased between 0.3° and 0.6°C during the last century.

As a result of these facts, the panel concluded that there is a reasonable chance that:

1. In the absence of greater human effort to the contrary, greenhouse gas concentrations equivalent to a doubling of the preindustrial level of CO_2 will occur by the middle of the next century.
2. The sensitivity of the climatic system to greenhouse gases is such that the equivalent of doubling CO_2 could ultimately increase the global mean temperature by somewhere between 1° and 5°C.
3. Because of the lag in heat transfer in the oceans, the global mean surface temperature at the time of CO_2 doubling may be as little as one-half the ultimate equilibrium temperature.

Possible Impacts of Greenhouse Warming

Because of the uncertainties in the projections, and because of the relatively wide range of possible temperature rises, the panel looked at the sensitivity of the affected human and ecological systems to the range of projected changes. On this basis the panel recommended actions that would help human and natural systems adapt to future climate change. It also examined the technical potential of various actions to slow or offset emissions of greenhouse gases and recommended several mitigation actions.

The study focused on the United States. It attempted to assess impacts on sectors of the domestic economy and on unmanaged ecosystems in this country. A few comparisons were made to other geographic regions where these were obvious, but no systematic analysis of other regions was made.

Assuming gradual changes and a global mean temperature rise of the range described above, the panel concluded that people in the United States will likely have no more difficulty adapting to future changes than to severe climatic conditions in the past such as the Dust Bowl. Needless to say, adjustments will likely involve some cost. There may also be substantial geographic differences in the impacts, with some regions winning and others losing. Overall, for the nation as a whole, winners and losers should balance out. Some panel members think this analysis, however, underestimates some of the adjustment costs.

Other countries are likely to have more difficulty adjusting to the projected changes, especially poor countries or those with fewer climate zones. They may face greater hardships in adapting, or have more difficulty taking advantage of favorable changes.

Many natural systems of plants and animals will change. Although key "ecosystem services " (e.g., absorption of CO_2 by plants) might be maintained on any given tract of land by replacement species, ecosystems will almost certainly be altered. Some people may find such changes unacceptable. The stronger this is felt, the greater the motivation to slow greenhouse warming.

The panel was unable to find credible assessments of major, sudden transformations with large consequences. For example, greenhouse warming conceivably could alter major ocean currents and thus regional climate, change precipitation in the far north and thus surface reflectivity and energy absorption, or release trapped methane from melting tundra and inject an additional surge of greenhouse gas. No one knows, however, just what might initiate such transformations or how they might proceed once started. Although there is no way to analyze their likelihood or impacts, the possibility that they might occur combines with the other uncertainties to lead the panel to conclude that investments to slow greenhouse warming would be good "insurance" against possible negative surprises.

Mitigation of Greenhouse Warming

The panel used a different approach to assessing ways of slowing, or mitigating, the onset of greenhouse warming than has been used in other studies. The panel developed a cost-effectiveness ranking in terms of reduction in annual emissions of CO_2 (or the equivalent in other gases, enhancement of CO_2 sinks, etc.). It is worth describing the major attributes of this approach.

First, it is a method that enable widely different options to be compared. Improvements in fuel efficiency in automobiles that last 10 years can be compared with combined cycle electricity generating plants that last 25 or 30 years. Replacement of incandescent light bulbs by high efficiency fluorescent tubes can be compared with reforestation of marginal land. This breadth of applicability is essential for a comprehensive approach.

Second, the priority ranking of options is based on current conditions. The ranking may not be the same in, say 2015. For example, among the most cost-effective options today are energy conservation in residential and commercial buildings. Further down the list are efficiency improvements in power plant heat conversion, and further still, electricity supply options. The full implementation of energy conservation measures, however, might well change electricity demand such that the cost-effectiveness ranking of energy supply options would be different by 2015. The panel made no attempt to analyze such interactions among options in its analysis, however.

Third, the reductions of greenhouse gas emissions are expressed in terms of 1990 emissions. The panel did not estimate how much implementing these options might reduce emission at any future point in time. The panel believes future emissions are highly dependent on economic conditions, technological progress, and success in implementing mitigation options. None of these can be predicted credibly. So the analysis does not project emissions reductions that could be achieved in, say, 2010. Rather, estimates are made of how much current annual emissions could be reduced if the decision were made now to implement these options. Putting the most cost-effective options in place, however, would be the wisest course of action whether future emissions follow a high or a low trajectory. The high priority options would be high priority under either scenario.

Fourth, the fact that this approach produces a ranking of options only under current conditions implies that the analysis should be repeated on a regular basis. After 5 or perhaps 10 years, costs and mitigation potentials may be sufficiently different to alter the rankings. If the most efficient options are to be used on the scale of decades involved in greenhouse warming, it would be necessary to repeat the analysis on a regular basis.

Fifth, the mitigation options are ranked in the study according to their cost-effectiveness. Those options that give the greatest reduction in emissions of carbon dioxide or its equivalent at lowest cost rank highest. The overall potential of that option is also considered. Hydroelectric dams, for example, may be very cost-effective, but there may be few remaining locations in this country where they could be built.

The cost-effectiveness ranking of the most attractive options can be found in Table 5. If the "low-cost" mitigation actions were implemented, the technical envelope[1] of emissions reduction would range from a little more than 10 percent of 1990 U.S. greenhouse gas emissions for pessimistic implementation assumptions to slightly less than 40 percent for optimistic assumptions about implementation. The analysis shows these reductions could be achieved at low cost, or even at a net savings if proper policies were implemented. Determining the cost of achieving the recommendations with any degree of precision, however, is extremely difficult. Thus, analysis in the study cannot be considered definitive. It is, however, a careful first cut. A

Table 5. Comparison of Selected Mitigation Options in the United States

Mitigation Option	Net Implementation Cost[a]	Potential Emission[b] Reduction (t CO_2 equivalent per year)
Building energy efficiency	Net benefit	900 million[c]
Vehicle efficiency (no fleet change)	Net benefit	300 million
Industrial energy management	Net benefit to low cost	500 million
Transportation system management	Net benefit to low cost	50 million
Power plant heat rate improvements	Net benefit to low cost	50 million
Landfill gas collection	Low cost	200 million
Halocarbon-CFC usage reduction	Low cost	1400 million
Agriculture	Low cost	200 million
Reforestation	Low to moderate cost[d]	200 million
Electricity supply	Low to moderate cost[d]	1000 million[e]

[a] Net benefit = cost less than or equal to zero
Low cost = cost between $1 and $9 per ton of CO_2 equivalent
Moderate cost = cost between $10 and $99 per ton of CO_2 equivalent
High cost = cost of $100 or more per ton of CO_2 equivalent

[b] This "maximum feasible" potential emission reduction assumes 100 percent implementation of each option in reasonable applications and is an optimistic "upper bound" on emission reductions.

[c] This depends on the actual implementation level and is controversial. This represents a middle value of possible rates.

[d] Some portions do fall in low cost, but it is not possible to determine the amount of reductions obtainable at that cost.

[e] The potential emission reduction for electricity supply options is actually 1700 Mt CO_2 equivalent per year, but 1000 Mt is shown here to remove the double-counting effect (see p. 61 for an explanation of double-counting).

NOTE: Here and throughout this report, tons are metric.

SOURCE: Committee on Science, Engineering, and Public Policy. Policy Implications of Greenhouse Warming. Washington, D.C.: National Academy Press. Table 6.2.

substantial reduction in greenhouse gas emissions can be accomplished at low cost, or even a net savings to the nation if proper policies were implemented.

Recommendations

The Synthesis Panel concluded that, "despite the great uncertainties, greenhouse warming is a potential threat sufficient to justify action now." It went on to point out that there are a number of effective and low-cost options available in the United States.

The panel developed recommendations in five areas: reducing or offsetting emissions of greenhouse gases, enhancing adaptation to greenhouse warming, improving knowledge for future decisions, evaluating geoengineering options, and exercising international leadership.

Three topics dominate the recommendations to reduce or offset emissions: eliminating emissions of halocarbons (principally chlorofluorcarbons, or CFCs), changing energy policy, and forest offsets.

Although the international agreements on ozone depletion mandate elimination of many halocarbon emissions, if they are not implemented the concentrations of greenhouse gases will be much worse. The panel thus recommended aggressive phaseout of CFCs and other halocarbon emissions and development of substitutes that minimize or eliminate greenhouse gas emissions.

In the are of energy policy, the panel recommended:

- Moving toward full-cost pricing[2] of energy
- Substantial improvement in energy conservation and efficiency
- Significant reduction in emissions from energy supply

In the area of forest offsets, the panel recommended:

- Reducing global deforestation
- Exploring reforestation programs

The great uncertainties surrounding virtually every aspect of policy about greenhouse warming puts information for future decisions at a premium. Policy decisions could be improved by careful gathering of data and targeted research in several areas:

- Field research on effects of carbon enhancement
- Data of all kinds on climate change
- Improved weather forecasts
- Research on climate feedback mechanisms
- Research on social an economic interactions with climate change

The panel also recommended research into what it called geoengineering options. These are ways of blocking solar radiation, enhancing natural sinks for greenhouse gases, or altering the reflectivity of the earth. They have the potential to substantially alter the radiative balance of the planet, and could be essential if major climatic changes occur. Research and development projects should be undertaken of both the potential of such options to offset global warming and their possible side-effects. However, their feasibility and especially the side-effects associated with them should be carefully examined before action is taken to implement them.

The panel recommended that the United States should participate fully and at an appropriate level in international agreements and programs to address greenhouse warming. It recommended that the U.S. resume full participation in international programs to slow population growth an contribute its share to their financial and other support.

Conclusions

The Synthesis Panel concluded that the threat of greenhouse warming justifies action now. It found a number of actions that could be taken to reduce the speed and magnitude of greenhouse warming. Others actions could prepare people and natural ecosystems for future adjustments to the conditions likely to accompany greenhouse warming. The fact that people can adapt, or even that they are likely to do so, does not mean that the best policy is to wait for greenhouse warming to occur and let them adapt. Waiting and adapting may sacrifice overall economic efficiency in the long run. The panel recommends pursuing low-cost actions employing currently available technologies, and concluded that options requiring great expenses are not justified at this time.

Notes

1. The term "technical envelope" refers to the performance limits on technologies performing a particular function. For example, the techncial envelop for the speed of jet aircraft is determined by the fastest and slowest models.

2. "Full-cost pricing" of energy refers to the ideal of including all social and environmental costs associated with the production of energy in the price to the consumer. In general, estimating the cost of what economists "externalities" is extremely difficult because it requires attaching value to items and conditions that are not bought and sold in markets. This recommendation is to strive to better reflect the indirect and often indeterminate costs to human health and environmental quality in energy prices.

OPTIONS FOR LOWERING U.S. CARBON DIOXIDE EMISSIONS

Rosina M. Bierbaum, Robert M. Friedman, Howard Levenson,
Richard D. Rapoport, and Nick Sundt

Office of Technology Assessment, U.S. Congress, Washington, DC 20510

ABSTRACT

The United States can decrease its emissions of carbon dioxide (CO_2) to as much as 35 percent <u>below</u> 1987 levels within the next 25 years by adopting an aggressive package of policies crossing all sectors of the economy. Such emissions reductions will be difficult to achieve and may be costly, but no major technological breakthroughs are needed. In this paper, we identify a "Tough" package of energy conservation, energy supply, and forest management practices to accomplish this level of emissions reductions. We also present a package of cost-effective, "Moderate" technical options, which if adopted, would hold CO_2 emissions to about a 15-percent increase over 1987 levels by 2015. In contrast, if the United States takes no new actions to curb energy use, CO_2 emissions will likely rise 50 percent during that time. A variety of Federal policy initiatives will be required to achieve large reductions in U.S. CO_2 emissions. Such policy actions will have to include both regulatory "push" and market "pull" mechanisms--including performance standards, tax incentive programs, carbon-emission or energy taxes, labeling and efficiency ratings, and research, development, and demonstration activities.

INTRODUCTION

This paper summarizes a recent study by the Office of Technology Assessment (OTA), *Changing by Degrees: Steps to Reduce Greenhouse Gases.*[1] The congressional committees requesting this assessment asked OTA to focus on a very specific question: "**Can the United States reduce carbon dioxide emissions in the near term?**"

To answer this question, OTA focused specifically on potential emissions reductions **in the next 25 years**. The analysis is structured around six key sectors of the U.S. economy: Buildings, Transportation, Manufacturing, Energy Supply, Forestry, and Food. To the extent possible, the report quantifies the potential for emissions reduction within each sector--areas where gains in efficiency, product substitution, conservation, or other technical options can ameliorate increases in CO_2 and other greenhouse gases. A selection of policy options that appear to offer the most promise for achieving these reductions in the United States is presented. OTA was charged to look abroad as well, so the special needs of Eastern Europe, the U.S.S.R., and developing countries--with respect to both energy and natural resource issues--are also addressed.

The assessment lays out three paths: a Base case ("business as usual"), a Moderate (essentially "no-cost") case, and a Tough case. Each of the Moderate measures that we identified typically require some initial capital investment but later save money through future fuel savings; in most cases savings more than compensate for initial costs. None of the measures are difficult to achieve technically, though inducing consumers to use them may not be easy.

The "Tough" measures would lower energy demand even further, but in many cases at a higher cost for the same level of convenience and comfort. All of the Tough measures analyzed are technically feasible, but most are not based on the best available prototypes or practices; OTA made judgments about what will be feasible for widespread use. Fully implementing the Tough measures would be challenging--politically, logistically, and perhaps economically.

Similar criteria were used to identify measures that apply to the forestry sectors (i.e., difficulty and cost) and estimates of CO_2 uptake over time were calculated. Data were not sufficient to calculate potential emissions reductions from the food sector.

Based on OTA's energy modeling analysis, under current trends and regulations carbon emissions by 2015 will be close to 50 percent greater than today's level--almost 1.9 billion metric tons per year (see figure 1). This Base

Figure 1--Carbon Emissions Under the Base Case, Moderate, and Tough Scenarios

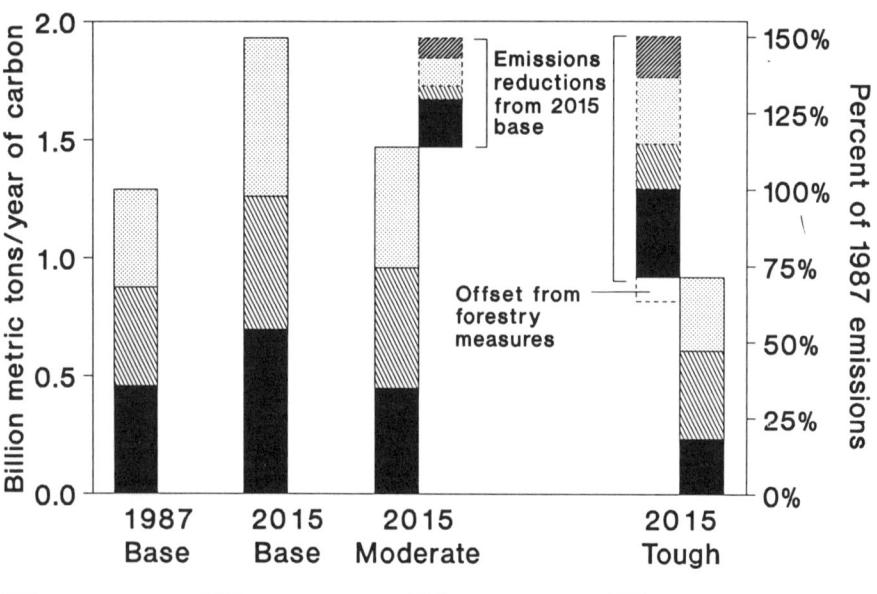

NOTE: The boxes outlined with dashed lines represent the *reductions* in carbon emissions associated with control measures applied in each of the three demand-side sectors (i.e., buildings, industry, and transportation) and electricity supply (electric utilities); additional carbon *offsets* afforded by forestry measures are also shown. The boxes outlined with solid lines represent *total emissions* from each demand-side sector. For the solid boxes, emissions associated with electricity generation have been allocated to the three demand-side sectors.

SOURCE: Office of Technology Assessment, 1991.

Case projection assumes that some efficiency improvements will occur even in the absence of new legislation. For example, by then we assume new homes will require 15 percent less heating, recently adopted appliance standards will have taken effect, and new cars will average close to 37 mpg.

By adopting all Moderate measures that lower energy demand, CO_2 emissions in 2015 could be held to about a 22-percent increase over 1987 levels. The emissions savings achieved by the Moderate measures are shown for each demand sector (buildings, transportation, industry) as well as for electricity supply in figure 1. Changes in the fuel mix used to generate electricity can lower emissions an additional 6 to 7 percent. The Moderate forestry measures provide about a 0.2-percent offset in carbon emissions by 2015. All Moderate measures together hold emissions to 15 percent above 1987 levels.

Finally, OTA's Tough scenario could lower net emissions by 2015 (excluding offsets from forestry measures) to 29 percent below 1987 levels--i.e., to about 0.9 billion metric tons per year. (For reference, this is about half of our Base case forecast for 2015.) Tough forestry measures could reduce emissions another 7 percent.

While we think the Moderate scenario is achievable at a net savings, nonetheless substantial shifts in the economy would have to occur. For example, energy expenditures, per se, would be 15 percent lower than they would be otherwise, but the cost of appliances, cars, and houses would be higher.

Many of the Tough scenario measures entail costs in excess of projected fuel savings; others are cost effective over their lifetime but are difficult to implement. A rough estimate of the cost range for the Tough scenario brackets a savings of $20 billion to a cost of about $150 billion per year (in 1987 dollars) by 2015, after subtracting fuel savings (assuming forecasted 2015 fuel prices). This range is equal to savings of a few tenths of a percent to a cost up to 1.8 percent of the Gross National Product (GNP) projected for 2015. For comparison, all environmental compliance costs today are about 1.5 percent of GNP; direct fossil fuel and electricity consumption purchases account for about 9 percent of GNP.

OPTIONS FOR REDUCING U.S. EMISSIONS

The major options available or likely to be available for **reducing** CO_2 emissions in the near-term fall into three categories:

1. increasing efficiency of end-use technologies,
2. changing use patterns to conserve energy, and
3. shifting energy supply away from high CO_2-emitting fuels.

Additional options to **offset** CO_2 emissions are primarily forestry-related or agricultural.

When choosing policy options Congress must consider **two** interdependent components: the universe of possible **technical changes** (or in some cases, behavioral changes) and the **policy instruments** (e.g., taxes, regulations, financial incentives) available to require or encourage the technical change. One policy option, for example, would be to reduce CO_2 emissions through regulations (i.e., a policy instrument) to require more fuel-efficient autos (i.e., a technical change).

An alternative or perhaps complementary policy option would be to use a high "gas guzzler" tax (i.e., policy instrument) to stimulate purchase of fuel-efficient autos.

Technical Options

OTA's report identifies a range of CO_2-reducing technical options available or likely to be available to the Nation over the next 25 years, and what their contribution might be. There are a large number of technical options to pick from and many targets of opportunity within each sector, as figure 2 shows. Significant progress in reducing U.S. CO_2 emissions will require that most of these options be pursued simultaneously.

Presently available energy "supply" options for achieving major CO_2 reductions over the 25-year timeframe of this assessment include: replacing high carbon-emitting fuels (e.g., coal) with lower carbon-emitting fuels (e.g., natural gas); using high-efficiency, electricity-generating technologies (e.g., high-efficiency gas turbines or cogeneration); and using nonfossil fuels.

While nonfossil energy offers the greatest long-term potential for achieving deep cuts in CO_2 emissions, we cannot count on **large-scale** use of nonfossil energy sources to replace fossil fuels within 25 years. These sources do not yet offer the performance, costs, or social acceptance needed to fully displace fossil fuels in such a relatively short period of time[2]. Only three nonfossil sources are presently being used on a significant scale in the United States: hydroelectric power, nuclear light-water reactors (LWRs), and biomass. Because of a combination of low baseload demand growth, cost, and environmental and social problems, no orders for new LWRs have been initiated in over a decade and there are no plans underway to build new reactors in the United States. Environmental factors set an upper limit on the number of potential new dam sites for hydroelectric facilities and on biomass production.

On the energy "end use" side, the technical options available today are primarily more efficient technologies or changes in energy use patterns. The first requires time and investment, whether for old equipment to be replaced or new equipment to be purchased. Changing energy use can include immediate (but reversible) changes such as fewer miles driven, lights dimmed, etc. In addition to the currently available technical options there is a large menu of additional options that could be developed over time. A diverse suite of energy R&D is ongoing, but what it will make available in the next quarter century depends greatly on Federal funding for demonstration.

Several technical options are available in the forestry sector to provide some **offsets** of CO_2 emissions. Increasing forest productivity and planting new trees can result in increased carbon storage that offsets fossil-fuel related emissions. Planting short-rotation tree crops for use as biomass fuels can partially replace the use of fossil fuels in some situations. These and other forestry options have attendant uncertainties and difficulties. For example, attempts to increase productivity focus on the timber component of forests (i.e., the commercially valuable portion). However, it is unclear whether increases in timber productivity actually indicate whether or not productivity in the entire forest has increased.

Policy Instruments

Policy instruments are the means government uses to require or encourage a desired technical or behavioral response. Many potential targets exist within each sector to achieve CO_2 emissions reductions (see figure 2). Whatever the CO_2

Figure 2--1987 U.S. CO_2 Emissions by Sector

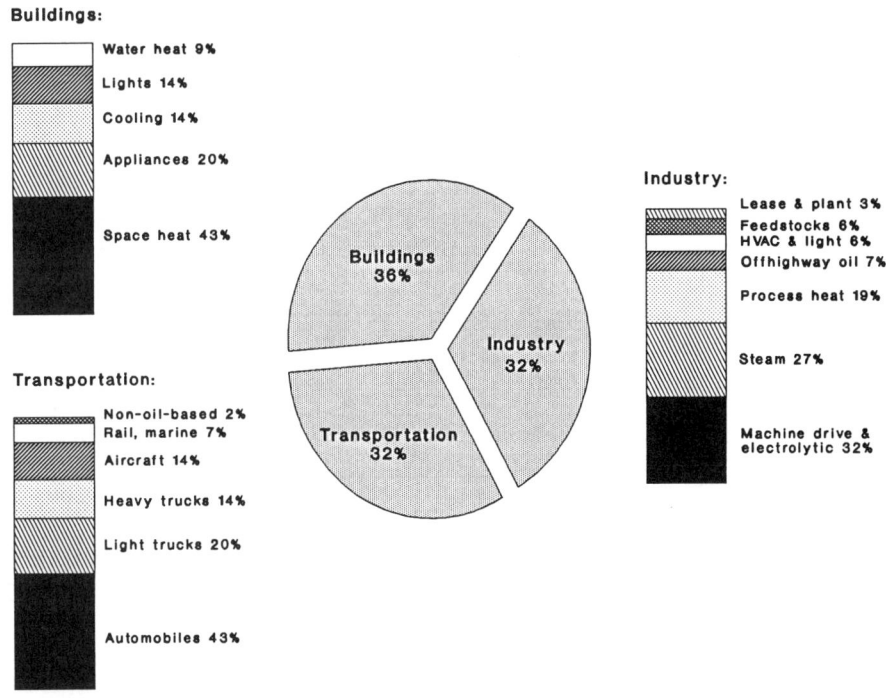

NOTE: Total carbon emissions from all three sectors equaled about 1.3 billion metric tons in 1987. Emissions have continued to increase since then.

SOURCE: Office of Technology Assessment, 1991.

reduction goal, Congress will have to use a variety of policy instruments to stimulate a diverse set of decisionmakers to use the appropriate fuels, technologies, and forestry and agricultural practices and to adopt energy use patterns that conserve energy. These instruments can be grouped into six generic categories:

1. taxes;
2. financial incentives;
3. marketable permits;
4. regulations;
5. research, development, and demonstration (RD&D); and
6. information and public education.

Just as there is no single technical option that is a cure-all, many policy instruments will be needed. The synergisms possible among taxation, regulation, incentives, information, and RD&D programs are key to significantly reducing emissions. Taxes, if properly set, can be used to adjust prices to tilt purchase decisions. Regulation (codes and standards) can be used to remove the least efficient equipment, appliances, and buildings from the market. Incentive and information programs can be used to clarify cost information and help create a market for improved energy performance. Education programs also provide consumers with the knowledge and information needed to make wiser energy choices. Government-sponsored RD&D can help provide producers and consumers with new technical options that can be used to reach national goals, as well as reduce, by cost-sharing, the risk to industry of developing these new options.

Taxes

Taxes offer a way to make high CO_2-emitting technical options more expensive than lower CO_2-emitting options. If Congress so desires, new tax monies could help fund incentive programs, offset the budget deficit, or replace other existing taxes.

Three possibilities include: 1) a general energy tax, 2) a carbon tax, and 3) initial purchase taxes. A general energy tax is levied on the energy (i.e., Btu) content of fuels. A carbon tax is set to reflect the fact that some fuels emit more carbon per unit of energy than do others. Both of these are thus "fuel" taxes. An initial purchase tax is levied on energy-consuming **technologies**, rather than fuels; the tax would be based on estimates of lifetime energy use or carbon emissions.

Energy and carbon taxes have the advantage of affecting **all emitters** simultaneously, rather than focusing on a few selected technologies. A carbon tax is a particularly effective way of targeting the heaviest economic sanctions against the worst emitters of CO_2.

Initial purchase taxes could have effects broadly similar to either an energy or carbon tax, depending on whether they were based on lifetime energy use or carbon emissions. Because consumers are often more concerned with the initial cost of a technology than with "life-cycle" costs (i.e., including fuel costs), purchase taxes can be more effective than either type of fuel tax in some situations.

Financial Incentives

Through financial incentives (e.g., tax incentives, low cost loans, and direct payment subsidies), the government pays part of the costs of utilizing desirable fuels, technologies, or practices. Tax incentives can be powerful instruments for stimulating desired actions by corporations and individual taxpayers looking for ways to reduce tax liabilities; however, tax incentives have little effect on those who pay low or no taxes.

Low-cost loans either defray some portion of loan interest or eliminate lender risk by insuring against loss. Low-cost loans can be effective policy instruments to stimulate utilization of CO_2-reducing technical options by both individuals and corporations. Direct payments for utilization of CO_2-reducing options (e.g., cash bonus for scrapping an old, fuel-inefficient car) are especially effective in stimulating the use of desired options by low-income or financially strapped decisionmakers.

Marketable Permits

Marketable permits and carbon taxes are closely related. Under a marketable permit system, policymakers fix the amount of carbon that can be emitted. The government then issues the allowed number of permits to emit a given amount of carbon. Permits can be bought and sold by energy users just like fuels. For example, for every 1 million Btu's of **coal** purchased, the user must also own (or purchase) permits to emit 57 pounds of carbon. To burn 1 million Btu's of **natural gas**, the user must own or purchase permits to emit 32 pounds of carbon. If demand for energy rises, the price of a carbon permit will rise to reflect the cost of lowering emissions. Some holders of permits will find ways to lower emissions (e.g., purchase more efficient equipment, switch from coal to natural gas) so that they can sell their permits (at a profit) to others.

A marketable permit system is the regulatory mechanism for limiting emissions of sulfur dioxide to control acid rain under the new Clean Air Act Amendments (Public Law 101-549).

Regulations

Regulations are policy instruments that can eliminate inefficient and/or high CO_2-emitting activities from the market. They can take the form, for example, of performance standards and building codes. Performance standards can be established for many diverse types of technologies (e.g., lighting standards) and applied nationwide; they are currently used for automobile efficiency and appliance efficiency. Building codes traditionally have been the province of local governments and their effective use depends on enforcement at that level.

Research, Development, and Demonstration (RD&D)

Through RD&D, government can search for and fine-tune technological fixes to the greenhouse gas emissions problem. **In fact, climate change can only be effectively addressed over the long-term with the development and worldwide use of better nonfossil energy sources.** Government can speed the process of testing and commercializing many energy-supply and end-use technologies. However, only about 5 percent of the $2.7 billion national budget for energy technology R&D in 1990 was devoted to renewables (including biomass energy) and only 7 percent to energy conservation. Fossil fuels had 25 percent of the research budget, nuclear fusion 12 percent, and nuclear fission 9 percent.

Information

Information as a policy instrument has the potential to change the awareness level and perceptions of decisionmakers. Information programs rest on the assumption that if decisionmakers are better informed they will make better decisions. The most common goal of information programs is to stimulate decisionmakers to opt for least cost (life-cycle) savings, as opposed to initial-purchase savings, in their energy decisions. For example, although the most efficient model of a household appliance often costs more initially, energy savings accrue over its useful life. Information can be supplied by Federal, State, or local governments, utility programs (see "Demand-Side Management" below), manufacturers, or nongovernmental organizations.

Information can be delivered to all decisionmakers in many ways, for example via label and rating systems and audits. Label and rating systems serve to provide purchasers with a basis for comparing front-end versus life-cycle costs at the time of purchase. Energy audits provide building owners and occupants with information they need when considering whether to purchase, rent, or retrofit alternatives. Energy audits can be effectively coupled with financial incentives to carry out retrofits that provide greater efficiency, and thus reduced CO_2 emissions.

Sectoral Policy Options

Buildings Sector

Table I summarizes the emissions reductions possible for each technical option modeled by OTA under both the Moderate and Tough scenarios. For buildings, improving shell efficiency and lighting are the two technical options with the greatest potential for lowering CO_2 emissions. Under the Base case, OTA assumes that by 2015 new homes and apartments will be designed such that they need about 15 percent less heating and 8 percent less cooling than current new homes. By adopting Moderate shell efficiency measures, such as thicker insulation and better windows, new homes will require an estimated 50 percent less heat and 25 percent less air-conditioning than today's average new home[3]. With Tough measures, homes can be built to require an estimated 85 percent less heat and 45 percent less air-conditioning[4].

As shown in table I, Moderate shell improvements in new residential buildings can reduce U.S. carbon emissions by 1.3 percent of current CO_2 levels by 2015. By implementing Tough improvements in the North and Moderate ones in the South, reductions of 2 percent in new residential buildings might be achieved. Tough measures for new commercial building shells can achieve reductions equal to 4 percent of 1987 levels by 2015.

Existing homes can also be made more efficient by installing more efficient heating and cooling equipment, insulation, windows, etc. The Base case assumes that existing homes will require 6 percent less heating by 2015 because of replacements and improvements that will happen anyway. Moderate measures boost this to 25 percent by 2015 and Tough measures boost it to 40 percent by 2015[4]. Tough measures in the North and Moderate ones in the South would reduce carbon emissions from existing buildings by 4 percent by 2000, but would have a declining effect thereafter as many of the older homes are replaced by new ones.

Improving the efficiency of lighting in new commercial buildings is another technical option that can yield substantial reductions. The Tough scenario measures together--a combination of high-efficiency fluorescent bulbs and ballasts, improved reflectors, and better use of daylight--would lower lighting energy needs by 60 percent in these buildings[5]. This achieves reductions equal to 3 percent of 1987 emissions by 2015.

Gains in commercial buildings can also be made by simply replacing existing bulbs with high-efficiency ones--without replacing fixtures--as shown under the "Operation and Maintenance" heading of table I. Replacing the most heavily used incandescent bulbs in homes with compact fluorescents and using high-efficiency fluorescents in commercial buildings can lower emissions by 1.3 percent under our Tough scenario.

The policies described below appear to offer the most promise to achieve these reductions. While a general carbon tax will certainly help, because there are so many different decisionmakers--some of whom may not be that responsive to price changes--a larger arsenal of policy instruments is needed. These include demand-side management (with the utilities as partners) as well as a series of targeted financial sanctions, incentives, and regulations.

Demand-Side Management (DSM)--DSM refers to electric utility programs designed to encourage customers to modify their patterns of energy use. Particularly promising--from a global warming perspective--are those situations where utilities allow energy conservation to compete with traditional supply technologies (e.g., power plants) to balance energy supply and demand. DSM can be an effective approach to reduce energy consumption by improving building shells as well as the equipment inside buildings. In some cases, utilities pay for rebate programs, give out high-efficiency light bulbs, or otherwise stimulate end-use efficiency improvements, and save energy at a fraction of the cost of new power supplies.

Demand-side management can result in greater investments in energy efficiency than customers would otherwise make. Utility programs have long time horizons and can capture the potential in both the new and retrofit markets, for both equipment efficiency and building shell improvements. There is already considerable support for DSM by many State energy offices, State legislatures, and public utility commissions.

However, in order for DSM to stimulate significant investment in conservation, incentive structures must be changed so that utilities can profit from demand-side investments. Any Federal legislation concerning DSM would need to be general enough to allow States flexibility in implementation and specific enough to have a genuine impact on conservation. Congress could provide funding to evaluate various incentive structures currently being examined by States and utilities. Should Congress wish to pursue more direct action, it could require States to formally consider demand-side resources in their planning, with oversight by the Federal Energy Regulatory Commission (FERC).

Further, the Federal Government could mandate that environmental externalities be considered in evaluating supply-side options (as New York State has done--i.e., penalizing polluting options based on estimates of the costs of environmental damage that would accrue)[6]. Congress has already mandated, in the 1980 Pacific Northwest Electric Power Planning and Conservation Act (Public Law 96-501), that the Northwest Power Planning Council adopt rate structures that give conservation measures a cost break over other, more traditional supply-side measures.

Technology-Specific Regulations--Congress can directly mandate efficiency improvements through appliance standards and building energy codes.

Appliance Standards--Appliance standards, by fiat, remove inefficient appliances from the market. The National Appliance Energy Conservation Act (Public Law 100-12), passed in 1987, are expected to lower residential energy use by up to 10 percent by the year 2000[7]. However, even stricter standards are possible. The law requires review of appliance standards twice during the 1990s, which provides an opportunity to obtain additional energy reductions through more stringent standards. Congress could also consider extending standards to other equipment such as commercial heating, ventilation, and air-conditioning equipment; light bulbs; and building components such as windows.

Standards could be even more effective if used in conjunction with other incentives. Policies such as utility programs, appliance labeling, and tax schemes provide incentives to do more than standards require.

Building Energy Codes--Building energy codes serve a function analogous to that of appliance standards by preventing the least efficient buildings from being constructed. Building codes have traditionally been under the jurisdiction of States and localities. Currently, there is little support from the States or the construction industry for a mandatory national building code. In 1976, Congress enacted legislation that required the development of the Building Energy Performance Standards (BEPS), a mandatory national code based on performance standards. In 1983, the law was modified to be a mandatory code only for Federal buildings.

Financial Measures--Congress can choose from among several sector-specific financial mechanisms, including building tax credits and subsidies and initial purchase taxes for appliances and other equipment.

Building Tax Credits and Subsidies--Tax credits and subsidies for using more efficient technologies can promote retrofitting of existing residential and commercial buildings. The Federal Government, for example, passed legislation that provided solar and conservation tax credits for the years 1978 through 1984. By 1983, 24 million households claimed a residential tax credit of up to $700 each for investments in energy conservation; however, no evaluation or monitoring of energy saved by this program was ever conducted.

Initial Purchase Taxes and Rebates for Appliances and Other Equipment--An initial purchase tax scaled to penalize inefficient equipment could accelerate the market penetration of efficient equipment. Examples include a lump-sum tax on appliances and equipment at the time of purchase. Taxes collected on the most polluting items could be used to provide a rebate on the least polluting items. However, although an initial purchase tax sends appropriate signals regarding consumer purchasing decisions, it would not--unlike an energy or carbon tax--change use of an appliance once it is purchased.

Consumer Information and Marketing Programs--Lack of information and uncertainty have been identified as key barriers to greater investment in energy conservation in the buildings sector. The large number of highly cost-effective investments in energy efficiency that are **not** chosen by consumers indicates that price alone does not stimulate optimal investment decisions. Therefore, information dissemination is a key element of several of the policy options discussed above, including the sector-specific financial measures and general energy and carbon taxes.

Home Energy Rating Systems--The Federal Government has been involved in home energy rating systems--which tell buyers how efficient their prospective homes or offices are--through its role in the mortgage market. In addition, the National Affordable Housing Act of 1990 requires HUD to develop a plan to make housing more affordable through mortgage financing incentives for energy efficiency. The Federal Government could play a further role by developing a uniform energy rating system for all residential and commercial buildings, making it easier and less expensive for lenders to include energy costs in their mortgage evaluations.

Energy Audits--The Federal Residential Conservation Service, created in 1978, mandated that gas and electric utilities provide their customers with on site energy audits. The program was implemented in 1981 and recently expired. There has been little evaluation of the program, and little reliable information has been kept on its success in reducing energy consumption. However, while it is unclear whether information from audits alone is enough to encourage conservation, it would certainly seem to be useful when combined with other measures.

Building Research, Development, and Demonstration--Major barriers to private investment in RD&D in the buildings sector include the fragmented structure of this sector and the short-term perspective of many of the decisionmakers (e.g., builders, renters). In addition, the U.S. Government currently spends a negligible amount on housing research. In contrast, in countries such as Sweden and Japan, RD&D spending has been part of a trend toward energy-efficient prefabricated housing. This spending has contributed to the energy efficiency of homes through standardization of energy-saving features and quality control in the design and manufacture of building components.

The Federal Energy Management Program (FEMP), administered by the Department of Energy, works with government agencies to implement cost-effective, energy-efficiency improvements. Congress could authorize FEMP to test and demonstrate performance, acceptance, and cost-effectiveness of new technologies in Federal buildings.

Transportation Sector

Urban passenger travel in cars and light trucks (i.e., light vehicles) in the United States requires the largest share of transport energy, consuming 15 percent of the world's oil production. The two main opportunities for reducing transportation's contribution to global warming are measures to increase the energy efficiency of light vehicles and measures to encourage urban passengers to drive less. Thus, under OTA's modeling exercise, the major reductions in this sector come from higher auto and truck efficiency, better control of traffic, and, under the Tough scenario, more use of public transit (see table I).

With respect to auto efficiency, our Base case assumes that new cars will average about 32 mpg by 2000 and about 37 mpg by 2010. Under the Moderate scenario, new car efficiency averages 35 mpg by 2000[8] and 39 mpg by 2010[9]. Under the Tough scenario, we assume a range of new car efficiencies. For example, efficiencies of 39 mpg by 2000 and 55 mpg by 2010 might be possible even if consumers maintain their current preferences for car size and performance[9]. If consumers are willing to buy smaller cars, new car fleet average

Table I—Measures To Lower U.S. Carbon Emissions (expressed as percentage of 1987 total emissions)[a]

	Reductions in 2015		Reductions in 2015	
	Moderate (in percent)	Tough (in percent)	Moderate (in percent)	Tough (in percent)
DEMAND-SIDE MEASURES			**Industry**	
Residential buildings			New investments:	
New investments:			Efficient motors	1.2	3.7 to 4.0
Shell efficiency	1.3	2.0	Lighting	0.5	0.7 to 0.8
Heating and cooling equipment	0.1	0.4 to 0.6	Process change, top 4		
Water heaters and appliances	1.2	1.5 to 2.3	industries	3.0	8.2
O&M, retrofits:			Fuel switch to gas	0.0	2.4 to 2.7
Shell efficiency	0.8	0.9	Cogeneration	0.8	5.2 to 5.8
Lights	0.6	0.8	O&M, retrofits:		
All residential measures together	4	5.6 to 6.6	Housekeeping	1.9	2.0
Commercial buildings			Lighting	0.1	0.2
New investments:			All industrial measures together	8	17 to 18
Shell efficiency	2.3	4.0	**ELECTRIC UTILITY SUPPLY-SIDE MEASURES**		
Heating and cooling equipment	1.0	1.2 to 1.9	Existing plant measures:		
Lights	2.1	3.0	Improved nuclear utilization	4.1	4.1
Office equipment	1.6	2.1	Fossil efficiency improvements	1.7	1.7
Water heaters and appliances	0.1	0.1	Upgraded hydroelectric plants	0.5	0.5
Cogeneration	0.2	1.5 to 2.3	Natural gas co-firing	—	3.7
O&M, retrofits:			New plant measures:		
Shell efficiency	0.8	0.8	No new coal; higher fraction		
Lights	0.5	0.5	of new nonfossil sources	—	0.0 to 4.7
All commercial measures together	8.5	13 to 15	CO_2 emission rate standards	0.4	0.0 to 0.1
Transportation			All utility supply-side measures together	6.6	9.9 to 14
New investments:			**FORESTRY MEASURES**		
New auto efficiency	0.8	3.5 to 3.8	Afforestation:		
New light truck efficiency	0.5	2.5 to 2.7	Conservation Reserve Program	0.2	0.2
New heavy truck efficiency	0.4	2.4 to 2.4	Urban trees	—	0.7
Non-highway efficiency	0.5	1.2	Additional tree planting	—	2.3
O&M, retrofits:			Increased timber productivity	—	3.1
Improved public transit	0.2	3.5	Increased use of biomass fuels	—	1.2
Truck inspection & maintenance	0.3	0.4	All forestry measures together	0.2	7.5
Traffic flow improvements/55 mph speed limit	1.2	1.4			
Ridesharing/parking controls	0.4	1.0			
All transportation measures together	4	14 to 15			

[a] 1 percent of 1987 emissions = 13 million metric tons C = 0.7 percent of 2015 emissions.

SOURCE: Office of Technology Assessment, 1991.

efficiencies of 42 mpg by 2000 and 58 mpg by 2010 might be achievable. Given this range of assumptions, reductions amount to about 3.5 to 3.8 percent of current CO_2 emissions by 2015 (see table I).

Reductions of about 2.5 to 2.7 percent from light trucks and another 2.4 percent from medium- and heavy-duty trucks are achievable under our Tough scenario, as well.

Traffic speed affects fuel consumption, too. By reinstating the 55 mile-per-hour speed limit and by reducing traffic congestion in urban areas in order to speed up travel, reductions of 1.4 percent by 2015 are possible under our Tough scenario.

Measures to move people out of their cars and into mass transit under the Tough scenario would yield reductions of about 3.5 percent by 2015. To achieve this, however, urban auto traffic would have to be reduced by 10 percent through urban light rail, sways, and improved urban design. Additionally, 5 percent of car travel between cities would have to shift to high-speed intercity rail.

The following four policy instruments will promote new car efficiency: gasoline taxes, vehicle taxes and rebates, fuel economy standards, and incentives for vehicle manufacturers. In addition, improved operation and maintenance practices will reduce energy use in existing cars. Two other measures, transportation control measures (TCMs) and controlling settlement patterns, can help reduce CO_2 by reducing vehicle miles traveled.

Gasoline Taxes--A gasoline tax would create incentives for both increased efficiency and reduced travel. Taxes would induce consumers to use less fuel while leaving them free to choose how they adjust their behavior. In concert with increasing fuel economy standards (see below), a long-term impact on the efficiency of the vehicle fleet could be achieved.

Although the effectiveness of taxes is hard to predict from studies of past responses to price changes, one might expect a 10-percent hike in gasoline prices to yield a 1- to 6-percent drop in gas consumption[10,11]. A 50-percent increase in price might reduce consumption 5 to 20 percent over the near term, even more over the longer term. A doubling or tripling in price (approaching the costs in Europe and Japan) might yield an immediate decrease of 13 to 20 percent and a longer term response of a 35- to 40-percent reduction in gasoline consumption. About half of this longer term adjustment to high price is expected from driving less, and the other half from more efficient vehicles. For example, consumers might choose to spend money on fuel-efficient technologies or to use mass transit, carpool, or simply travel less.

Vehicle Taxes and Rebates--Taxes on inefficient vehicles can create incentives to choose better fuel economy and forego large size and extra power. Such a program would be most effective if accompanied by rebates for highly efficient cars. In a "revenue neutral" program, the money taken in from the taxes would be redistributed through the rebates. The Federal Gas Guzzler Tax already applies to cars with fuel economies below certain thresholds; the Omnibus Budget Reconciliation Act of 1990 recently doubled the tax for cars getting less than 22.5 mpg.

An expanded program of **auto purchase taxes and rebates** could complement **fuel economy** standards and taxes, but it could also pose serious trade difficulties as long as the high-efficiency end of the auto market is dominated by imports. If implemented suddenly, such measures would put domestic manufacturers at a disadvantage; on the other hand, measures designed to protect domestic manufacturers might conflict with General Agreement on Tariffs and Trade (GATT) rules.

Fuel Economy Standards--Standards influence the tradeoffs among cost, performance, size, and efficiency that underlie the development and introduction of new models. The current fuel economy standards for cars, in place since 1978, have helped to increase auto fuel economy[12]. More stringent standards can both lower CO_2 emissions and reduce our dependence on imported oil. Redesigned standards that vary with vehicle volume can help minimize the burden on U.S. manufacturers that offer a full range of car sizes[13,14].

Incentives for Vehicle Manufacturers--One incentive, aimed at producers instead of consumers, is the use of government-sponsored competitions to induce manufacturers to develop high-efficiency or alternate-fueled cars. A variant of the incentive scheme injects competitive elements into a high-efficiency rebate program. For example, the government could identify a few classes of vehicles most in need of fuel economy improvement and offer a competitive reward in the form of consumer rebates on a large (e.g., 200,000 units) production run of a new vehicle achieving the best fuel economy above a specified threshold.

Efficient Vehicle Operating Practices--Changes in vehicle operating practices offer small potential reductions individually but often have short startup times and do not require large, up-front capital investment. They include reimposing (and enforcing) the 55-Mph speed limit; requiring efficiency inspections for trucks; and charging efficiency-promoting parking fees at Federal offices.

Transportation Control Measures (TCMs)--TCMs include a wide variety of measures to reduce the number of vehicle miles traveled (VMT) and lower congestion. They are attractive because they typically have short startup times and low capital costs, and can reduce energy use and greenhouse emissions even within existing settlement and employment patterns. In aggregate, TCMs appear to hold modest promise for reducing VMT. These include such measures as: ridesharing (promotion and matching services); employer-based transportation management (high parking charges or transit or vanpool subsidies; High-Occupancy Vehicle (HOV) lanes (restricting lanes on freeways to cars with three or four occupants and to buses); Park and Ride (intercept drivers near their origins); and mass transit improvements (bus service expansion, operational changes, and fare changes).

Controlling Settlement Patterns--Long-term reductions in emissions can be achieved by changing patterns of settlement to reduce the need for travel. This can be accomplished through higher densities, or through mixing uses so that residences, jobs, and services are roughly balanced at a local scale. When more destinations are close to home, more trips can be made by foot; when densities are higher, public transit can serve more people effectively.

In the United States, except possibly for some high-growth areas in the South and West, efforts to change the shape of settlement in major cities may meet local resistance. Nevertheless, some changes are feasible in suburban areas nationwide. Stringent suburban restrictions on development--sometimes only on commercial and industrial development, sometimes on new residential development as well--have been attempted in some regions of the United States[15].

Transportation RD&D--American automakers lag behind their Japanese, and to a lesser extent their European, counterparts in moving research results to the market[16]. In the 1980s, a program to support more aggressive research and development in the American auto industry--the Cooperative Automotive Research Program--was briefly attempted by the Department of Transportation. A revived, combined government/industry program could be successful if domestic automakers, their suppliers, and innovative research companies all are key players. The program could target important efficiency areas such as continuously variable transmissions, energy-storage systems, new engine designs for heavy trucks, improved safety for lighter vehicles, and innovations to permit increased intermodal freight.

An area of longer term research that deserves special attention is development of truly clean, economically acceptable, alternative fuels and a supporting infrastructure. Fuels with the greatest potential--electricity or hydrogen from noncarbon energy sources (e.g., solar and nuclear) and woody biomass fuels grown on a sustainable basis--are the furthest from large-scale technical viability. Research in these areas could be expanded, with parallel programs to assess and demonstrate the actual performance of a variety of fuels.

Manufacturing Sector

For manufacturing, as shown in table I, three types of technical improvements offer the greatest promise. The first area is "process changes." The top four manufacturing energy consumers (paper, chemicals, petroleum, and primary metals)--which account for more than 75 percent of energy use in this sector--improved their energy efficiency by between 2.3 and 4.3 percent per year between 1980 and 1985[17]. If this pace can be maintained, as we assume in our Tough scenario, reductions equal to about 8 percent of current emissions (by 2015) will result.

Cogenerating electricity and steam for industrial processes is a second promising option. When utilities generate electricity, about two-thirds of the energy from burning the fuel is released as heat. If electricity is generated at industrial sites where the heat can be used, the efficiency of fossil fuel use can be increased dramatically. Under our Tough scenario, we assume that 90 percent of new and replacement industrial steam boilers will cogenerate electricity. Such measures can lead to reductions equivalent to about 5.8 percent of current total U.S. emissions.

More efficient motors are a third technical improvement that can lead to substantial improvements. Moderate and Tough measures might improve motor efficiencies by 10 percent and 30 percent[18], respectively, yielding reductions of about 1.2 percent by 2015 under the Moderate scenario and 4 percent under the Tough one.

The following policy instruments could encourage these technical measures: carbon taxes, DSM, efficiency standards, marketable permits, tax incentives, informational policies, and RD&D.

Carbon Tax--A carbon tax would levy economic penalties against the highest industrial emitters of CO_2. Under such an approach, the tax would be highest on coal, low for natural gas, and zero for noncarbon sources (e.g., wind, solar, geothermal, or nuclear). For industries where the cost of energy is particularly important, carbon taxes should encourage energy efficiency, fuel switching and cogeneration.

Using several econometric models, the Congressional Budget Office estimated that a carbon tax of $100 per ton would lower CO_2 emissions from industry by between 10 and 35 percent by the year 2000. The higher reduction estimate reflects a 70-percent reduction in coal use.

Demand-Side Management--DSM programs--joint programs between electric utilities and their customers discussed previously--can help lower electricity use in the industrial sector. The major programs include: 1) rebates to customers who install approved equipment, 2) low-interest loans to customers for conservation installations, and 3) installation of conservation equipment at utility cost[19].

Standards--A more traditional regulatory policy is to require efficiency standards for common energy-using equipment, similar to those existing for automobiles and some appliances. Motors would be the most likely candidate for this approach.

Marketable Permits--CO_2 emissions can be regulated by requiring permits for emissions; manufacturers could be issued permits based, for example, on some percentage of their 1990 emissions. Reductions might be accomplished by installing energy-efficient technologies and fuel switching; offsets could result from approved reforestation/afforestation projects. It would be up to the manufacturer to choose the most cost-effective strategy. Marketable permits would allow firms to trade their unused carbon rights to a firm that is exceeding its budget.

Manufacturing Tax Incentives--Much industrial equipment is old and energy-inefficient compared to the best available technology. In many cases, replacing old equipment improves energy efficiency by 10 to 50 percent. Financial policies, such as tax credits or accelerated tax depreciation schedules, aimed at stimulating rapid replacement of older equipment have the potential to stimulate improvements in energy use. Such policies have a precedent: the Energy Tax Act of 1978 provided a 10-percent added "energy investment tax credit" for certain energy-conservation investments (as well as tax credits for some energy-supply investments). The tax credits were available until 1985 and applied to a specific list of technologies. However, rather than specify which technologies qualify, Congress could foster innovation by offering similar--or greater--tax breaks for company-chosen conservation technologies.

Informational Policies--A barrier to reducing emissions in the manufacturing sector is lack of information about how to improve energy use--especially for smaller, less energy-intensive industries. Informational policies can include performance goals, the collection of performance data, labeling of the energy performance of equipment, training, and performance audits.

Renewed support for cooperative government/ industry information-sharing programs could help. For example, DOE's Energy Analysis and Diagnostic Center program funds faculty and students at several universities to perform free energy audits for small and medium-sized manufacturers in more than 30 States. Because costs saved by manufacturers translate to increased taxable income, the program can provide additional tax revenues to the Federal Government. The biggest cost savings have come from efficiency improvements associated with cogeneration, space heating, lighting, and process equipment maintenance and replacement (in descending order of savings[20]). This program could be expanded or new programs could be modeled after it.

Manufacturing R&D--Research and development sponsored by DOE's Office of Industrial Programs in waste energy reduction and industrial process efficiency, if funded, are projected to save more than 3 to 4 percent of energy used by industry per year over the next decade. Research areas identified by Oak Ridge National Laboratory as particularly promising are: improved use of catalysts in chemical production; intelligent sensors and controls; and heat recovery and cogeneration[21]. R&D in nonenergy areas, such as materials science, also holds promise for partial replacement of energy-intensive materials like steel and aluminum. Likewise, research and development to improve the quality of products made with recycled materials could help reduce energy use by increasing the demand for recycled materials such as paper, steel, and aluminum.

Electricity Generation

About one-third of U.S. carbon emissions come from generating electricity; by 2015 under our Base case this may be as high as 45 percent. Thus measures that lower the rate of carbon emissions per kilowatt-hour (kWh) of electricity generated would translate into substantial reductions.

Table I shows OTA's estimate of the technical potential for emissions reductions in the electric utility sector depending on the demand for electricity and the stringency of policies. Moderate utility supply-side measures can lower emissions by about 6.6 percent. The two with the greatest reduction potential are: 1) increasing the efficiency of fossil fuel-fired plants (by about 5 percent) through improved maintenance[22]; and 2) operating existing nuclear power plants 70 percent of the time (similar to Western Europe and Japan[21] and extending their useful life to 45 years.

Our Tough measures eliminate coal use wherever possible. A combination of renewable energy sources, nuclear plants with improved designs that may be available after 2005, and high-efficiency gas turbines are the only new utility plants built under the Tough scenario. However, if all the Tough demand-side measures in the buildings and industrial sectors are implemented, growth in demand for electricity is so low that very few **new** plants are needed through 2015. Thus, the only way to lower emissions under this scenario is to either cofire existing coal plants (e.g., with 50 percent natural gas), or retire existing coal plants after 40 years of operation (rather than the typical 60 years) and replace them with renewable or nuclear fuels or natural gas. The former measure would reduce emissions by about 3.7 percent by 2015; the latter, by about 4.7 percent of current levels by 2015.

The following policy options could be used to encourage these technical measures: carbon taxes, marketable permits, subsidizing noncarbon sources, emissions limits and standards, and RD&D.

Carbon Taxes--A carbon tax, if set high enough, would encourage fuel switching and conservation. A carbon tax in the range of $75 to $150 per ton would make natural gas a more economic choice than coal at many facilities. A carbon tax would also provide added motivation to develop more noncarbon energy sources.

Marketable Permits--Utilities could be issued marketable permits for CO_2 emissions allowed from their coal-fired units, based on their generation in a historic year (e.g., 1990) multiplied by an allowed emission rate. Under this approach some utilities could curtail coal use more than necessary to meet their limits and sell permits to others exceeding their limits.

Subsidize Noncarbon Sources--Any of the general financial instruments, such as a carbon tax or fossil fuel energy tax, will serve to encourage use of nonfossil sources for electricity generation. According to one estimate[2], a 2 cent-per-kWh subsidy or its equivalent for **only** renewable sources of electricity might double the contribution of renewable sources of electricity by 2010--i.e., allow them to supply 40 percent of new demand under a Base case growth scenario. Under our Tough scenario, we assume nonfossil sources can provide between 30 and 45 percent of new demand (depending on the success of other demand-side measures).

CO_2 Emission Limits and Efficiency Standards--Congress could **mandate** reductions by setting CO_2 emission limits or efficiency standards. For example, an emission rate limit of 0.55 pounds carbon per kWh (lbs C/kWh) would require a typical Midwestern plant burning Illinois coal to burn between about 10 and 30 percent gas, depending on its efficiency. At 0.55 lbs C/kWh, the most efficient **new** coal burning technologies (e.g., integrated coal gasification combined cycle, or IGCC) would just qualify burning coal alone.

Two somewhat different strategies could be pursued to set CO_2 emission limits for **new** plants. If the intent is to force development of ultra-efficient coal technologies, then a standard in the range of 0.35 to 0.40 lbs C/kWh would be appropriate. Molten carbonate fuel cells, if successful, might be able to achieve such emission rates using bituminous coals. If the intent is to limit new fossil fuel-fired generation to the cleanest sources only--advanced combined cycle turbines burning gas--then a new source performance standard of about 0.25 lbs C/kWh would be more appropriate. To **speed up replacement** of old plants with new, lower emitting ones, Congress could mandate the retirement of existing fossil-fuel-fired plants earlier than their expected lifetime of 60 years.

Energy RD&D Funding--Over the last decade, Federal funding for renewable energy, conservation, and nuclear (fission) R&D fell rapidly. The 1990 combined energy technology R&D budgets (in 1990 dollars) for these three categories were 82 percent lower than they were in 1980. To reinstate the funding levels of 10 years ago would require adding about $2.6 billion. By doing so, the

Federal Government could hasten the development and demonstration of supply technologies that would reduce greenhouse gas emissions. The most promising of these technologies include: commercial fuel cells; storage technologies for solar electricity; biomass-driven turbines; variable-speed wind turbines; and better designs for nuclear power plants. Many experts estimate that these technologies could be commercially available within the next few decades.

The government could also play a role in reducing the perceived risk of new technologies and integrating renewable energy sources in existing energy systems by conducting demonstration projects or, perhaps, providing government-backed loans. To encourage new nuclear energy sources, a two-track process appears best: the Department of Energy could help fund full-scale demonstrations of both new "evolutionary" light water reactors and "revolutionary" design changes such as a modular high-temperature gas reactor.

For existing nuclear power plants, the goal should be to increase the number of hours of operation, rather than to increase efficiency of fuel use. A Department of Energy demonstration program (coordinated with the Nuclear Regulatory Commission) might bring U.S. hours of nuclear plant operation from well below to above the average for Western Europe and Japan. Key elements of such a program would include improving preventive maintenance; installing automated controls to improve reactor operation; and speeding up time spent refueling.

Forests

Forestry-related measures with the greatest potential to offset carbon emissions include increasing the productivity of existing forests, planting trees in new areas, and growing tree crops for biomass energy; we consider these to be Tough measures, with the exception of ongoing tree planting in the Conservation Reserve Program. OTA estimates that the increased carbon uptake from increasing productivity on about 60 million hectares of timberland might be equivalent to annual emissions reductions of about 3 percent of current levels by 2015. Planting new trees (i.e., afforestation) on farmland and other nonforested areas and in cities might result in carbon storage equivalent to emissions reductions of about 2 percent of current levels by 2015. Planting trees for biomass energy might result in an additional reduction of about 1 percent by 2015.

There are several caveats to this potential for offsetting emissions. Trees planted today can continue to store carbon beyond this report's 25-year timeframe. But this carbon eventually will be released to the atmosphere, either when trees die and decompose naturally, when they are harvested and burned, or when products made from wood eventually decompose. Unless the wood is used to displace fossil fuel use or is permanently stored under conditions that do not allow decomposition, carbon offsets in later years will dwindle. These estimates also assume that increasing the productivity of a forest's **commercial timber** component is equivalent to increasing the productivity of the **entire** forest ecosystem, but this assumption needs to be tested. Finally, forests--and the feasibility of using forestry practices to offset emissions--are likely to be affected by future climate changes. Therefore, forestry options in industrialized countries such as the United States cannot be considered a substitute for reducing total energy use, but rather as a way of "buying" time while developing alternative nonfossil fuel sources and improving the efficiency of energy use in general.

Congress could promote management practices that increase carbon storage or offset CO_2 emissions by augmenting existing forest management and tree planting programs of the U.S. Forest Service (USFS) and the Agricultural Stabilization and Conservation Service, and by enhancing the biomass energy research program of the Department of Energy. In addition to direct support for such programs, Congress also could consider using financial mechanisms (e.g.,

changing income tax policies to encourage more investments in forest management; imposing a tax on fossil fuels to make biomass fuels more competitive).

Incentives To Increase Carbon Storage on Forest Lands--Incentives to increase productivity--i.e., net carbon storage--will differ for publicly and privately owned forests. On public lands, which are located mostly in the West, management objectives are determined by planning processes legislated by Congress. Government investments in these lands are likely to focus on reforestation and timber stand management. Congress could direct the USFS and Bureau of Land Management to, for example, increase reforestation activities and to conduct research on the ability of "new forestry" practices that proponents contend might help to both maintain higher levels of diversity and allow commodity production.

Privately owned forests are most extensive in the East and South. For **nonindustry** private forests, Congress could continue to increase assistance to States and private landowners under programs such as the Forestry Incentives Program and the Agricultural Conservation Program. These programs currently reach only about 2 percent of nonindustry private owners[23], even though these owners undertake over 40 percent of all reforestation. The Interior and Related Agencies Appropriations Act for fiscal year 1991 almost doubled funding for the USFS's State and private forestry programs, which include tree planting and management. The 1990 Food, Agriculture, Conservation, and Trade Act (Public Law 101-624), known as the 1990 Farm Bill, also authorized a forestry stewardship program in which the USFS would work with State and local governments, land grant universities, and the private sector to improve resource management on privately owned forest land.

For **industry**-owned timberland, investments might be stimulated through changes in capital gains provisions (e.g., restoring preferential tax rates or providing a partial exclusion from taxable income for timber held longer than 20 years) or allowing full annual deductions for expenses, as well as by increasing funding for Federal assistance programs. One possibility for increasing support of such programs is to use funds that would accrue if below-cost timber sales in National Forests were eliminated.

Incentives for Growing New Trees on Unforested and Urban Lands--Mechanisms to promote afforestation include the Conservation Reserve Program (CRP), the President's proposed America the Beautiful program, and financial incentives such as tax credits for carbon storage. In general, any tree-planting program needs to consider the costs of maintaining trees in a healthy state once planted; this will be even more critical as climate changes occur.

Incentives for Biomass Energy To Offset CO_2 Emissions--Growing short-rotation woody crops on nonforested land for use as an energy source shows some promise. Congress could increase funding for Department of Energy research on uncertainties regarding long-term productivity, including effects on nutrient availability, and costs. Increasing fossil fuel taxes would make biomass fuels more competitive. Even then, farmers wishing to invest in biomass crops may be limited by loss of base acreage in commodity support programs and by lack of revenues for several years. Thus, changes in support programs or provision of some subsidy may be needed to stimulate investments in biomass crops on current cropland.

Food Sector

In the other U.S. sectors, CO_2 is the primary focus of OTA's analysis, although both CFCs (e.g., in buildings and transportation) and methane (e.g., from natural gas production and distribution) also are assessed. The food sector, though, differs in two important aspects. First, the relative importance of methane (CH_4) and N_2O emissions is greater than in other sectors. Although estimates are uncertain, the food sector may account for one-third of global CH_4 emissions and anywhere from one-tenth to one-fifth of current global N_2O emissions. Its contribution to total U.S. CH_4 emissions is roughly 9 percent (its contribution to U.S. N_2O emissions is uncertain, though).

Second, fossil fuel-related CO_2 emissions (i.e., from farm machinery, irrigation equipment, fertilizer manufacturing, food transport, processing and packaging, and cooking) and CFC emissions (primarily from refrigeration) are subsumed in the transportation, industry, and buildings analyses summarized earlier. Further, CO_2 emissions from agricultural-related deforestation in the United States are very small (although they are very important in developing countries). To place the food sector in perspective, though, we estimate that it accounts for at least 8 percent of total U.S. CO_2 emissions and about 5 percent of U.S. CFC-11 and CFC-12 emissions (worldwide, it may account for one-fifth of global CO_2 emissions and up to 15 percent of global CFC-11 and CFC-12 emissions).

In the past, congressional concern about agriculture largely has focused on farm production, promotion, and income. With the passage of the 1985 Food Security Act, Congress began dealing with some of the environmental impacts of U.S. agriculture. Although the 1990 Farm Bill expanded these efforts, including extending the CRP until 1995, additional steps can still be taken, as discussed below for methane and nitrous oxide emissions.

Some of the opportunities discussed earlier for the buildings, industry, and transportation sectors also can affect food sector activities (e.g., more efficient cooking, processing and packaging, etc.). In addition, fossil fuel-related CO_2 emissions from the U.S. food system could be reduced by making fertilizer manufacture, farm machinery, and irrigation more energy efficient.

Reducing Methane Emissions--U.S. methane emissions from the food sector are primarily from ruminant animals (e.g., cattle, sheep). Congress could direct the U.S. Department of Agriculture (USDA) to determine the potential for techniques such as improved nutrient management, feed additives, and manure management to reduce methane emissions. To limit future growth in, or even reduce, livestock populations in the United States, Congress could consider reducing or removing price supports for feed grains, which might make beef and dairy products more expensive (although it is unclear if the costs would rise or fall over the long term).

Reducing N_2O Emissions--To reduce nitrous oxide emissions, Congress could modify commodity program policies, which now encourage monocropping and heavy fertilizer use, to give farmers more control over the types of crops they plant without losing program crop base acreage and support payments. Congress could provide cropping flexibility only to those farmers who adopt environmentally sound cropping patterns. Congress also could make implementation of best management practices (BMPs) a prerequisite for receiving Federal price and income supports. BMPs, designed by the Soil Conservation Service (SCS) to reduce soil degradation and water contamination from agricultural activities, include more efficient fertilizer use, water impoundments, permanent vegetative cover, and manure storage.

Food RD&D--The development of an accurate emissions database for the food sector is perhaps the most critical research priority. Increased research is needed to quantify CO_2 emissions from agricultural land-clearing activities, CH_4 emissions from ruminant animals (and from rice cultivation, particularly in the developing world), and N_2O emissions from nitrogenous fertilizers. The emissions reduction potential of different alternative practices must also be investigated; for example, support is needed for research on methane-reducing techniques, especially for livestock in confined and range-management systems. Congress also could increase funding for RD&D efforts to develop new alternative practices, especially those that simultaneously increase crop yields and reduce greenhouse gas emissions per unit of food output.

U.S. Policy Options To Help Limit Greenhouse Gas Emissions Abroad

Earlier sections in this summary set forth specific policy options that the United States could pursue to reduce or offset its own greenhouse gas emissions. By taking such actions to reduce its own emissions, the United States can provide leadership through example. In the broader context, the United States also can work towards the adoption of international conventions and protocols regarding climate change, similar to those developed for phasing out CFCs and halons.

The United States also can attempt to help developing countries, Eastern Europe, and the U.S.S.R. to minimize their greenhouse gas emissions, without hindering the prospects for needed economic development. Indeed, strategies to lower greenhouse emissions can simultaneously help these nations become more economically efficient. Numerous existing programs and organizations in the United States and on the international scene directly influence development and indirectly can affect greenhouse gas emissions. The United States, for example, provides direct bilateral assistance through the U.S. Agency for International Development (A.I.D.). Numerous other U.S. agencies--such as the State Department, the Commerce Department, the U.S. Trade Representative, the Treasury Department, the Agriculture Department, and the Environmental Protection Agency--support technology transfer and development assistance in certain areas. Through these U.S. and international organizations, the United States currently contributes about $9 billion annually in foreign aid assistance (including bilateral aid, food aid, security-related economic support funds, and multilateral aid) to developing countries.

The United States can continue to work through its own bilateral assistance programs and international organizations, as well as through NGOs, to increase the development and transfer of technologies and policies related to energy, family planning, and land use and management practices that provide sustainable alternatives to deforestation and dependence on fossil fuels.

REFERENCES

1. U.S. Congress, Office of Technology Assessment, Changing By Degrees: Steps to Reduce Greenhouse Gases, OTA-O-482 (Washington, DC: U.S. Government Printing Office, February 1991).

2. Solar Energy Research Institute, Idaho National Engineering Laboratory, Los Alamos National Laboratory, Oak Ridge National Laboratory, and Sandia National Laboratories, The Potential of Renewable Energy, An Interlaboratory White Paper, prepared for U.S. Department of Energy, Office of Policy, Planning and Analysis, SERI/TP-260-3674 (Golden, CO: Solar Energy Research Institute, March 1990).

3. Lawrence Berkeley Laboratory, "Buildings Energy Use Compilation and Analysis Project" (Berkeley, CA: Mar. 6, 1986).

4. Goldman, C.A., "Measured Results of Energy Conservation Retrofits in Residential Buildings," paper presented at 1986 ASHRAE Winter Meeting, LBL-20950 (San Francisco, CA: Jan. 19-22, 1986).

5. Geller, H.S., "Commercial Building Equipment Efficiency: A State-of-the-Art Review," contract prepared for U.S. Congress, Office of Technology Assessment (Washington, DC: American Council for an Energy-Efficient Economy, February 1988).

6. Ottinger, R.L. et al, "Environmental Costs of Electricity," Oceana Publications, 1990.

7. Geller, H.S., "Residential Equipment Efficiency: A State of the Art Review," contract prepared for U.S. Congress, Office of Technology Assessment (Washington, DC: American Council for an Energy-Efficient Economy, 1988).

8. DiFiglio, C., K.G. DuLeep, and D.L Greene, "Cost Effectiveness of Future Fuel Economy Improvements," submitted to Energy Journal, 1989.

9. DiFiglio, C., presentation at OTA workshop on transport and global warming, Apr. 6, 1989.

10. Bohi, D.R. and M.B. Zimmerman, "An Update on Econometric Studies of Energy Demand Behavior," Annual Review of Energy 9:105-154, 1984.

11. Dahl, C.A., "Gasoline Demand Survey," The Energy Journal 7(1):67-82, 1986.

12. Greene, D.L., "CAFE or Price? An Analysis of the Effects of Federal Fuel Economy Regulations and Gasoline Prices on New Car MPG, 1978-89," Energy Journal, 11(3):37-57, 1990.

13. McNutt, B. and P. Patterson, "CAFE Standards--Is a Change of Form Needed?" SAE Technical Paper Series, #861424 (Warrendale, PA: Society of Automotive Engineers, 1986).

14. U.S. Congress, Office of Technology Assessment, S. Plotkin, testimony before the Senate Committee on Commerce, Science and Transportation, May 2, 1989, "Increasing the Efficiency of Automobile and Light Trucks--A Component of a Strategy to Combat Global Warming and Growing U.S. Oil Dependence."

15. Downs, A., "The Real Problem With Suburban Anti-Growth Policies," Brookings Review, pp. 23-29, Spring 1988.

16. Bleviss, D.L., The New Oil Crisis and Fuel Economy Technologies (Quorum Books, New York: 1988).

17. U.S. Department of Energy, Energy Information Administration, Indicators of Energy Efficiency: An International Comparison, EIA Service Report, SR/EMEU/90-02 (Washington, DC: July 1990).

18. Baldwin, S., "Energy-efficient Electric Motor Drive Systems," in T.B. Johansson et al. (eds.), Efficient End Use and New Generation Technologies, and Their Planning Implications (Lund, Sweden: Lund University Press, 1989).

19. Ross, M., "Improving the Energy Efficiency of Electricity Use in Manufacturing," Science 244:311-317, Apr. 21, 1989.

20. Kirsch, F.W., Energy Conserved and Costs Saved by Small and Medium-Sized Manufacturers: 1987-88 EADC Program Period (Philadelphia, PA: University City Science Center, March, 1988).

21. Fulkerson, W. et al., Energy Technology R&D: What Can Make A Difference?, Volume 2, Supply Technology, ORNL-6541/V2/P2 (Oak Ridge, TN: Oak Ridge National Laboratory, 1989).

22. Electric Power Research Institute, Power Plant Performance Monitoring and Improvement, EPRI report CS/EL-4415, vol. 3 (Palo Alto, CA: February 1986).

23. National Association of State Foresters, "Global Warming and Forestry in the United States," background paper (Washington, DC: April 1990).

OPTIONS FOR REDUCING CARBON DIOXIDE EMISSIONS

Arthur H. Rosenfeld
Department of Physics
University of California-Berkeley, Berkeley, Ca.
Center for Building Science
Lawrence Berkeley Laboratory, Berkeley, Ca.

Lynn Price
Center for Building Science
Lawrence Berkeley Laboratory, Berkeley, Ca. 94720

ABSTRACT

Improvements in energy efficiency can significantly reduce the annual growth in greenhouse gas emissions. Such improvements occur when energy intensity is reduced; no reduction in energy services is required. Using the concept of "cost of conserved energy" to develop conservation supply curves similar to resource supply curves, researchers consistently find that electricity and natural gas savings of nearly 50% of current consumption are possible for U.S. buildings. Such reductions in energy consumption directly reduce emissions of greenhouse gases.

To capture these savings, we must continue to develop energy-efficient technologies and strategies. This paper describes three recent energy-efficient technologies that benefitted from energy conservation research and development (R&D) funding: high-frequency ballasts, compact fluorescent lamps, and low-emissivity windows. Other advanced technologies and strategies of spectrally selective windows, superwindows, electrochromic windows, advanced insulation, low-flow showerheads, improved recessed lamp fixtures, whitening surfaces and planting urban trees, daylighting, and thermal energy storage are also discussed.

INTRODUCTION

Worldwide combustion of fossil fuels produces enormous emissions of greenhouse gases. Annually, 5.7 gigatonnes of carbon (GtC), equivalent to one ton of carbon (C) for every living person on earth are pumped into the atmosphere to fulfill our energy needs. Carbon dioxide (CO_2) can be accounted for in either units of C or CO_2; one kilogram (kg) of C is equivalent to 3.667 kg of CO_2 [3.667 = $m(CO_2)/m(c)$ = 44/12]. For this paper we have chosen to use CO_2.

Ideally, to reduce CO_2 emissions, we would cease energy production by fossil fuel burning facilities and switch to non-fossil fuel sources. However, the technical and economic barriers of the non-fossil sources must first be resolved. In the meantime, we can reduce about half of the annual growth in greenhouse gas emissions through increased energy efficiency in buildings, industry, and transportation.

Improvements in energy efficiency occur when energy intensity is reduced. Energy intensity is the ratio of energy consumed to the products and services produced, defined as:

$$\text{Energy Intensity} = \frac{\text{Energy (E)}}{\text{Gross National Product (GNP)}} \quad (1)$$

Energy efficiency does not mean a reduction in energy services; indeed the exact same services of heat, light, power, etc. are provided with technologies and processes that use less energy. In many cases, improvements in energy efficiency cost dramatically less than building new power plants or generating expensive peak power at existing facilities.

In the U.S., energy consumption is divided almost equally by three sectors: 1) industry, 2) transportation, and 3) commercial and residential buildings. After describing consumption trends of all of these sectors, we will focus on the building sector. We will discuss conservation supply curves that estimate the overall potential for energy and CO_2 savings and will describe specific energy-efficient technologies and strategies for this sector.

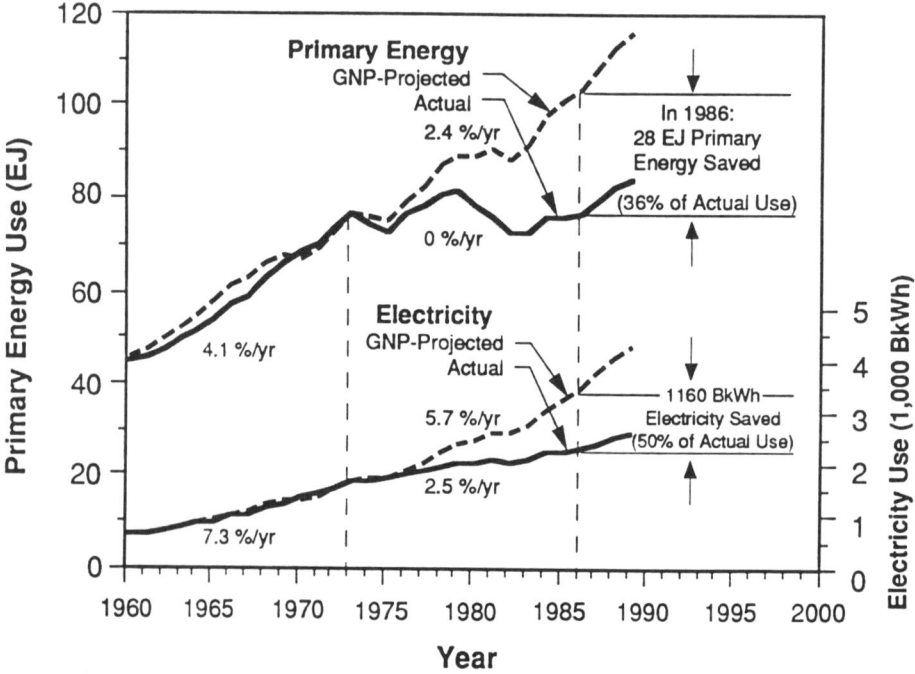

Figure 1. Total U.S. Primary Energy and Electricity Use: Actual vs. GNP Projected (1960-1989). GNP-projected energy values are based on 1973 efficiency and GNP. The electricity projections include an additional 3% per year to account for increasing electrification. Electricity use is given in terms of total equivalent primary energy input - exajoules (EJ) - on the left-hand scale, and net consumption - 1,000 billion kilowatt-hours (BkWh) - on the right-hand scale.

Source: Energy Information Administration, U.S. Department of Energy.

RECENT ENERGY CONSUMPTION TRENDS

During the past 30 years, primary energy consumption in the U.S. has fluctuated dramatically. These fluctuations are divided into three recent energy eras in **Figure 1**: the initial frozen energy efficiency era (1960-1973), the energy conservation era (1973-1986), and the current frozen energy efficiency era (1986-1989).

Primary Energy

Between 1960 and 1973, primary energy use and U.S. GNP were inexorably linked and climbed about 4% per year. In 1973, the Organization of Petroleum Exporting Countries (OPEC) oil embargo provided a powerful incentive to conserve energy; during the 13 years of high oil prices and progressive energy policies from 1973 to 1986, national energy use stayed constant, while U.S. GNP grew by a total of 35%, or 2.4% per year. Efficiency measures implemented during this period avoided an increase of approximately 50% in U.S. greenhouse gas emissions. By 1986, projected primary energy use was 36% higher that actual use, indicating a savings of 28 exajoules. One-third of this savings is attributed to structural changes in the economy and the remaining two-thirds is attributed to improved energy efficiency.[1] In 1986, when OPEC's oil prices collapsed, gains in energy efficiency nearly stopped. Now, primary energy consumption is climbing again at a rate of about 2% per year and it is feared that energy use and GNP could return to the lockstep relationship experienced prior to 1973, directly contributing to increased emissions of CO_2.

Electricity

Even more impressive than the past reductions in primary energy is the conservation experienced in electricity as shown in **Figure 1**. Since buildings consume two-thirds of total U.S. electricity, improvements in this sector contributed significantly to total electricity savings. Until 1973, total electricity use was growing at a rate of 7.3% per year (3% faster than the GNP). During the energy conservation era, electricity use grew only as fast as GNP, for an annual savings of 3.2%. In 1986, projected electricity use was 50% higher than actual electricity use. This savings, of 1160 billion kilowatt-hours (BkWh) per year, is equivalent to the annual output of 230 baseload (1000 megawatt) power plants. Using the 1989 all-sector average price of electricity of 6.4¢ per kilowatt-hour (kWh), this is a savings worth over $50 billion per year.

Residential and Commercial Buildings

Figure 2 shows the performance of residential and commercial buildings. During the energy conservation era between 1973 and 1986, energy use in this sector demonstrated the same remarkable "gaps" between actual and GNP-projected energy use. Actual energy use during this period was about 2% less per year than projected by GNP for both residential and the commercial buildings. In 1986, 8 exajoules (EJ), or 28% of the primary energy consumed in buildings, were saved (1 EJ = 10^{18}J). Using the 1989 cost of energy to buildings of $6.70 per gigajoule ($200 billion for 30 EJ consumed), this energy is worth over $50 billion and equal to four million barrels of oil per day (Mbod) of oil equivalent.

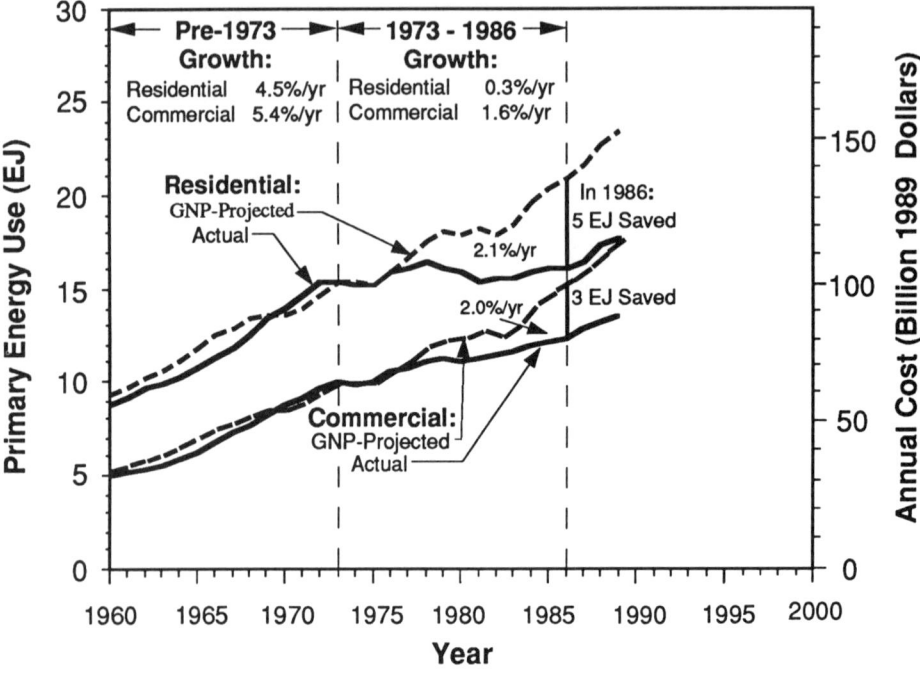

Figure 2. Primary Energy Use in U.S. Buildings (1960-1989). Before 1973, total primary energy use in residential and commercial buildings was growing at 4.5 and 5.4% per year respectively. From 1973 to 1986, energy use in residential buildings leveled off at about 16 exajoules (EJ) per year, while commercial building energy use grew only 1.6% per year from 10 to 12.4 EJ. Since the collapse of oil prices in late 1985, total primary energy use for both residential and commercial buildings has grown about 10%, or greater than 3% per year.

The residential GNP-projected curve is straightforward:

$$\text{Projected Energy (t)} = \frac{\text{Energy(1973)}}{\text{GNP(1973)}} \times \text{GNP (t)}. \qquad (1)$$

But for the commercial sector, the pre-1973 trend was for energy use to grow 1% faster than residential energy use (or GNP). Accordingly, the commercial GNP-projected curve has been tilted up by 1% per year to reflect this trend, i.e. (1) has been multiplied by the factor $1.01^{(t-1973)} \approx 1 + .01 (t-1973)$.

Source: Energy Information Administration, U.S. Department of Energy.

CONSERVATION SUPPLY CURVES

A conservation supply curve relates energy savings achieved by implementing a given efficiency measure, to that measure's "cost of conserved energy" (CCE). Thus, for electricity:

$$\text{Cost of Conserved Energy (CCE)} = \frac{\text{Annualized Investment (\$ per year)}}{\text{Annual Energy Saved (kWh per year)}} \quad (2)$$

The initial investment in an efficient technology or program is annualized by multiplying it by the "capital recovery rate" (CRR):

$$\text{CRR} = \frac{d}{1-(1+d)^{-n}} \quad (3)$$

where "d" is the real discount rate and "n" is the number of years over which the investment is written off, or amortized.[2] We use "real" discount rate (i.e. corrected for inflation) in order to compare the CCE with the price of energy, excluding inevitable inflation from both measures.

For example, an energy-efficient refrigerator that consumes 690 kWh/year (or 240 kWh/year less than the 1990 average consumption of 930 kWh/year), has an incremental cost of about $66.[3] Assuming a 6% discount rate and a 20 year amortization period, the CRR is:

$$\text{CRR} = \frac{.06}{1-(1+.06)^{-20}} = .09 \quad (4)$$

and the CCE is calculated as:

$$\text{Cost of Conserved Energy (CCE)} = \frac{6 \ (\$ \text{ per year})}{240 \ (\text{kWh per year})} = 2.5 \ \cancel{c}/\text{kWh}. \quad (5)$$

This cost can then be compared to the price of electricity to determine whether the investment should be made.

On a conservation supply curve, each measure or step (such as "efficiency improvements to residential refrigerators") is defined as follows:

>Height = CCE (cents/saved kWh)
>Width = annual kWh saved
>Area under the step = total annualized cost of investment ($)

The steps are ranked in order of ascending CCE, with the cheapest options plotted first, causing the curve to be upward-sloping.

Although conservation supply curves all have the same general shape, there are a number of underlying assumptions that can make them appear more or less attractive. These include the level of technology saturation assumed, the baseline and analysis period chosen, the number of new buildings included, whether existing efficiency is frozen or increases naturally, economic considerations such as retail vs. wholesale prices and discount rates, and whether fuel switching is included.

Traditionally, conservation supply curves have assumed one of two technology saturation levels: "technical potential" or "achievable." The "technical potential" saturation level is based on engineering and economic calculations without concern for the probability of successful implementation. "Achievable" saturation scenarios are based on actual experience; typical utility conservation programs have captured only about 50% of the technical potential. However, with recently adopted profit incentive mechanisms, some utilities can now earn up to 15% of avoided costs. Given this powerful profit motive to sell efficiency, the "achievable/technical" ratio will probably increase.

Conservation supply curves assume a specific baseline year and energy consumption level. They also address a specific analysis period (typically 10 to 20 years) and, depending upon the length of time, may or may not include new buildings. If new buildings are included, then the number of new buildings must be estimated. Also, existing levels of energy efficiency are either assumed to stay constant ("frozen efficiency") during the analysis period or to grow at a "naturally occurring" rate. In addition, economic assumptions, such as the discount rate, must also be made. An important economic assumption is whether retail or wholesale prices are used. Many utilities are now involved in promoting energy efficient technologies and are supplying products at wholesale prices, significantly reducing initial costs and payback periods. Finally, some conservation supply curves include fuel switching options to conserve electricity.

Electricity Conservation Supply Curves

LBL U.S. Residential Electricity Conservation Supply Curve

Analysts at Lawrence Berkeley Laboratory (LBL) have recently completed a comprehensive electricity conservation supply curve for U.S. residential buildings.[4] This curve was derived using a thorough database of appliance efficiency and costs developed for the U.S. Department of Energy (DOE) and a detailed analysis of thermal integrity measures in single-family dwellings. The LBL conservation supply curve evaluated the technical (versus achievable) potential for electricity efficiency improvements and assumed a 7% real discount rate, an analysis period of 1990 to 2010, and frozen efficiency. New buildings have been included. Conservation costs are those for consumer installation; utility or government administrative costs are not included.

Figure 3 shows the LBL technical potential conservation supply curve for residential electricity savings in 2010. For those measures costing less than the price of electric power to residential customers, or 7.6¢/kWh in 1989, the technical potential for residential electricity savings in all buildings in 2010 is about 40%, or 404 BkWh of 2010 baseline use of 1028 BkWh.

The LBL conservation supply curve is based on an analysis of 214 residential electricity conservation measures. This curve includes better equipment for space conditioning, appliances, and lighting. Fuel switching from electricity to natural gas for water heaters, ranges, and clothes dryers is also included. Further, engineering estimates for certain advanced technologies such as "superwindows," spectrally-selective glazings, evacuated panels for refrigerators, heat-pump water heaters, and heat-pump dryers are included. The LBL supply curve does not include "promising" technologies for which there are no data.

LBL Supply Curve Compilation

Analysts at LBL have also recently compiled and adjusted nine potential conservation supply curves that depict the technical potential for electricity savings for both U.S. residential and commercial buildings (which consumed 1627 BkWh or 64% of all 2630 BkWh sold in 1989) by about the year 2000.[5] LBL adjusted all curves to a real discount rate of 6%, to frozen efficiency, and to technical potential energy savings. All of these studies were based on available technologies; technologies that only exist as

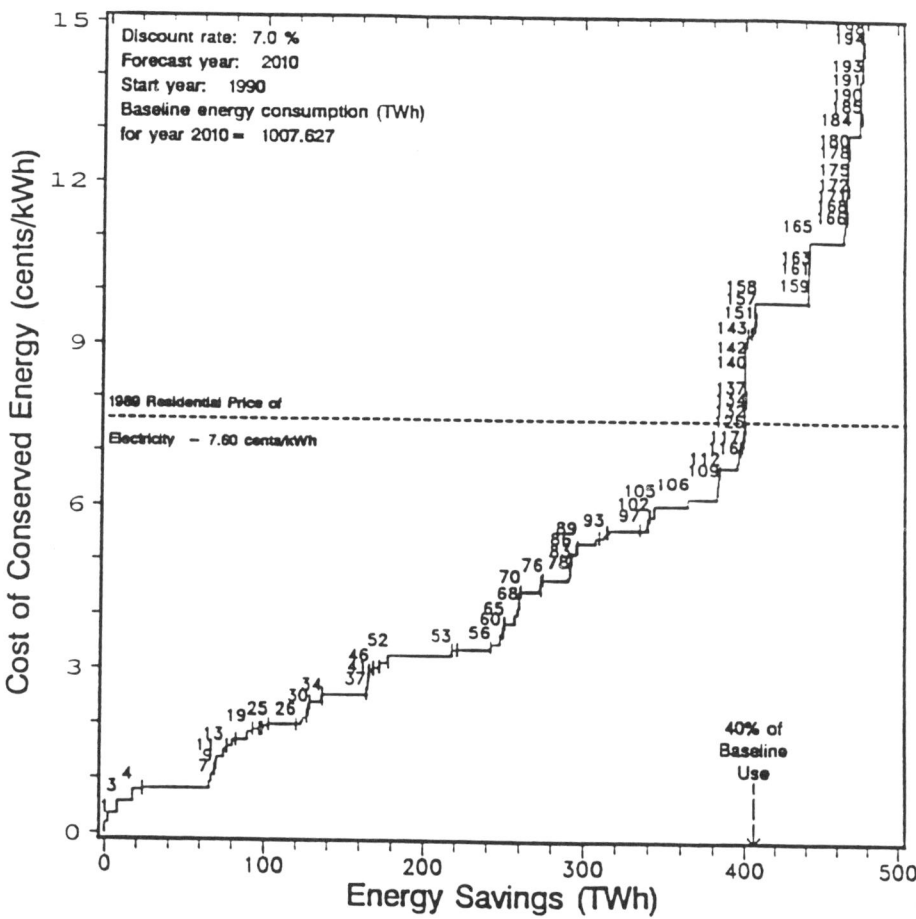

Figure 3. Supply Curve of Conserved Electricity for the United States Residential Sector - Maximum Technical Potential in 2010. Each step represents a conservation measure (or a package of measures). For example, step 36 is the conservation measure "improve refrigerator to 1993 standard," step 90 is "switch electric clothes dryer to gas," and step 154 is "improve windows in existing single family homes, North." The width of the step indicates the nationwide electricity savings from the measure and the height of the measure indicates the cost of conserved electricity. The end uses include space conditioning, water heating, refrigeration, lighting, and miscellaneous.

Source: J. Koomey, C. Atkinson, A. Meier, J. McMahon, S. Boghosian, B. Atkinson, I. Turiel, M. Levine, B. Nordman, and P. Chan, *The Potential for Electricity Efficiency Improvements in the U.S. Residential Sector*, LBL-30477, (Lawrence Berkeley Laboratory, Berkeley, 1991).

prototypes were excluded. Cumulative electricity savings of the conservation supply curves range between 35% and 55%. Other conservation supply curves that include technologies that are now only prototypes will undoubtedly result in larger technical potential savings.

Figure 4 presents the 12-step Electric Power Research Institute (EPRI) curve which represents the approximate mid-range of the compiled supply curves. (The EPRI curve actually includes only 11 steps; an additional first step for white surfaces and urban trees has been added by LBL.) The EPRI curve represents a cumulative savings of 734 BkWh, which is 45% of 1989 U.S. building sector electricity use.[6] The EPRI curve is also consistent with a new National Research Council study[7] which is too coarse to compile as a supply curve. That study (in its Table 4-12) estimates a near term retrofit potential savings of 30% and a long term retrofit potential savings of 50%. These potential savings nicely bracket the EPRI potential savings of 45%.

Natural Gas Conservation Supply Curves

LBL has also reviewed two supply curves of conserved natural gas for the residential sector. One is a study of the U.S. residential sector by the Solar Energy Research Institute[8] and the other is a study of the California residential sector.[9] **Figure 5** presents these two supply curves and shows savings of about 50% are possible for U.S. residential natural gas use at less than the 1989 average price of $5.63 per million Btu. Extrapolating this estimate to cover all gas and oil use in U.S. buildings yields savings of about 5.2 quads. (Natural gas and oil are interchangeable for many utilities and industries, so we combine these two fuels together to estimate potential fuel savings.)

From these studies it appears that potential natural gas savings are slightly larger than electricity savings. However, natural gas savings are not well studied, presumably because less is spent annually in the U.S. on natural gas (about $60 billion versus about $140 billion for electricity) and because even during high energy prices natural gas use stayed constant while electricity use grew.

Fuel Switching

Assuming that electricity is generated from the mix of fuels burned by U.S. power plants (including coal with its high C content), then fuel switching from electricity to natural gas represents another method to reduce CO_2 emissions. In the building sector, fuel switching involves replacing electric resistance heat with on-site combustion of natural gas and replacing electric appliances with gas appliances (mainly water heaters and clothes dryers). Fuel switching is the least well studied U.S. potential conservation option. Even so, we estimate that U.S. buildings electricity use could be reduced by 10% through fuel switching.[10]

As an example of fuel switching we will discuss residential water heaters which represent the largest single U.S. potential switch. In Michigan, 400,000 homes had gas heat but electric resistance water heaters, and could switch with a simple payback time of 2 years. In the 1988 Michigan Electric Options Study, the switching potential was about 20% of residential electricity. Because Michigan seems to have a higher fraction of homes with gas available than does the rest of the U.S., we have picked a symbolic 10% reduction in U.S. building electricity, and used the data for water-heaters as a proxy for all fuel switching.[11] In this example, 163 BkWh of electricity (10% of the building sector 1989 consumption of 1627 BkWh) are replaced by 0.7 quads of natural gas. Such a fuel switch ultimately reduces CO_2 emissions because natural gas contains emits less CO_2 than the mix of fuels used to produce electricity.

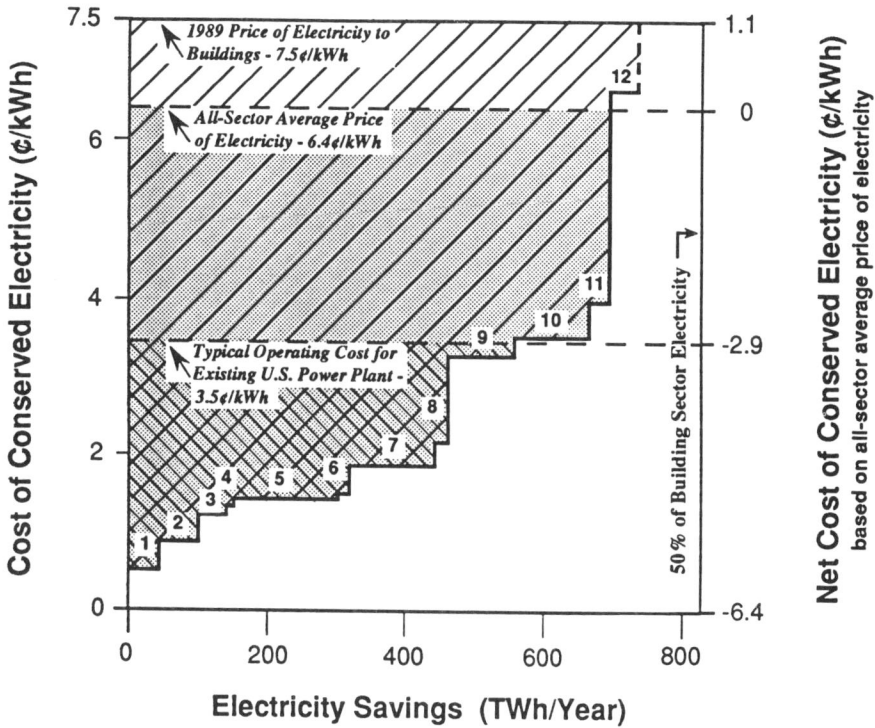

Potential Net Savings:

▨ **$ 37 B/yr.** - CASE 1: Below 7.5¢/kWh 1989 Price of Electricity to Buildings

▧ **$ 29 B/yr.** - CASE 2: Below 6.4¢/kWh All-Sector Average Price of Electricity

▩ **$ 10 B/yr.** - CASE 3: Below 3.5¢/kWh Typical Operating Cost for Existing U.S. Power Plant

Figure 4. Cost of Conserved Electricity (CCE) for Buildings. This figure is the EPRI curve with a discount rate of 6%. The full X-axis corresponds to 813.5 TWh, which is half of the total 1989 U.S. buildings electricity use of 1627 TWh and which cost $140 B. The Net CCE scale is displaced by 6.4 ¢/kWh - the all-sector average price of the avoided electricity. All recommended measures that have a CCE of less than 6.4¢/kWh have a negative cost, i.e. save money.

Areas between the CCE and a price line represent annual dollar savings. Case 1 (lightly hatched area) shows this potential annual net savings of $37 B, based on the average price of the avoided electricity of 7.5¢/kWh. Case 2 (shaded area) represents the potential annual savings of $29B, based on the all-sector average price of 6.4¢/kWh (defined as Net CCE of 0 on the right hand scale). To be extremely conservative, the net CCE can be referenced to the avoided cost of merely operating an existing plant - about 3.5 ¢/kwh at the meter. Case 3 (heavily hatched area) represents this most conservative estimate of savings of $10B/year.

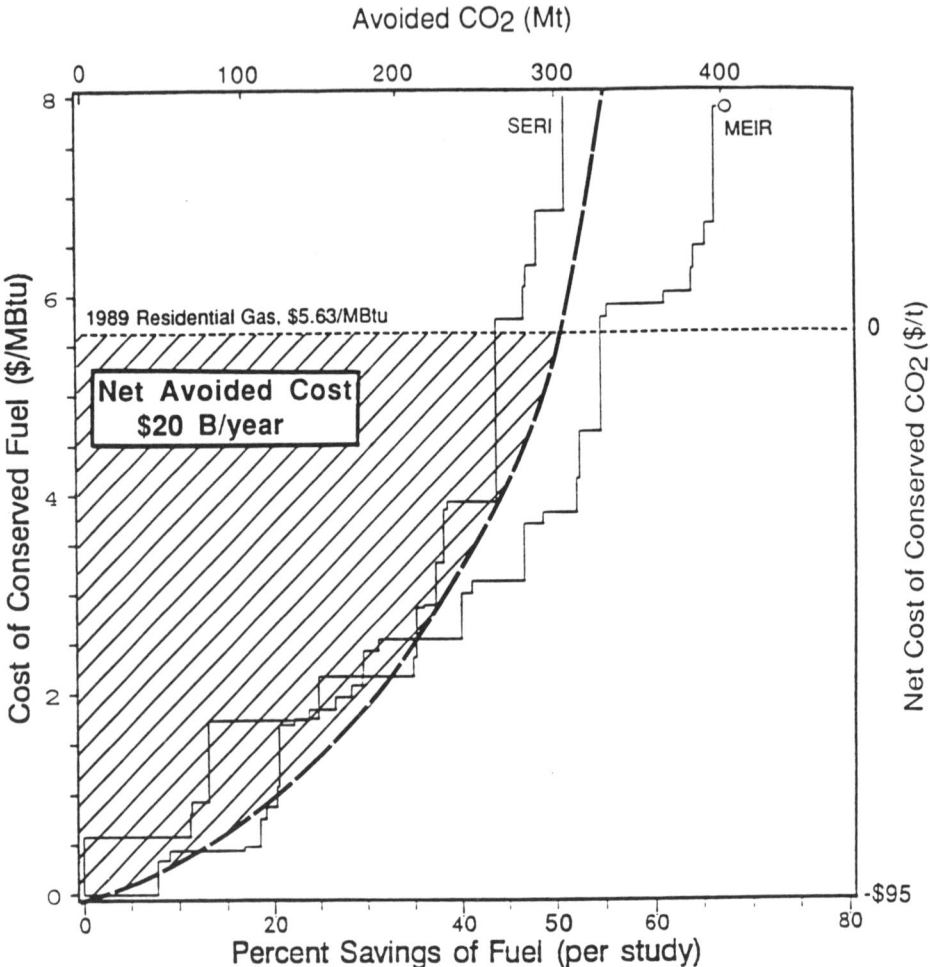

Figure 5. Cost of Conserved Energy (CCE) and Net Cost of Conserved Carbon Dioxide (CC CO_2) for the Residential Sector. The figure shows two supply curves of conserved natural gas for the residential sector: a study of the U.S. residential sector by the Solar Energy Research Institute and a study of the California residential sector by Meier, et al. The smooth curve between them goes through 50% savings at the current average price of $5.63/MBtu for natural gas. Extrapolating this to natural gas and oil use for the entire buildings sector yields a potential savings of about 5.2 quads of "fuel." The right-hand y-axis scale shows the net CC CO_2 running down from 0 to -$95/t CO_2, and the average CC CO_2 is -$70/t CO_2. The potential CO_2 savings are about 300 Mt CO_2.

Sources: Solar Energy Research Institute (SERI), *A New Prosperity: Building a Sustainable Energy Future* (Brickhouse Publishing, Andover, MA, 1981) and A. Meier, J. Wright, A. Rosenfeld, *Supplying Energy Through Greater Efficiency: The Potential for Conservation in California's Residential Sector*, (University of California Press, Berkeley, 1983). CCEs for SERI curve are based on 6% discount rate and 10-year lifetime and are approximated from the original CCEs that were based on a 3% discount rate and slightly different lifetimes. CCEs for MEIR curve are calculated using known lifetimes.

Table 1 shows that, for this example, 114 Mt CO_2 produced by electricity are replaced by 40 Mt CO_2 produced by natural gas, with a net savings of 74 Mt CO_2.

Fuel switching as an electricity conservation policy is now attracting interest in several locations. In Vancouver, Canada, BC Hydro gave subsidies to consumers to replace electric water heaters with gas units. In Vermont, winter-peaking utilities have begun fuel switching programs that promote switching from electricity to natural gas or propane. These utilities audit electric customers to determine if they are eligible for fuel switching and, if so, arrange and oversee the switch. Because of this program, alternative fuel dealers are offering financial incentives to attract new customers. The utilities are also offering incentive payments and have completed low-income rehabilitation projects.[12] Other New England utilities are being encouraged to establish fuel-switching programs. In Wisconsin, the Wisconsin Public Service Commission has directed utilities to fuel switch where economic.

Table 1. A fuel switching example of saving 10% of building sector electricity by switching water heaters from electric resistance to gas heat.

	1989 Use	1989 Potential Savings
Electricity		
BkWh (10% of Buildings BkWh)	1627	163
Mt CO_2 (1kWh$_e$ = 0.7 kg CO_2)	1139	114
$ (6¢/kWh x 163 BkWh)	--	$9.8B
Gas		
Quads (.0043 MBtu replaces 1kWh$_e$)	10.4	-0.7Q
Mt CO_2 (1Q = 57 Mt CO_2)	600	-40
$ (at average $4.20/M Btu)	--	-$3.0B
Net		
Mt CO_2	1739	74
$	$170B	$6.8B

OVERALL ENERGY AND CARBON DIOXIDE SAVINGS POTENTIAL

The studies described above indicate large potential savings in the U.S. building sector of about 45% of electricity and about 50% of natural gas. Further electricity savings are possible through fuel switching.

Table 2 characterizes these energy savings as dollar savings to the U.S. economy. First, using the 1989 all-sector average price of 6.4¢ per kWh, the potential electricity savings using the EPRI residential and commercial buildings estimate of 734 BkWh have a net value of $29 billion (taking the cost of the efficiency measures into account). Second, using the 1989 residential average price of natural gas of $5.63 per MBtu, the potential natural gas savings of 5.2 quads have a net value of $20 billion. When fuel switching savings of $6.8 billion are added, the total technical potential energy savings is valued at $56 billion.

Potential energy savings may also be characterized as savings of CO_2 emissions. For electricity we make this conversion using the CO_2 produced by the mix of fuels burned by U.S. power plants,[13] which is estimated to be 500 million tonnes (Mt) of carbon (C) for 1990 electric sales of 2610 BkWh, or 0.19 Mt C/BkWh. One kilogram (kg) of carbon is equivalent to 3.667 kg of CO_2, so 0.19 Mt C = 0.7 Mt CO_2. Thus:

$$1 \text{ kWh} = 0.7 \text{ kg } CO_2 \qquad (6)$$

and

$$1 \text{ BkWh} = 0.7 \text{ Mt } CO_2. \qquad (7)$$

Using (6), the cost of electricity is then converted as follows:

$$1¢/\text{kWh} = \$14.3/tCO_2. \qquad (8)$$

Using (8), net CCE can be converted to the net cost of conserved CO_2, or CC CO_2.

In order to transform fuel savings to CO_2 savings we add the CO_2 that oil and natural gas contribute to our base case fuel use in buildings. In 1989, natural gas accounted for 7.7 quadrillion Btu (quads) and oil accounted for 2.7 quads. Weighting these fuels by their respective carbon content, assuming natural gas contains 14.5 kgC/MBtu and oil contains 20.3kgC/MBtu,[13] yields:

$$\begin{array}{c} 1 \text{ MBtu "fuel"} = 16 \text{ kg C} = 59 \text{ kg } CO_2 \\ \text{or} \\ 1 \text{ quad "fuel"} = 59 \text{ Mt } CO_2 \end{array} \qquad (9)$$

Overall potential savings from electricity and fuel efficiency improvements along with fuel switching are summarized in Table 2. Total net CO_2 savings are 890 Mt, or slightly over 50% of 1989 emissions from this sector of the U.S. economy.

Table 2. Summary of the Potential Savings of Electricity, Fuel (Gas and Oil), and Carbon Dioxide for Existing Buildings.

	1989 Use	1989 Potential Savings
Electricity		
BkWh	1627	734
Mt CO_2 (1kWh$_e$ = 0.7 kg CO_2)	1139	513
$ (at 7.5¢/kWh)	$112B	--
Net $ (at 6.4¢/kWh)	--	$29B
CC CO_2 ($/t)	--	-57
Fuel (Gas and Oil)		
Quads	10.4	5.2
Mt CO_2 (1M Btu - 57 kg CO_2)	600	300
$ (at $5.63/M Btu)	$58B	--
Net $ (at $5.63/M Btu)	--	$20B
CC CO_2 ($/t)	--	-70
Fuel Switching (from Table 1) Net Savings from Switching 10% of Electricity to Gas		
Mt CO_2 (Electricity and Gas)	1739	74
Net $	--	$6.8B
Net CC CO_2 ($/t)	--	-92
Total		
Mt CO_2	1739	890
Net $	$170B	$56B
Net CC CO_2 ($/t)	--	-63

ENERGY EFFICIENCY RESEARCH AND DEVELOPMENT

Need For Increased Funding

Conservation supply curves indicate that the technical potential exists for energy and CO_2 savings of close to 50% of current consumption levels in the U.S. building sector. But how can such large savings be realized? First, we must continue to develop energy-efficient technologies and strategies to capture these savings. Then we must ensure widespread adoption of existing and new technologies and strategies through development of effective energy policies. Adequate funding for energy efficiency research and development (R&D) is an *essential* element of this picture.

Current energy conservation R&D funding, however, is wholly inadequate. In 1989 U.S. public domain R&D for efficiency was approximately $200 million, less than 10% of public domain R&D spending for all U.S. energy technologies. **Table 3** sets the scale:

Table 3. R&D Spending Comparison

R & D	Spending	Comments
Total U.S.	3% of U.S. GNP	
Non-military	2% of U.S. GNP	
Mature industries	1% of revenues	
Public domain energy efficiency	0.04% of U.S. energy sales	(2/5000 of bills of $500 B/yr.)
Building energy efficiency	0.025% of utility revenues	(1/4000 of bldg energy of $200 B/yr.)

Figure 6 shows a time-series of public domain R&D spent on energy technologies in fields including efficiency, renewables, fossil fuel, nuclear power, and magnetic fusion.[14] In 1978, public energy efficiency R&D reached $250 million, only 5% of all energy R&D, and it remained at this level throughout fiscal year 1991. This level of spending is .04% of U.S. energy sales from annual energy expenditures of $500 billion. In 1991, under the Bush administration, energy efficiency R&D spending did grow from $200 million to $265 million, still extraordinarily out of step with the dramatic economic contribution of energy efficiency.

Figure 7 compares the energy performance of various energy sciences to their DOE funding. Between 1973 and 1989, 22 exajoules (EJ) of primary energy were saved as a result of energy efficiency improvements. When divided into the 1991 budget for energy efficiency of $200 million, energy efficiency gets only $9 million per EJ that it has contributed. In comparison, fossil fuel-related R&D, with its strong political muscle, receives $328 million per EJ supplied, and nuclear power, with its devoted administrative and congressional support, gets $78 million per EJ supplied. Energy efficiency R&D is probably neglected because there is no perception of political muscle (apart from transportation) either by the fragmented building sector and industry or by the Congress. An added problem is that many policymakers honestly feel that since efficiency is such a success compared to fossil fuel, nuclear, and renewables, it can take care of itself and needs little help from the government.

Of the total $200 million per year for efficiency shown in Figures 6 and 7, buildings-related R&D gets only $50 million per year. Compared with annual energy expenditures of $200 billion, that's only 1/4000 of revenues, despite the dramatic economic growth achieved by efficiency improvements, the escalating threat of global warming, and the huge successes of simple, affordable energy technologies such as those we will discuss later in this chapter. Furthermore, the highly-fragmented industry invests very little in private R&D -- the majority of builders and component

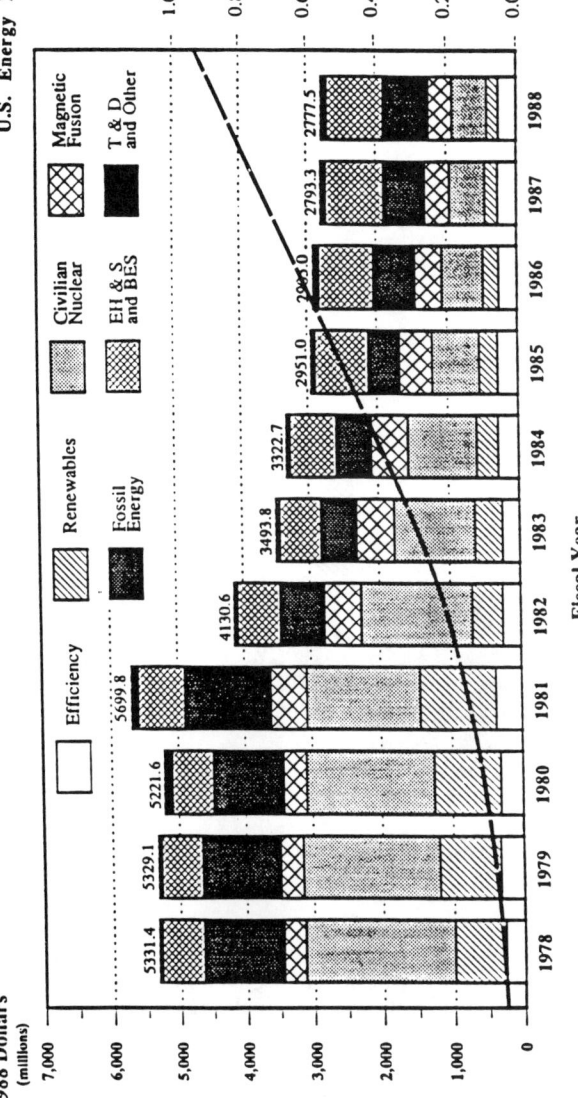

Figure 6. Combined energy technology R&D budgets (DOE, EPRI, GRI, and USNRC), in millions of 1988 dollars, for efficiency improvement; various energy sources; environment, health, and safety (EH&S) research; basic energy sciences (BES); and "T&D and other." The latter includes GRI and EPRI funds for transmission, distribution and planning and management functions.
Sources: DOE, FY 1988, derived from summaries of the House-Senate Conference Report on the DOE Budget, which appeared in Inside Energy, Jan. 4, 1988. DOE, FYs 1978-87, Appendix to the Budget of the U.S. Government, 1980-1989; Department of Energy Congressional Budget Request; Department of Energy Budget Highlights; Department of Energy Budget Formulation Office, personal communication. EPRI, Annual Reports of the Electric Power Research Institute; and Research and Development Plans. GRI, Five-Year Research and Development Plans and Program; and Gas Research Institute Annual Reports. USNRC, Appendix to the Budget of the U.S. Government, 1980-1989.

Source: Oak Ridge National Laboratory, Energy Technology R&D: What Could Make A Difference? (ORNL-6541/V1). May 1989.

Dashed line shows how energy efficiency R&D spending should have increased to about 1% of U.S. energy sales by 1988, comparable to R&D spending levels in mature industries.

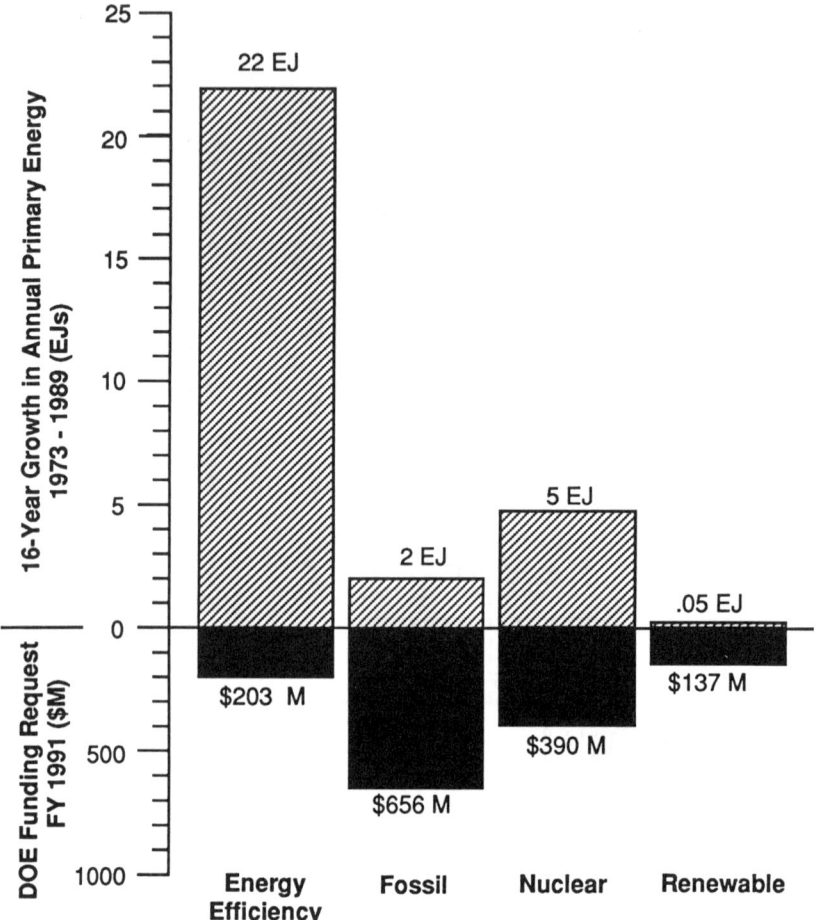

Figure 7. Growth in Annual Primary Energy Supplied or Saved by Energy Efficiency (1973-1989) vs. DOE FY 1991 Funding Request. 22 exajoules (EJ) of primary energy were saved as a result of energy efficiency improvements between 1973 and 1989. If efficiency is measured as the difference between actual 1989 energy use and energy use projected by a constant 1973 E/GNP, then full efficiency gain would be approximately 33 EJ. However, structural change is credited with 1/3rd of the change, resulting in an energy efficiency value of 22 EJ. Despite the fact that 22 of the 29.5 EJ of primary energy were provided through energy efficiency, this source is allocated only $200 million of the $1.4 billion budget.

Sources: Funding Data: DOE/MA-0400, *U.S. DOE Posture Statement and FY 1991 Budget Overview*. Energy Data: DOE/EIA-0384, *Annual Energy Review* 1989. Structural Change: L. Schipper, R.B. Howarth, H. Geller, *Annual Review of Energy*, Vol. 15, 1990 and U.S. Office of Technology Assessment, *Energy Use and the U.S. Economy*, June 1990.

manufacturers are simply not large enough to do research or to lobby effectively for increased governmental R&D funding.

In order to optimize the potential benefits of energy efficiency in buildings, the U.S. federal government should follow the example of Sweden which has a nationally-funded council for buildings research whose funding is equivalent to $1 billion U.S. The results of such an energy efficiency commitment are twofold. First, Sweden leads the world in energy-efficient buildings and second, its building sector runs an annual international trade surplus equivalent to $60 billion U.S. The U.S. building sector's performance is dismal in comparison, with a trade deficit of $6 billion per year.[15]

Research and Development Successes

During the past decade, significant strides have been made in the development of energy-efficient technologies. **Table 4** provides a summary of the characteristics and economics of three energy-efficient technologies that will be described in this section: high frequency electronic ballasts, compact fluorescent lamps, and low-emissivity windows.[16] These technologies, which were developed with DOE R&D funding, illustrate the remarkable benefits-to-R&D-cost ratios that can be realized with energy-efficient technologies.

Lighting: Fluorescent versus Incandescent

The visible light output of lamps is measured in units of "lumens" and the "efficacy" of lighting is measured in lumens/watt. Thus, 40, 60, or 100 watt incandescent lamps have efficacies of 13, 15, or 17 lumens/watt. Fluorescent lamps, by contrast, are 4 to 5 times more efficient. Thus, 34 or 40 watt fluorescent lamps (four feet long) are 60 to 80 lumens/watt (including ballast losses) depending on whether their ballasts (which start them and limit their current) are of the older "core-coil" type or of the recently developed high-frequency, electronic type.

Now we can understand U.S. lighting in 1973, just before the OPEC oil embargo provoked the sudden rise in electricity prices. With their 5-fold advantage in efficiency, fluorescent lamps had taken over commercial buildings, most frequently with "cool white" (sunlight colored) phosphors. A typical fixture, with 2 lamps of 40 watts each, yielded about 6000 lumens, equivalent to 5 incandescents of 75 watts each. The smallest fluorescent on the market at the time was the 22 watt "circline" (890 lumen) equivalent to a 60 watt fluorescent, but (with end losses) its efficacy was down to 40 lumens/watt. These fluorescents were fiercely resisted by homeowners and decorators who preferred the nostalgic, reddish, 15 lumen/watt incandescent.

Thus, in 1973, about 200 BkWh out of U.S electric sales of about 2000 BkWh, went to fluorescents (for 80% of the lumens) and 200 more BkWh went to incandescents (for most of the remaining lumens). As electricity prices shot up, two developments became inevitable: first, improve fluorescents (with high-frequency ballasts), and second, develop 20 watt compact fluorescents (with "warm red phosphors if necessary) to screw into the sockets then filled with 50-75 watt incandescents. This compact fluorescent lamp development was jump-started when high-frequency ballasts were shown to economically cut end losses and ballast losses in half.

Electronic Ballasts

Fluorescent lamps typically have 6 milli-torr of mercury (Hg) vapor, "buffered" by 1-3 torr of noble gas (argon or krypton). The Hg is about 1% ionized during operation, and emits mainly ultraviolet photons, which excite a "phosphor" coating on the inside of the glass tube, which in turn radiates visible light. In the middle of the

Table 4. Economics of Three New Energy Efficiency Technologies and Appliance Standards. An update of Tables 1 & 4 of Geller et al., *Ann. Rev. of Energy* 12, 1987.

	RESEARCH & DEVELOPMENT				STANDARDS
	HIGH FREQUENCY BALLASTS VS. CORE COIL BALLASTS	COMPACT FLUORESCENT LAMPS [1] VS. INCANDESCENTS	LOW-E (R-4) WINDOWS VS. DOUBLE GLAZED WINDOWS Per small window (10 ft^2)	TOTAL	REFRIGERATORS AND FREEZERS '76 base case vs. '85 CA Stds.
1. UNIT COST PREMIUM					
a. Wholesale	$8	$5	$10		
b. Retail	($12)	($10)	($20)		($100)
2. CHARACTERISTICS					
a. % Energy Saved	33%	75%	50%		66%
b. Useful Life	10 years	3 years	20 years		20 years
c. Simple Payback Time (SPT)	2 years	1 year	2 years		1 year
3. UNIT LIFETIME SAVINGS					
a. Gross Energy	1330 kWh	440 kWh	10 MBtu		24,000 kWh
b. Gross $	$100 [2]	$33 [2]	$70 [2]		$1800
c. Net $ [3b-1a]	$92	$28	$60 [3]		$1700
d. Gross Equivalent Gallons [4]	100	40	80		1920
e. Miles in 25 mpg car	2500	1000	2000		48,000
4. SAVINGS 1985-1990					
a. 1990 Sales	3M	20M	20M		not
b. Sales 1985 through 1990	8M	50M	50M		ramping
c. Cum. Net Savings [4b x 3c]	$750 M	$1.4B	$3B	$5B/5yr	up
5. SAVINGS AT SATURATION					
a. U.S. Units	600M	750M	1400M		100M
b. U.S. Annual Sales	60M	250M	70M		6M
c. Annual Energy Savings [5b x 3a]	80 BkWh	110 BkWh	0.3 Mbod		144 BkWh
d. Annual Net $ Savings [5b x 3c][5]	$5.5B	$7B	$4B	$16.5B/yr	$10 B
e. Equivalent power plants [6]	16 "plants"	22 "plants"			29 "plants"
f. Equivalent offshore platforms [7]	45 "platforms"	60 "platforms"	30 "platforms"		78 "platforms"
g. Annual CO$_2$ savings [8]	55 Mt	80 Mt	18 Mt	153 Mt	100 Mt
6. PROJECT BENEFITS					
a. Advance in Commercialization	5 years	5 years	5 years		5 years
b. Net Project Savings [6a x 5d]	$27.5 B	$35B	$20B	$82.5B	$50 B
7. COST TO DOE FOR R&D	$3M	$0 [9]	$3M	$6M	$2M
8. BENEFITS/ R&D COST [6b/7]	9000:1		6500:1	14,000:1	25,000:1

From: "The Role of Federal Research and Development in Advancing Energy Efficiency," Statement of Arthur H. Rosenfeld before James H. Scheuer, Chairman, Subcommittee on Environment, Committee on Science, Space, and Technology, U.S. House of Representatives, April 1991.

---

[1] Calculations for CFLs based on one 16-watt CFL replacing thirteen 60-watt incandescents, burning about 3300 hours/year, assuming that a CFL costs $9 wholesale, or $5 more than the wholesale cost of thirteen incandescents. For retail we take $18 - $8.
[2] Assuming price of 7.5¢/kWh for commercial sector electricity and a retail natural gas price of $7/MBtu (70¢/therm).
[3] For hot weather applications where low-e windows substantially reduce cooling loads, air conditioners in new buildings can be down-sized, saving more than the initial cost of the low-e window.
[4] Assuming marginal electricity comes from oil or gas at 11,600 BTU/kWh, thermally equivalent to 0.08 gallons of gasoline.
[5] Net annual savings are in 1990 dollars, uncorrected for growth in building stock, changes in real energy costs, or discounted future values. See Geller et al., Table 1.
[6] One 1000 MW baseload power plant supplying about 5 BkWh/year.
[7] One offshore oil platform = 10,000 bod. To convert "plants" burning natural gas to "platforms": 1 "plant" = 27,000 bod = 2.7 "platforms." Alaska National Wildlife Refuge, at 0.3 Mbod, is equivalent to about 30 "platforms."
[8] 1989 U.S. emissions of CO$_2$ were 5000 Mt.
[9] Descended from high-frequency ballasts (only DOE assistance was in testing).

discharge column, the conversion of plasma energy to ultraviolet is very efficient, but at each end there are voltage drops (anode and cathode "falls") adding up to about 15 volts, as ions and electrons drift into the electrodes to produce heat and not light. By raising the exciting frequency from 60 Hz (core-coil ballast) to 20-50 kHz (electronic ballast), the ion and electron drift distances are greatly reduced, the 15 volt end losses drop to about 8 volts and the efficacy of a 4 foot lamp rises 10 to 15%. In addition, the electronic ballasts are themselves much more efficient than core-coil, so the system efficiency rises another 10%, for a total gain of about 25%. Specifically, the heat dissipations of ballasts which operate *pairs* of 40 watt lamps are as follows: an outmoded "standard" core-coil ballast is 16 watt (i.e. an additional 20%), an "efficient" core-coil ballast is down to 10 watts, and an electronic ballast is only 4 watts.[17,18,19]

A further benefit of electronic ballasts are that they are easier to control electronically, permitting "daylighting," i.e. the practice of dimming lighting to save electricity when daylight is available. This raises the system efficiency of electronic ballasts, averaged over an entire floor of an office building, easily 30 to 40% above undimmed "standard" ballasts.

The energy-efficient electronic ballasts described above were developed through DOE-sponsored research at LBL in the late 1970s. Electronic ballasts are now commercially available for about $15 each wholesale and, over their 10 year lifetime, save 1330 kWh and $100 (See Table 4). These savings are equivalent to 100 gallons of gasoline, enough to drive 2500 miles in a 25 mpg car. Between 1985 and 1990, 8 million electronic ballasts were sold in the U.S. Based on the net lifetime savings of $85 per ballast, cumulative net lifetime savings for these 8 million ballasts is $680 million. It is expected that 600 million electronic ballasts, saving production of 80 BkWh, emissions of 55 Mt CO_2, and expenditures of $5.5 billion annually, will have been sold when market saturation is reached. The initial DOE project to develop electronic ballasts cost $3 million and is estimated to have advanced commercialization by 5 years, for a net project savings of $25 billion. This represents over an 8000:1 return on DOE's investment.

Compact Fluorescent Lamps (CFLs)

The economics of CFLs are shown in **Table 5**. An individual CFL rapidly pays for itself through reduced energy bills. For example, one 16-watt CFL replaces a series of about one dozen 60-watt incandescents since it burns 12 times longer than the incandescents. This CFL would save 440 kWh and about $33 in electricity costs over its 40 month life in a commercial building.

A modern, automated CFL production plant costs $7.5 million and can produce six million lamps annually, each of which will save 440 kWh over its service life, for a total savings of approximately 2.5 BkWh per year, equivalent to the sales of a 500 MW intermediate or baseload power plant that costs up to $1 billion to construct and $200 million per year to operate.

Within a decade CFLs will penetrate enough of the U.S. market to save over half of the 200 BkWh used annually by incandescents. As shown in Table 4, when CFLs have saturated the market they will save production of 110 BkWh, emissions of 80 Mt CO_2, and expenditures of $7 billion annually. DOE spent no R&D money on CFLs -- they descended directly from the development of electronic ballasts -- but because of the electronic ballasts, commercialization of CFLs was also advanced by 5 years for a net savings of $35 billion. Thus, DOE R&D expenditures of $3 million for electronic ballasts actually resulted in total savings from advancing commercialization of electronic ballasts and CFLs of $62.5 billion, an incredible 20,000:1 benefit-to-R&D-cost ratio.

Table 5. Economics of a 16-Watt Compact Fluorescent Lamp (CFL) Assuming 1 kWh Costs 7.5¢.

	Prices			
	Wholesale[1]		Retail	
	Low	High	Low	High
Price of 1 CFL	$7.00	$10.00	$9.00	$20.00[2]
Price of Initial Incandescent	$ .75	$ .75	$ .75	$ .75
Net First Cost of CFL	$6.25	$9.25	$8.25	$19.25
Monthly Savings Using CFL (Electricity Savings plus Avoided Incandescent Cost)	$1.06	$1.06	$1.06	$1.06
Simple Payback Time (months)[3]	6	9	8	18
Lamp Life (months)	40	40	40	40

Assumptions

- **CFL/Incandescent Ratio:** One 16-watt CFL replaces thirteen 60-watt incandescents. One CFL lasts approximately 10,000 hours; thirteen incandescents at 750 hours each last 9750 hours.

- **Lifetimes:** We estimate that the lifetime of approximately 10,000 hours is spread over 40 months at 250 hours/month for typical usage in a commercial or office space.

- **Monthly Electricity Savings:** Replacing a 60-watt incandescent with a 16-watt CFL saves 44 watts at the meter. Over its lifetime of 10,000 hours, the CFL saves 440 kWh. Using the average price of electricity of 7.5¢/kWh, the CFL saves 440 x 7.5¢ = $33.00. This results in a monthly savings of $ .83.

- **Monthly Avoided Incandescent Cost:** The initial incandescent costs $ .75 and the remaining 12 incandescents are calculated to cost $ .23/month (12 x $ .75 = $9.00 divided by 40 months).

---

[1] "Wholesale" is included above because innovative methods are being used by some utilities to make CFLs available to customers at wholesale prices, allowing the customers to realize large savings.
[2] For less developed countries, like India, exorbitant import fees make this cost $35.00.
[3] Simple payback time (STP) is the interval needed to recoup the money invested in an energy-efficient technology through reduced energy bills. STP ignores discount rates.

Windows From a Physics Perspective

Heat losses and gains through windows are responsible for 25% of all heating and cooling requirements in U.S. buildings. The fossil fuel equivalent of the heat loss alone is the 1.8 million barrel per day (Mbod) output of the Alaskan pipeline, or of Kuwait before 1991. If we understand how windows work thermally, we can easily see how to save half or all of this 1.8 Mbod.

Heat flow is typically measured in watts per square meter (W/m²) in SI (Systeme Internationale, a subset of metric) units, and if linear in temperature, is written:

$$q = U \Delta T = \frac{1}{R} \Delta T, \qquad (10)$$

where U is the conductance (W/m²K) and R is the resistance (m²K/W).

In the U.S., where the IP system (inch, pound, Btu, etc.) is used, U_{IP} is expressed in units of Btu/hr ft² °F. The conversion factors are: $U_{SI} = 5.68\ U_{IP}$, and $R_{SI} = R_{IP}/5.68\ R_{IP}$. As an example, 4-inch stud/fiberglass insulated walls are R-11, i.e. have $R_{IP} = 11$, and 9 inch ceiling insulation is R-19. Converted to SI, R-11 becomes $R_{SI} = 2\ \Omega$, using Ω as a shorthand for m²K/W.

Figure 8 shows the heat leak between a warm indoor room at T_i (at right) and a cold outdoors at T_o (at left).[20] (The convention that indoors is at the right comes from a more complete description of a window, with the sun on the left, shining through the window from left to right.) Glass itself is a poor thermal insulator; 1/8" window glass typically has a resistance of only 4 milliohms. Glass is also nearly "black" to heat at room-temperature (T_o or T_i), i.e. its emissivity, ε, is 0.84, so that heat radiates easily from all glass surfaces. Thus, the thermal resistance of a window is determined almost entirely by the resistance of air and by Plank's constant, σ.

Radiation across a gap is given by:

$$q_{rad} = \frac{\sigma (T_2^4 - T_1^4)}{1/\varepsilon_1 + 1/\varepsilon_2 - 1}, \qquad (11)$$

where ε_1 and ε_2 are the emissivities shown in Figure 8. This is linearized by writing $T_2 = T_1 + \Delta T$ to get:

$$q_{rad} = \frac{3\sigma (T_1^3) \Delta T}{1/\varepsilon_1 + 1/\varepsilon_2 - 1}, = \frac{\Delta T}{R_{rad}}. \qquad (12)$$

Setting $T_1 = 255$ K, we get:

$$R_{rad} = 0.2\ (\frac{1}{\varepsilon_1} + \frac{1}{\varepsilon_2} - 1). \qquad (13)$$

For uncoated glass, $\varepsilon_1 = \varepsilon_2 = 0.84$, and:

$$R_{rad}\ (uncoated) = 0.2\ (1.4) = 0.28, \qquad (14)$$

which is "worse" than the parallel R(conduction and convection) shown in Figure 8 as Rc+c = 0.5 Ω.

Low-Emissivity (Low-E) Windows

Low-E windows follow the thermos bottle approach by using a thin, metallic mirror on one of the gap surfaces. As we shall see below, there are many semiconductors (like tin oxide) which have a high enough electron density so as to act nearly like a mirror to heat ($\varepsilon=0.1$) but transmit visible light. The technology of depositing low emissivity films on plastic was perfected in a collaboration between LBL and Southwall Technologies, which trade-marked the nice term "heat mirror," leaving the rest of the industry to use the words "low-E."

If ε_1 or $\varepsilon_2 = 0.1$, equation (13) becomes:

$$R_{rad} = 0.2 (10.2) = 2\Omega, \qquad (15)$$

which is now 4 times as good as Rc+c. By coating *both* surfaces one could achieve Rrad = 4 Ω, but it's better to put the extra expense into filling the air gap with a heavier gas. Gas conduction is proportional to $1/\sqrt{m}$, where m is the atomic number. Argon, for example, will raise the gap resistance by about 1/3rd. One chooses a monatomic gas to avoid the heat capacity associated with rotational states.

To complete our discussion of Figure 8 we must still address the heat transfer to the outer surface. From the room to the inner glass, we have labelled $R_i = 0.13$ Ω. For uncoated glass this heat transfer is about half radiative, half convective. Outdoors is windy, so conduction overwhelms radiation, and Figure 8 shows $R_o = 0.03$ Ω. Now we can calculate R for an air-filled, low-e window. From equation (15), $R_{rad} = 2$, which in parallel with Rc+c = 0.5 gives:

$$R_{gap} = \frac{2 \times 1/2}{2 + 1/2}, = \frac{2}{5} = 0.4. \qquad (16)$$

Then:

$R_{window} = R_{outer} + R_{gap}$, (i.e. $R_{window} = 0.03 + 0.13 = 0.4 = 0.56$ Ω),

and in IP units $R_w = 3.2$ Ω_{IP}, called "R-3.2." This is significantly better than single glazing ($R_{SI} = 0.16$, $R_{IP}=1$), but still poor compared to a 4 inch wall at $R_{IP} = 11$. An Argon fill adds about 30% to the total resistance of the window and is becoming standard with the major window manufacturers. The latest trend is to go to "triple glazing," by stretching two thin films of low-E plastic inside the gap. Triple glazing with gas fills produce "superwindows" with $R_{IP} = 6 - 9$ which is nearly as good as a stud wall. But a wall can only insulate, while a window admits solar heat during the day. The result is that superwindows are net energy gainers facing in any direction in any part of the U.S.

Economics of Low-E Windows

The economics of low-E windows are impressive: their payback time is only two years and they are rapidly saturating the market. As shown in Table 4, at saturation, 70 million one square meter low-E "windows" will be sold in the U.S. every year. The <u>net</u> annual savings from these windows will be $4 billion. In the past five years, 50 million of these windows were sold in the U.S.; they have already saved $3 billion in cumulative avoided energy bills. One of these low-E windows costs $10 wholesale (or $20 retail) more than a typical thermopane window, but saves 10 to 15 million Btu over its 20 to 30 year lifetime, worth approximately $70 in avoided energy bills.

Spectrally Selective Windows - Plasma Frequency

The previous discussion focused on low-E coatings for cold weather, where all that was needed was for ε to be small (≤ 0.1) for "heat" (with wavelength λ > 5 μ), but approach 1.0 for light (λ < 0.8μ). But windows can also be made spectrally selective, creating the opportunity to use them more effectively in hot climates. The energy in sunlight is about half visible (λ < 0.8μ) and about half in visible near-infrared heat. In winter, this near-infrared is welcome, but in hot weather it must be reflected along with the far infrared. To "take the heat out of sunlight" [21] the transition in ε must be moved very close to 0.8μ as shown in **Figure 9**.[22] Not only does this save air conditioning bills, but it also reduces the first cost of a new building because the designer can down-size expensive chillers.[23] A visually transparent but selective window is more more desirable than the conventional reflective "solar control" glazing used universally on commercial buildings because they do not darken the interior space and thus avoid the need for artificial lighting even near the windows.

Also, in hot climates vernacular architecture often relies on vertical and horizontal overhangs to block incoming sunlight to reduce solar gains and air conditioning needs. Because spectrally selective windows solve the problem of solar heat gains, these overhangs are no longer essential and greater application of daylighting principles is possible. Daylighting saves even more electricity by reducing demand for lighting. In fact, the effect of using spectrally selective windows in hot climates is so dramatic that it calls for a new "least cost" approach to building design that adequately addresses these interactions.[24]

The basic physical idea behind a low-E or spectrally selective coating is the optical response of conduction electrons in a semiconductor or metal. This can be approximated by the dielectric function :

$$\varepsilon(\omega) = \varepsilon_\infty [1 - (\frac{\omega_p}{\omega})^2] = (\tilde{n})^2 \qquad (17)$$

where ñ is the index of refraction which governs wave propagation.[25] For frequency ω greater than the plasma frequency ω_p, ε is positive, the refractive index ñ is real, and waves can propagate in the material. For $\omega < \omega_p$, ε < 0, the refractive index is imaginary, so a wave incident on the material is reflected.

The most familiar example of this transition is the difference in propagation of electromagnetic waves in the ionosphere. Low-frequency radio has $\omega < \omega_p$, and ñ is imaginary, so the waves are reflected and will bounce between the earth's surface and the ionosphere, all around the world. Short-waves (fm band, tv, and microwave) have ñ real, easily penetrate the ionization, and are lost; hence to receive these high frequencies, we have to be within line of sight of the transmitter.

The plasma frequency ω_p depends on the conduction electron concentration n through:

$$\omega_p^2 = \frac{4\pi n e^2}{m\varepsilon_\infty} \qquad (18)$$

Here e is the electronic charge, m is the effective mass, and ε_∞ is the background dielectric constant from the bound charges. In a metal, n is typically $10^{22} cm^{-3}$, and ω_p falls in the ultraviolet. In a heavily-doped semiconductor, n can now be 10^{20} to $10^{21} cm^{-3}$, with ω_p in the near-infrared. This is shown for a tin-doped indium oxide

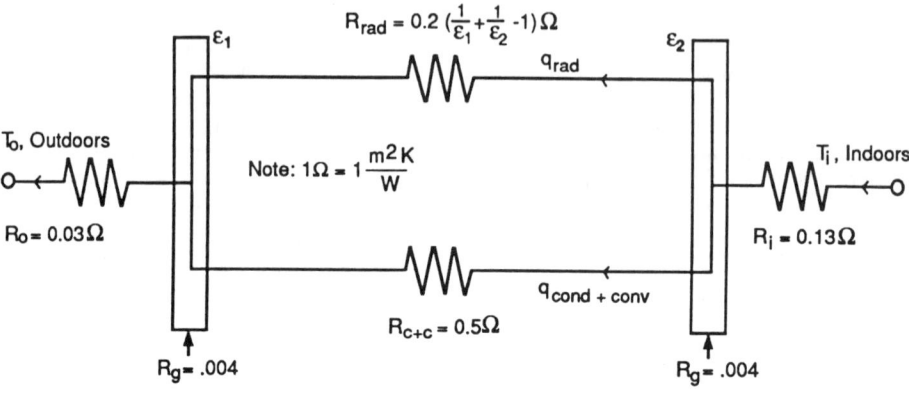

Figure 8. Thermal Circuit for a Double-Glazed Window.

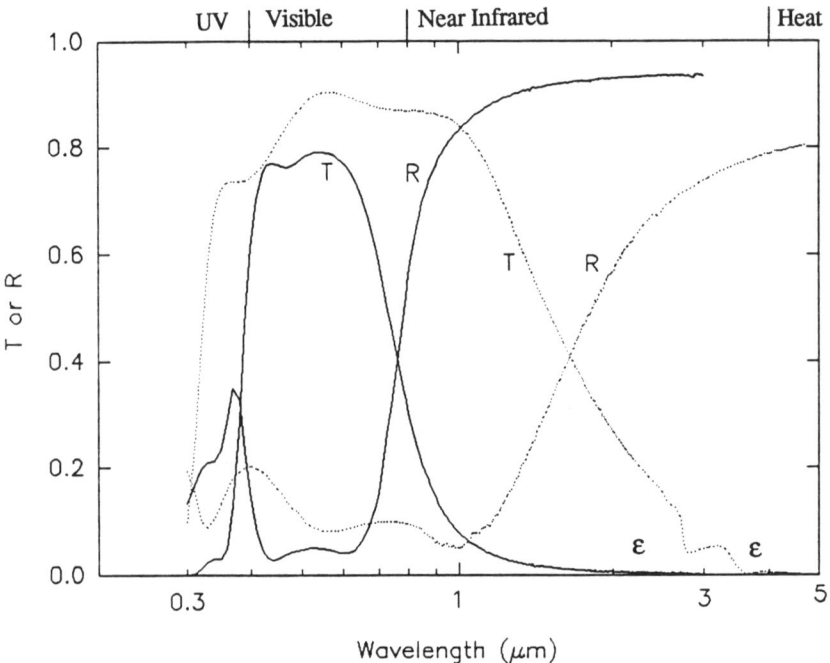

Figure 9. Transmittance (T or ε) and reflectance (R), for normal incidence, of two samples of coated glass. Dotted line: In_2O_3:Sn-coated glass manufactured by Donnelly Corporation, Holland, Michigan. Solid line: Multilayer-coated glass manufactured by Cardinal IG, Minneapolis, Minnesota 55426.

coating in Figure 9. The reflectance changes over a range of a few μm near the plasma frequency due to scattering and trapping of the electrons.

For a sharper roll-off and better spectral selectivity, multi-layer coatings are used. One layer is a very thin (~ 10nm) metal film, often Ag. In this case ω_p is in the ultraviolet, but the magnitude of ε changes fairly slowly near ω_p. As a result, for a thin metal film alone, the reflectance changes slowly from nearly 0 in the ultraviolet to nearly 1 in the near-infrared. When the metal film is sandwiched between dielectric layers, thin-film interference effects can sharpen the transition from high transmittance to high reflectance. A five-layer coating can give a close approximation to a step at the visible-near infrared boundary, as illustrated in Figure 9.

Appliance Standards

Along with R&D funding, legislatively-enacted standards also perform the function of advancing technology development. An example of the benefits of such standards in improving the energy efficiency of appliances is provided in the last column of Table 4. This column illustrates the energy and economic savings attributable to the 1985 California refrigerator and freezer appliance standards when compared to 1976 base case appliances. Manufacturers complied nationwide with the California standards; annual energy saving are now over 140 BkWh, valued at $10 billion and equivalent to the electricity produced by 29 baseload (1000 MW) power plants.

In many cases, appliance standards are the easiest way to remove or discourage inefficient products from the market. A recent study compared various policy options and found that standards result in more savings than other methods, including tax credits, rebates, and consumer education.[26] Currently, the Federal government has set energy efficiency standards for many home appliances and fluorescent ballasts. Significant additional energy and CO_2 savings may be achieved by standards for other products such as motors, lamps and lighting fixtures, office equipment, windows, and commercial HVAC equipment.

Building Energy Performance Standards

Performance standards for new buildings have yielded amazing results.[27] In 1975, the American Society of Heating, Refrigerating and Air-Conditioning Engineers (ASHRAE) announced its ASHRAE 90 series of voluntary standards. By 1978 California had adopted tougher mandatory standards that have about 3-year payback times and which are enforced. Most states have adopted the ASHRAE standards, but enforcement is inconsistent, with notable offenders being the federal government and some state governments.

The savings are remarkable. In 1975 a typical new office tower annually consumed 170 kBtu of natural gas and 30 kWh of electricity per square foot. This consumption has now dropped to 10 kBtu (a savings of 95%) and 10 kWh (a savings of 2/3rds) and the buildings are better designed and more comfortable. The California Energy Commission estimates that annual savings from new California buildings alone will reach $1 billion by 1995, so we can significantly spur the U.S. economy by updating and enforcing standards in all states and for federal buildings.

LATEST DEVELOPMENTS AND DIRECTIONS FOR FURTHER ENERGY EFFICIENCY EFFORTS

Advanced Windows

As discussed above, a new generation of "superwindows," rated from R-6 to R-10, is advancing the performance of windows even farther. Superwindows increase a window's performance from R-4 to R-8 by stretching two plastic films with low-E coatings between the double glazing of an R-4 window and filling the air gaps with argon or krypton. When the window frame is also improved, the advanced window's performance is increased to R-8. In field tests, superwindows outperform 6-inch thick R-19 walls because they let sunlight into the building during the day and block heat loss to the outdoors at night. The superwindow is a net energy gainer, whereas the surrounding wall only prevents heat loss.[28] In addition, superwindows minimize interior condensation and reduce the damage to furnishings by blocking ultraviolet rays. Although these windows can cost 20% to 50% more than conventional windows, their initial cost is repaid in about four years of avoided energy bills.

Electrochromic windows that control the flow of radiation are now under development. These windows switch from clear to white under electronic or thermal control. Initially, these windows will appear in automobile applications, but eventually the new windows will penetrate the buildings market.

Advanced Insulation

Optimum U.S. homes have walls with 6 inches of insulation (R-19) contained within an outer insulating sheath (for a total of R-24) in contrast to standard homes which use only a scant 4 inches of insulation (R-11). Some builders, such as Bigelow Homes in Chicago, have so much confidence in the performance of optimum and super-insulated residences that they offer to pay all owners' annual energy bills that exceed $100 for townhouses and $200 for private homes. In 1989, the contest for the homeowner with the lowest heating bill was won by a customer who paid only $24 for heating for an entire year.

Concern over both ozone depletion and global warming has led to regulations banning the manufacture of chlorofluorocarbons (CFCs), whose most widely recognized use is as a refrigerant. Less well known is the fact that a typical U.S. refrigerator contains about 1/2 lb. of CFCs in the compressor, and twice that much in the CFC-blown highly-insulating foam that fills the shell of the box. Hence, there are now many parallel R&D efforts underway to replace this foam.

At least four kinds of advanced super-insulation are currently being developed with the support of the Department of Energy. One form of superinsulation is an outgrowth of advances in low-E window technology. Gas-filled panels (GFPs), which have been developed specifically in response to the need to replace CFC blown foam insulation, are an assembly of reflective foils that simultaneously minimize radiative, conductive, and convective heat transfer. Multiple layers of highly reflective metallized polymer film compartmentalize the interiors of the GFPs, virtually eliminating radiative heat transfer. Much thinner, crumpled film is inserted between these parallel layers to minimize convection and further decrease radiative heat transfer. In addition, to reduce conduction, a low thermal conductivity gas (such as argon or krypton), or gas mixture, is encapsulated at atmospheric pressure within sealed panels. These GFPs may someday be applied in HVAC insulation, hot water heaters, swimming pool and spa covers, refrigerated transport walls, airplane bodies and even homes. Compared with fiberglass at R-3.5 per inch, GFPs have tested at R-7 to R-13 per inch, and an R-15 performance is anticipated.[29]

"Aerogel" is a transparent or translucent insulating material also developed at LBL. Aerogel can be used in windows and skylights as well as in appliances. In its opaque, evacuated form, it has reached R-30 per inch.

Low-Flow Showerheads

The typical showerhead flows at 5 gallons per minute (gpm) and the typical shower is 5 minutes long. (The showerhead industry claims that normal showerheads use 6 to 10 gpm. A recent study conducted in Yakima, Washington found that the average showerhead used 3.1 gpm.[30]) The fuel required for heating the water from 55°F to 110°F is 600 Btu/gallon; annually, this is the equivalent of 44 gallons of oil. The annual cost of natural gas for this typical shower is $32.

Low-flow showerheads use 2.5 gpm, half the amount of conventional showerheads, so they save $16/year, with a present value (20 years, 6% real interest) of nearly $200. Based on initial costs for a replacement showerhead of between $5 and $10, simple payback times are well under a year.

Extrapolating to the entire U.S., these water- and energy-conserving showerheads will reduce annual U.S. fuel consumption from the equivalent of 0.3 million barrels of oil per day (MBod) to 0.6 MBod, equivalent to the expected oil production rate from the Alaskan National Wildlife Refuge.

Improving Recessed Lamp Fixtures

In U.S. commercial buildings there are about 300 million recessed fixtures for incandescent lamps. The advantages of replacing incandescent lamps with compact fluorescents have been discussed above. But a problem remains: 100-200 million of the incandescent fixtures are still equipped with all-too-familiar black "microgroove" collars that were initially designed to keep a bright filament out of sight. Such an absorber allows omni-directional light to escape only downward in a 45 degree cone, and necessarily absorbs three-quarters of that light. Thus, if the lamp is 60 watts, the black collar probably absorbs about 40 watts. If the lamp burns 3000 hours per year, the collar absorbs 120 kWh, worth about $10 per year. Multiplied by 100-200 million, the annual cost of this wasteful technology is $1-2 billion per year.

These collars persist even though the direct view of a filament disappeared in the 1940s, because lighting designers still emphasize style over energy costs. If designers were trained in energy efficiency, 3 to 4 baseload power plants would be liberated from generating the light and heat absorbed by such fixtures.

Whitening Surfaces and Planting Urban Trees

Most U.S. cities are 3 to 6 degrees Fahrenheit (°F) hotter on summer afternoons than they were 50 years ago. These summer "urban heat-islands" arise because heat-absorbing asphalt and buildings have replaced trees and fields. Downtown Los Angeles is 7 °F hotter than in 1940, and is heating up by 1 °F every 8 years.

An unshaded U.S. home with a dark or terra-cotta colored roof typically needs 2 to 3 kilowatts (kWs) for air-conditioning. This costs the average homeowner between $100 and $300 annually. In contrast, a shaded home with a light-colored exterior generally uses only half this amount of electricity.

Such simple, low-technology mitigation measures as planting urban shade trees and whitening surfaces of roofs, streets, and sidewalks can garner tremendous savings when their impact is aggregated for an entire city, state, or region. In Los Angeles alone $100,000 per hour in peak summer air-conditioning costs would be saved.[31] Lowering temperatures in urban heat islands by 5 °F could save U.S. rate payers about $1 billion per year in avoided air-conditioning. And since smog "cooks" slower at reduced temperatures, eliminating heat islands would significantly lower levels of smog. Of course, such temperature reductions would also reduce (by 1/2%) the power plant CO_2 emissions associated with generating electricity.

Daylighting

Designing buildings that maximize sunlight to illuminate interiors is a low-technology approach to reducing building energy demand while meeting building lighting needs. This practice is widespread in Europe, where every office has an outside window, but is still under-utilized in the U.S. Just one square foot of direct sunlight can actually illuminate 200 square feet of interior space, if evenly distributed through a skylight or clerestory window.[32]

Thermal Energy Storage

In a well-insulated building with low-E, well-managed windows, thermal storage can economically save most of the energy now used for heating and cooling. For example, a super-insulated home with exposed masonry floors and perhaps with a south-facing "Trombe" wall will hold heat for days. If its windows are concentrated towards the south, and are well-managed, it can get through the winter comfortably on the free heat from sun, appliances, and occupants, plus space heat corresponding to about 1 Btu/ft^2/degree day (°F) compared with 8 Btu for a typical pre-1975 U.S. home, or 4-5 Btu for newer homes. Also, in most of the U.S., the same super-insulated home, with a white roof, well-managed windows, and a whole-house fan to draw in cool night air, can remain comfortable all day with no air-conditioning or with one or two small window units.

The same thermal energy storage techniques apply to offices, where in fact there is more free heat than in homes. In Sweden, a modern office building is designed to store free heat during a winter week, cool slightly over the weekend, and use the stored heat to warm itself up on Monday morning. Accordingly, many offices no longer request a connection to the Stockholm district heating system. Similar strategies of night cooling are used during the summer, and greatly reduce the demand for peak power.[33]

CONCLUSION

Numerous studies show that energy-efficient technologies and strategies such as these can save an incredible amount of energy and slow the annual growth in CO_2 emissions that contribute to global warming. Even more impressive, these important environmental savings can be realized at a net economic benefit. Adequate R&D funding is imperative to continue to develop technologies and strategies to capture these savings. Then, effective energy efficiency policies that promote widespread technology transfer and implementation are essential.

ACKNOWLEDGEMENT

This work was supported by the Assistant Secretary for Conservation and Renewable Energy, Office of Building Technologies, Buildings Division, of the U.S. Department of Energy under Contract No. DE-AC03-76SF00098.

We would like to thank Dariush Arasteh, Susan Reilly, and Dave Wruck, all of the LBL Windows and Daylighting Program, for help with the section on the physics of windows.

REFERENCES

1. L. Shipper, R.B. Howarth, H. Geller, *Ann. Rev. Energy,* 15, 1990 and U.S. Office of Technology Assessment, *Energy Use and the U.S. Economy,* (U.S. GPO, Washington, D.C., 1990).

2. A. Meier, J. Wright, A. Rosenfeld, *Supplying Energy Through Greater Efficiency: The Potential for Conservation in California's Residential Sector,* (University of California Press, Berkeley, 1983).

3. Personal communication, I. Turiel, Lawrence Berkeley Laboratory.

4. J. Koomey, C. Atkinson, A. Meier, J. McMahon, S. Boghosian, B. Atkinson, I. Turiel, M. Levine, B. Nordman, and P. Chan, *The Potential for Electricity Efficiency Improvements in the U.S. Residential Sector,* LBL-30477, (Lawrence Berkeley Laboratory, Berkeley, 1991).

5. A. Rosenfeld, C. Atkinson, J. Koomey, A. Meier, R. Mowris, and L. Price, *A Compilation of Supply Curves of Conserved Energy for U.S. Buildings* (Lawrence Berkeley Laboratory, Berkeley, 1991). This paper contributed to the National Academy of Science *Policy Implications of Greenhouse Warming: Report of the Mitigation Panel.* (National Academy Press, Washington, D.C., 1991).

6. A. Faruqui, *Efficient Electricity Use: Estimates of Maximum Energy Savings,* prepared by Barakat & Chamberlin, Inc. for the Electric Power Research Institute, CU-6746, (Electric Power Research Institute, Palo Alto, 1990).

7. National Research Council, *Confronting Climate Change: Strategies for Energy Research and Development* (National Academy Press, Washington, D.C., 1990).

8. Solar Energy Research Institute (SERI), *A New Prosperity: Building a Sustainable Energy Future* (Brickhouse Publishing, Andover, MA, 1981).

9. A. Meier, J. Wright, A. Rosenfeld, *Supplying Energy Through Greater Efficiency: The Potential for Conservation in California's Residential Sector* (University of California Press, Berkeley, 1983).

10. A. Rosenfeld, C. Atkinson, J. Koomey, A. Meier, R. Mowris, and L. Price, *A Compilation of Supply Curves of Conserved Energy for U.S. Buildings* (Lawrence Berkeley Laboratory, Berkeley, 1991).

11. F. Krause, A. Rosenfeld, M. Levine, *Analysis of Michigan's Demand-Side Electricity Resources in the Residential Sector*, LBL-23025, (Lawrence Berkeley Laboratory, Berkeley, 1988).

12. F. Krause, J. Busch, J. Koomey, *Incorporating Global Warming Risks in Power Sector Planning: A Case Study of the New England Region*, LBL-31019, Draft, (Lawrence Berkeley Laboratory, Berkeley, 1991).

13. J. Edmonds, W. Ashton, H. Cheng, M. Steinberg, *A Preliminary Analysis of U.S. CO2 Emissions Reduction Potential from Energy Conservation and the Substitution of Natural Gas for Coal in the Period to 2010*, DOE/NBB-0085 (U.S. DOE Office of Energy Research, Washington, DC, 1989).

14. Oak Ridge National Laboratory, *Energy Technology R&D: What Could Make A Difference? Part I: Synthesis Report,* ORNL-6541/V1 (Oak Ridge National Laboratory, Oak Ridge, May 1989).

15. A. Rosenfeld, "Energy-Efficient Buildings in a Warming World," *Proceedings of the Massachusetts Institute of Technology Conference: Energy and the Environment in the 21st Century,* March 1990, (MIT, Cambridge, 1990).

16. A. Rosenfeld, "The Role of Federal Research and Development in Advancing Energy Efficiency," *Hearing on DOE Conservation Budget Request, Before James H. Scheuer, Chairman, Subcommittee on Environment, Committee on Science, Space, and Technology, U.S. House of Representatives*, Washington, D.C, 1991.

17. S. Berman, "Energy and Lighting," *American Institute of Physics Conference Proceedings* No. 135 (AIP, New York, 1985).

18. J.W.F. Dorleijn and A.G. Jack, "Power Balances for Some Fluorescent Lamps," *Journal of the Illuminating Engineering Society*, (Fall 1985).

19. R. Verderber, *Status and Applications of New Lighting Technologies*, LBL-25043, (Lawrence Berkeley Laboratory, Berkeley, 1988).

20. D. Arasteh, M. Reilly, and M. Rubin, *A Versatile Procedure for Calculating Heat Transfer Through Windows*, LBL-27534, (Lawrence Berkeley Laboratory, Berkeley, 1989). The authors have also prepared a personal computer program for window designers: *WINDOWS 3.1*, LBL-25686, (Lawrence Berkeley Laboratory, Berkeley, 1988).

21. B. Davids, "Taking the Heat Out of Sunlight - New Advances in Glazing Technology for Commercial Buildings," Presented at the *American Council for an Energy-Efficient Economy 1990 Summer Study on Energy Efficiency in Buildings*, August 26-September 1, 1990.

22. S. Berman, and S. Silverstein, "Infrared-Reflecting Selective Surface Materials Which Can Be Useful for Architectural and/or Solar Heat Collector Windows," *American Institute of Physics Conference Proceedings* No. 25.(AIP, New York, 1975). For more information on spectrally selective glazings, see also: Howson, R., *Solar Optical Materials*, (Pergamon, Oxford, 1988) and C. Lampert, *Solar Energy Mater.* $\underline{6}$,1 (1981).

23. A. Gadgil, A. Rosenfeld, D. Arasteh, and E. Ward, *Advanced Lighting and Window Technologies for Reducing Electricity Consumption and Peak Demand: Overseas Manufacturing and Marketing Opportunities,* LBL-30389 (Lawrence Berkeley Laboratory, Berkeley, 1991).

24. A. Rosenfeld and L. Price, "Making the World's Buildings More Energy-Efficient," Presented at the *Technologies for a Greenhouse-Constrained Society Conference,* June 11-13, 1991, Oak Ridge, TN.

25. C. Kittel, *Introduction to Solid State Physics,* 5th edition, (Wiley, New York, 1976).

26. U.S. Congress, Office of Technology Assessment, *Changing By Degrees: Steps to Reduce Greenhouse Gases* (U.S. GPO, Washington, D.C., 1991).

27. A. Rosenfeld and D. Hafemeister, "Energy Conservation in Large Buildings," *American Institute of Physics Conference Proceedings* No. 135 (AIP, New York, 1985).

28. R. Bevington and A. Rosenfeld, "Energy for Buildings and Homes," *Scientific American,* 263:3 (September 1990).

29. B. Griffith, D. Arasteh, S. Selkowitz, "Gas-Filled Panel High-Performance Thermal Insulation," *Insulation Materials: Testing and Applications, Volume 2, ASTM STP1116,* (ASTM, Philadelphia, 1991). Also LBL-29401 (Lawrence Berkeley Laboratory, Berkeley, 1991).

30. B. Manclark, "Low Flow Showers Save Water? Who Cares?" *Home Energy,* (July/Aug., 1991).

31. H. Akbari, A. Rosenfeld, and H. Taha, "Summer Heat Islands, Urban Trees, and White Surfaces," Presented at *American Society of Heating, Refrigerating and Air-Conditioning Engineers January 1990 Meeting,* Atlanta Georgia. Also, LBL-28308 (Lawrence Berkeley Laboratory, Berkeley, 1990).

32. R. Bevington and A. Rosenfeld, "Energy for Buildings and Homes," *Scientific American,* 263:3 (September 1990).

33. A. Rosenfeld and D. Hafemeister, "Energy Conservation in Large Buildings," *American Institute of Physics Conference Proceedings* No. 135 (AIP, New York, 1985).

SCIENCE AND DIPLOMACY:
A NEW PARTNERSHIP TO PROTECT THE ENVIRONMENT

Ambassador Richard E. Benedick
World Wildlife Fund

INTRODUCTION

I am ever more intrigued by the intersection between two vital aspects of human affairs: science and diplomacy. It is an intersection that is not intrinsically obvious, but one that is becoming increasingly important in the real world of international relations. The efforts, for example, of the international community to protect the planet's environment rest upon modern scientific discoveries, theories, and understanding.

Environment and Foreign Policy

In a way this new element in foreign affairs is symbolized by my own personal situation as a senior fellow at the World Wildlife Fund on assignment from the State Department. Traditionally, the Department of State has sent foreign service officers as diplomats-in-residence to places like Georgetown, Brookings, Hoover Institution, or the Kennedy School, where they pursue traditional diplomatic preoccupations in politics, arms control, and economics. My assignment, however, represented the first time that a senior officer in the foreign service was assigned to an environmental organization: perhaps this may be a symbol of the coming of age of the environment in our foreign policy.

The experience outside the Department has been a rather exhilarating period of activity and reflection, of speaking and consulting and participation in conferences and symposia, and of considerable writing -- including my recent book which, even in its title Ozone Diplomacy, symbolizes the combination of science and foreign affairs.

We all know in fact that environmental issues are rising to the top of the foreign policy agendas of world leaders. Some developments have been surprising: former Prime Minister Thatcher of the United Kingdom, President Mitterand of France, and Pope John Paul II have in the last three years come out with significant declarations on the importance of protecting the environment. These are people who have not customarily been at the forefront of these issues. The leaders of the seven major industrialized countries, at their 1989 Paris Economic Summit meeting, actually devoted one-third of their communique to environmental affairs. I can remember when I was responsible for environment issues in the State Department. I would consider it a triumph to get a sentence or two in a Summit communique. I am happy to say that my immediate successor, Bill Nitze, was able to make one-third of the communique. My hat is off to him. Environment has even penetrated the bastions of economics at the Summit.

Gorbachev and the Environment

Perhaps the most surprising symbol of this growing importance in foreign policy agendas has been the commitment of President Gorbachev. With all the preoccupations that face him in presiding over the disintegration of the Soviet empire, he has taken time on several occasions to enunciate very strong concern about what is happening to the world's environment, including some eloquent statements at the United Nations' General Assembly.

For me personally, the most memorable demonstration of Gorbachev's interest in the environment occurred when he hosted an international conference of religious and spiritual leaders, parliamentarians, scientists, and environmental policy people such as myself. This meeting took place in January 1990. Usually these conferences in the winter are in places like Acapulco, so when I was invited to Moscow in January I said to myself this must be a serious conference. I was glad that I had attended, not only because of Gorbachev but because it turned out to be a very interesting and colorful assemblage: there were grand muftis and patriarchs, rabbis and mullahs, swamis and medicine men, with robes and turbans and feathers. All in all, an unusual conference.

As evidence of the seriousness with which he regarded this environmental group, Gorbachev invited over a thousand participants to the Kremlin on the concluding day for a reception. And there he gave a serious and eloquent speech that was full of substance and specific recommendations and also an admission of past errors of Soviet policy in the environmental field. It was a very impressive performance that made him look like a genuine environmental president.

Even more impressive, however, was that all of this occurred at a time of virtual civil war in the Soviet Union. The reception and speech were given on the night before troops struck in Azerbaijan to put down serious civil unrest. Gorbachev arrived for his speech at the Kremlin somewhat late and breathless and, of course, none of us knew at the time about the plans for Azerbaijan.

Recently I found out more about the circumstances of his late arrival and I am happy to be able to share with you how Gorbachev got to the Kremlin that night. As it happened, he had been meeting across town at military headquarters with his general staff to plan for the next morning's action in Azerbaijan. Time slipped past, and all of a sudden an aide slipped him a message that these parliamentarians and spiritual leaders were awaiting him in the Kremlin. Gorbachev dashed out to his limousine and ordered his driver to speed back to the Kremlin at once.

Now a little-known fact is that new regulations had been passed as part of perestroika, providing that Moscow traffic regulations were to be strictly enforced -- even for VIPs. As Gorbachev was late, he urged his driver to drive faster and faster. Finally the poor man said, "I'm sorry Mikhail Sergeyevich, I can't drive any faster. I will lose my license. I will lose my job." Gorbachev was by now so exasperated that he exclaimed: "All right, move over. I'll take full responsibility, I'll drive!" And he got behind the wheel of that limousine and pressed the accelerator to the floor.

As luck would have it, he sped through an intersection where two of Moscow's famous tough motorcycle cops were posted. They looked first at the limousine, then at each other, and then grinned broadly and set off in hot pursuit. But they were just a little bit tense. One motorcycle cut off the limousine and the other policeman stayed behind it with hand on pistol. He saw his colleague look into the car and when, a few seconds later, the colleague returned, he snapped, "What happened? Why didn't you give him a ticket?" And the first policeman replied sheepishly, "I couldn't do it. I was afraid to. This was a real big shot." "What do you mean?" asked the second, "Who was it?" "Well," the first replied, "I didn't recognize him but it must have been somebody really stupendous because Gorbachev was his driver!"

REFLECTIONS ON GLOBAL ENVIRONMENTAL CHANGE

Things are indeed happening in Eastern Europe concerning attitudes toward the environment. As I reflected on these developments, it struck me that it was no coincidence that the social and political upheavals in Eastern Europe and the Soviet Union have occurred at the same time as a great wave of ecological consciousness is sweeping across the world. In point of fact, in many countries ecological movements were at the cutting edge, as a catalyst and a spearhead in demands for reform. It seems to me that both of these trends -- the ecological awareness and the demands for political and economic and social reform -- share common roots: a desire by millions of people around the world to regain control of their own destiny.

A New Generation of Environmental Dangers

These trends are a reflection of a broadening time perspective. There is a growing realization that the preoccupation of modern society with short-term objectives, whether they are political or ideological, military or economic, is fundamentally not only unsatisfying to the human soul but also downright dangerous to the planet on which we depend.

This realization has been fostered by the work of modern science. Scientists are demonstrating that the often inadvertent byproducts of the unprecedented economic growth experienced since the industrial revolution, and especially since the end of World War II, are beginning to threaten basic natural cycles upon which life on Earth depends. These developments are linked to our energy policies and our industrial and agricultural practices, to wasteful consumption patterns in the West and burgeoning population growth in the developing countries. All of these factors are bringing new dangers.

In effect, what we are facing now is a new generation of global environmental challenges; an assault on the planet by land, by sea, and in the air. The list of these global problems is topped, of course, by global warming. In addition, there is depletion of the stratospheric ozone layer, mass extinction of plant and animal species, the destruction of tropical forests, the spread of deserts and soil erosion, and the pollution of wetlands, lakes, freshwater, and oceans. All of these represent something new in their scope and potential consequences.

These are not the environmental issues of twenty years ago, which were basically local and could be addressed by local measures. This new generation of environmental dangers shares a number of unique characteristics. They are very slow in developing, and they are long-term in their impacts. Because of this, their effects can be virtually irreversible over any reasonable time-frame. They are global in scope rather than local or even regional. And they are almost always characterized by a considerable degree of scientific uncertainty. The proof isn't all there.

These are not the kinds of issues that politicians like to deal with. The concepts are not obvious. For example, it is not intuitively obvious that using a perfume spray in Paris propelled by chlorofluorocarbons (CFCs) can lead to deaths by skin cancer and extinction of species half a world away and several generations into the future.

In a sense, we are now in the situation of a boat on a misty night, steaming full speed ahead and confronting a blinking light directly in its way. The commander of this ship orders his first mate to signal a warning light to the approaching object as follows: "You are in my course, please change your direction five degrees south." After a few seconds there is an answering light flashing through the mist: "Have received your message. Request you turn five degrees north." This response makes the commander most impatient and he directs his mate to send another message: "I am repeating my request: change your course five degrees south. This is Admiral Jones, who are you?" The answering light flashes again: "This is Mr. Brown. Repeat urgent request to change your course five degrees north." The admiral is now incensed and signals: "This is my last message. Change your course five degrees south at once. I am a battleship." And the reply flashes back: "Please turn five degrees north. I am a lighthouse."

We are on that ship and the planet is trying to send us some urgent messages -- via science. But there is a lot of fog and the messages aren't always clear and they are not always definitive. There is not always the last shred of evidence that some people seem to need.

Redefining National Security

What this means to me, in looking at our traditional concerns of diplomacy, is that we need to redefine our concept of national security to reflect these new ecological dangers. Arguably, the risk to civilization from a nuclear war is now less acute than the possibility of some kind of ecological collapse unforeseen by the scientists. Such a collapse could ensue from the gradual cumulative effects of hundreds of millions of casual decisions made by people around the world every day -- decisions like using a hairspray that is propelled by chlorofluorocarbons, or building a road through a forest, or driving a few extra blocks when we could have walked.

Addressing these new environmental challenges will require international cooperation of a type and on a scale that has never before been achieved. No single nation or group of nations, however powerful they are, can do it alone. Military power and economic clout are irrelevant in the face of these global environmental threats. What is required is a new kind of diplomacy and new modes of relations among sovereign states.

THE OZONE PROTOCOL: LESSONS FOR GLOBAL COOPERATION

It is against this background that I'd like to try to draw some lessons from an unprecedented example of international cooperation to protect the global environment: an international agreement so unusual that it has been characterized as one of the most significant international accords in diplomatic history; an accord so unusual that The Washington Post uncharacteristically praised the Reagan administration for its leadership on this issue. And President Reagan himself, in urging prompt ratification of the treaty by the U.S. Senate, called it "a monumental achievement ... an extraordinary process of scientific study ... and of international diplomacy."

Obstacles to an Ozone Treaty

The response of the international community to the threat to the ozone layer, as manifested in the September 1987 Montreal Protocol and its subsequent revision in London in June 1990, offers hopeful and useful lessons to a world facing these additional global dangers. We now have some guidelines for the kind of international cooperation that is needed.

Of course, the overarching issue now is global warming, which is admittedly much more complex than the ozone problem. Some observers, in considering the contrasts, conclude that it was easy to control CFCs and hence that the Montreal Protocol has little relevance to climate change. With the benefit of hindsight, the Montreal Protocol has almost achieved an aura of inevitability; it all seems so simple in retrospect.

Well let me tell you, as someone who was there, that memories are short. I am reminded of the historian who said, "All revolutions seem impossible before they occur -- and inevitable afterwards." Most knowledgeable observers back in 1985 or 1986 would not have bet very much that the world would achieve any meaningful international agreement to eliminate an important part of the chemical industry, chlorofluorocarbons and halons that had become nearly ubiquitous in modern society, useful products that were almost synonymous with modern standards of living.

It was not only the chemical producers but also a broad range of industries that were affected by the treaty: food processing, air conditioning, refrigeration, housing construction, oil drilling, transportation, the defense industry, telecommunications, electronics, health care, plastics. Billions of dollars of investment worldwide, and hundreds of thousands of jobs, were at stake. The industrial sector was implacably opposed to international regulations, arguing that the costs of change would be unacceptable and that alternative technologies and chemicals were completely unavailable.

There was also extreme hostility to the idea of international controls from countries that together accounted for two-thirds of the world's production of CFCs, namely, the twelve-nation European Community, Japan, and the Soviet Union. For their part, the developing countries, whose cooperation was also essential because of their large populations and their potential use of these chemicals, was, at best, indifferent to an international treaty. As an Indian diplomat remarked to me, "rich man's problem -- rich man's solution."

And finally, the whole argument for a treaty rested on unproven scientific theories. I'll remind you that at the time we negotiated and signed the Montreal Protocol there was no measurable evidence that CFCs were causing loss in the ozone layer. It all rested on abstract and disputed theories based on arcane computer models at the frontiers of modern science. (In that respect, incidentally, global warming is very much in the same category today.) In fact, each time the scientists went back to their models and refined them, putting in new data and new theories, contradictory predictions for future ozone layer depletion came out. The situation was almost tailor-made for industry and others who wanted to prevent action against CFCs.

In sum, the odds were heavily stacked against any kind of a strong treaty coming out of all this. And I would submit to you that against this background there is really no precedent in international environmental law or in diplomacy for the kind of agreement that ensued.

I'd like to pay tribute in this context to another State Department officer, Dick Smith. Dick carried the protocol into the very necessary next step after Montreal and is responsible for the strengthening of the treaty that took place last June in London. Without Dick's persistence, building on my work and that of others who had been involved, I don't think we would be at the current stage where these chemicals will actually be phased out. With the latest scientific evidence of even more rapid ozone loss than predicted, it seems clear that the Treaty will probably be even further strengthened in the not-distant future.

The Crucial Role of Science

Science played an extraordinary role in these historic environmental negotiations. From the very beginning my State Department colleagues and I steeped ourselves in the science of atmospheric chemistry and atmospheric physics to a degree that would have astonished my high school science teachers. I found it fascinating. I discovered that the scientists themselves were only too happy to explain their still evolving theories to a naive but curious outsider.

Furthermore, this solid understanding, or at least solid superficial understanding, on the part of the diplomats, combined with our continual interaction with the scientists, proved of enormous advantage to the U.S. negotiating team vis-a-vis our protagonists from the European Community. We had the scientists at our elbow while the European diplomats often wouldn't listen to their own scientists. I believe this gave us a real advantage in addressing the uncommitted countries and in presenting a reasonable and well-founded U.S. position.

Let us not forget that without modern science and technology, the world would have been unaware of the danger to the ozone layer. It was almost serendipity that scientists came up with a theory that this almost perfect chemical, CFCs -- nontoxic, noncorrosive, nonflammable, and so very useful in thousands of products and processes -- could in fact cause damage which would threaten human life: damage to a layer of gas thirty miles above the earth's surface that protects animal and plant cells from harmful radiation. It seemed an almost unbelievable presumption.

Scientists and Diplomats: Forging an International Consensus

In arriving at an international political agreement it was extremely important that an international scientific consensus first be developed: a commonly accepted body of data and a narrowing of the range of uncertainties. It was essential to get scientists from other countries involved in this effort, even though it was spearheaded by our own National Aeronautics and Space Administration (NASA) and National Oceanic and Atmospheric Administration (NOAA). The scientific conclusions could not, for obvious political reasons, come only from American scientists.

It was a very interesting and unusual situation. What happened in effect was the gradual development of an international network of atmospheric scientists that had previously worked together only minimally. Dedicated to scientific objectivity, this community was able, working with concerned government officials, to transcend narrow national commercial interests. Before this powerful collaboration of scientists and diplomats, it had been largely the commercial interests that were dictating policy in many governments, especially in Europe and Japan.

The scientists, however, had to get out of their laboratories and interact with us and share some responsibily for the policy options. The negotiators would turn to the scientists and ask them to test on their models the consequences of different policy options -- and to explain and defend their findings. It was uncomfortable for some of them because they were used to staying in their laboratories representing the pure science. Others, however, blossomed under this new scrutiny and limelight and found the policy arena stimulating and challenging.

Scientists participated in the interagency deliberations as we formulated the U.S. negotiating position, and were also on the U.S. delegation. In addition, I held a trump card, Dr. Bob Watson of NASA; he was not on our delegation, but rather I persuaded him to serve as a special advisor to the United Nations Environment Programme, UNEP. In this way, he could function as an objective contributor and not be tied to the U.S. position. He could say what he wanted to in that position and thus had enhanced credibility as a scientist.

In general, the extremely fruitful collaboration between the diplomats and the scientists was an important element in the success of the Montreal Protocol and an important lesson as we approach global warming and similar issues.

Diplomatic Strategy and Scientific Objectivity

In our own diplomatic strategy for these difficult negotiations, we placed great emphasis on the science. We admitted the uncertainties as we approached other governments to explain our position favoring tough controls. We realized that when you have a seventy percent case, it is unwise to try to stretch it to a hundred percent. We were careful not to exaggerate the degree of certainty, because to do so would risk losing valuable credibility and providing ammunition to those who wanted to block the treaty or have weak or no controls.

This reliance on the science and this commitment to objectivity made us seem reasonable to the many uncommitted countries who were involved in these negotiations. And it made the chief opponents of a strong treaty, the European Community, seem as if they were defending narrow commercial interests rather than looking at the broader global threat.

So I believe that the history of the Montreal protocol negotiations presents an extremely useful paradigm for the importance of collaboration between science and diplomacy, as well as for the use of objective scientific data in establishing the rationale for international cooperative action.

Public Opinion and the Media

There were additional important lessons from the ozone experience that are relevant in addressing other global environmental challenges. One is the importance of educating and mobilizing public opinion, which can play a critical role in applying pressure on often cautious politicians. In the United States, these functions were assumed by the media, by environmental organizations, and by congressional hearings.

One should never make the mistake of underestimating the power of consumers when they are educated and motivated. In the United States, for example, between 1975 and 1977 the market for aerosol products dropped by two-thirds before there was any formal government ban on CFCs in aerosols. This was the result of tens of millions of individual decisions by American consumers, when they went to their supermarkets and drugstores, not to purchase products that were propelled by aerosol sprays. Today your aerosol products are fine, but at that time CFCs were the propellants, and what consumers were reading about and seeing on television persuaded them to buy something different. And this provided a very powerful signal to industry that they should come up with substitutes, which they soon did.

Mindful of the power of public opinion, an important part of the U.S. diplomatic strategy was directed at the media. We tried to influence foreign public opinion, reaching out to get beyond the governments and to reach directly to foreign consumers and environmental groups. We scheduled many overseas press conferences, briefings, and radio and television appearances.

In fact, Charles Wick, who was director of USIA at the time, featured the ozone issue on his WorldNet, which is a televised real-time interactive interview program. We would be in a studio in Washington and there would be an audience in studios in Europe, Japan, or South America that could then ask us questions. The program was beamed to particular capitals and through our embassies' efforts the discussions were directly plugged into foreign media. The usual participants were Dr. Bob Watson of NASA and myself -- again science and diplomacy -- and we appeared so often that USIA staff dubbed it the "Bob and Dick Show." If any of you people had insomnia a couple of weeks ago you would have seen our reunion: Bob and Dick reunited to discuss implications of the latest scientific evidence on the faster-than-expected depletion of the ozone layer. It was on CBS Nightwatch at 4 a.m. and fortunately for Bob and Dick they tape it at a different time.

National Leadership and Multilateral Diplomacy

A major lesson from the Montreal Protocol is the crucial importance of national leadership. On the ozone issue, the United States took a prominent leadership role. Leadership in an international endeavor is, of course, important not only in the realm of the environment. Recent events in the Middle East are a clear demonstration that when the United States government finally puts its mind to something, it can really get things done. The same kind of leadership will be needed in effectively addressing the other new environmental threats, particularly global warming.

U.S. leadership in the ozone negotiations required an interesting synthesis of bilateral diplomacy and multilateral diplomacy. It became clear that you can't just show up for a complex multilateral negotiation without having undertaken a lot of advance bilateral negotiation and preparation. You have got to know and understand the other countries' positions and concerns before they arrive at the formal negotiation with fixed positions. The ozone negotiation was almost a classic study of the use of our overseas embassies in engaging host country governments in a constructive dialogue aimed at influencing and gradually winning over other delegations.

In this connection, I would also like to acknowledge the crucial role of the United Nations. The United Nations Environment Programme, UNEP, a relatively little known and unsung specialized agency, was the model of an effective multilateral organization in action. UNEP provided an objective forum and it stuck to the substantive issues. Unlike some UN bodies, it was able to avoid the temptation of debating extraneous political issues (such as apartheid or the West Bank) that belong in other fora. The head of UNEP, Mostafa Tolba, an Egyptian, is himself a scientist. Tolba was instrumental in providing the right atmosphere for discussion of these scientific and technical issues, tough enough as they were, without getting into political debates. UNEP was also responsible for persuading countries to join in the process. Through cajoling and pressuring and pleading, Tolba gathered around the bargaining table representatives of governments that may not have been initially interested.

Industry and the Market Mechanism

Another element that should not be forgotten is the role of private industry and of the market mechanism. Industry is not just a source of environmental problems; it is a major part of the solution. And the operations of the market mechanism are an important factor as well.

The market, of course, is basically neutral with respect to the environment. Sometimes, through policy, through regulations, through taxes, it is necessary to give the market a nudge in the right direction order to protect the environment. Through the market mechanism, one can provide a clear signal to industry that it is worth their while to invest in environmentally beneficial substitutes and alternative technologies. In the absence of that signal they'll simply say that it is impossible or too costly.

The Montreal Protocol tried to harness these powerful forces -- the intelligence and the financial resources of industry -- to environmentally beneficial ends. In effect, the protocol was a technology forcing instrument. We imposed deadlines on industry to phase down CFCs in full knowledge that the substitutes for these substances did not yet exist. The treaty thus served to stimulate competition and the kind of entrepreneurship that would bring forth new technologies.

When certain ideologues within the Reagan administration accused me of being some kind of a radical because I was arguing for strong controls and tried to remove me as chief U.S. negotiator, I reminded them that I had graduated from Harvard Business School. As Lincoln Gordon (who was my mentor at Harvard) can testify, my position on the Montreal Protocol was rooted in a genuine faith in the market system: namely, that if you send industry the right kind of signal, Yankee ingenuity will come up with the right answers.

And this is precisely what has happened. Almost every week we now read of some new product or technology coming onto the market to substitute for CFCs or for processes that used CFCs. A spokesman for DuPont, which was the prime producer of CFCs, actually admitted that without the right market constellation, meaning prices and regulations, it just didn't pay them to do the research. In fact, when the Reagan administration entered office in 1981, DuPont stopped the small amount of research they had been doing on CFC substitutes, and they only resumed five years later when it became clear that there would be negotiations for international regulation. So sometimes you have to set the market in motion and use the amazing forces of our wonderful market system to a good end beyond the time-frame of the quarterly profit-and-loss statement.

Unprecedented Population Growth

The developing countries didn't cause these problems, but with their huge populations and continuing unprecedented population growth, their potential contribution to exacerbating environmental degradation problems could be enormous. Even though population growth rates have declined slightly, in terms of absolute numbers each year shows greater additions than the preceding year; this is because the base population is so large, and even if today's young parents have fewer children per couple, there are simply more parents producing more babies in total. The United Nations in the last few years has had to repeatedly revise upwards its projections of population levels for the coming decades.

And I am not reluctant to say this at a Jesuit university: when the Pope came out last year with his otherwise excellent statement on protecting the environment, he did not address this crucial population dimension. Shortly after the papal document was released, I was invited to speak at the U.S. Catholic Bishops Conference. I praised the Pope's statement but I also reminded the audience that it left out a very important component of the ecological threat which he had otherwise aptly identified. And I was gratified when some prominent churchmen came up afterwards and told me that they were happy that I had made this point. I guess I may have been uncustomarily diplomatic in my formulation of the population issue that day.

It is increasingly clear that the industrialized nations cannot ignore population growth in the Third World, but they must also, by leadership and by direct contributions, assist developing countries in dealing with the problem. These countries aspire to emulate the lifestyles and the energy and industrial policies that have made us wealthy. And if they do that, with the numbers of people involved, it will make the kind of ecological collapse that I indicated earlier a real possibility.

Equity and the Third World

In effect, we in the West have grown wealthy while inadvertently polluting the global commons. And now the time has come to use some of that wealth to help the developing countries assume a fair share of responsibility for addressing the new global environmental dangers. Developing countries must themselves undertake environmentally sensible policies. But in doing so, they cannot be asked to depress their own already low standards of living or to jeopardize their prospects for future economic growth.

Thus, they need our help, and on balance it will be a good investment. Because, otherwise, we will all lose together.

Another pertinent innovation of the Montreal Protocol was the first global environmental fund. This mechanism was designed at the Meeting of Parties in London last year to provide financial assistance and transfer new technologies to enable developing countries to achieve the required phaseout of ozone-depleting substances. To solve the global warming issue we may well need to go further. We are going to need creative approaches, new modes of aid and technology transfer, and enhanced scientific cooperation.

ACTING UNDER SCIENTIFIC UNCERTAINTY

A final and singularly important lesson from the Montreal Protocol experience that I would like to bring to your attention is the necessity for governments to act under conditions of scientific uncertainty.

We simply cannot wait for complete evidence on these issues -- especially global warming. Political leaders must learn to avoid the temptation to heed the self-serving claims of entrenched economic interests who continually stress that the threats are remote and the costs of adaptation or change are astronomical. Somehow it seems to me that we have a natural tendency to overestimate the costs and underestimate the possibilities for social and technological change once we put our minds to it.

The Danger of Thresholds

Here again science can be a big help to policymakers. What we are finding in many of these issues is that nature doesn't provide us with convenient early-warning signals. This is what the scientists characterize as non-linearity: there is not necessarily a smooth and continual build-up of signals. What happens instead is that the Earth's systems absorb human abuse up until a certain threshold and then there is a sudden collapse that surprises us all. And one cannot readily predict when the threshold will be crossed.

A good example of this phenomenon of non-linearity is the Antarctic "ozone hole," of which we all have heard so much. I would remind you that this seasonal ozone collapse had not been definitively linked with CFCs at the time that we negotiated and signed the Montreal Protocol. It was only six months after the protocol was signed that the evidence became clear that the Antarctic ozone hole was caused by chlorine from CFCs. The governments thus acted before there was complete evidence.

As the scientists came to better understand the ozone chemistry, they realized that for years the amount of chlorine in the atmosphere over Antarctica, and in fact over the whole world, had been slowly but steadily increasing. Chlorine is the chemical that has the ability to destroy the ozone layer. The scientists discovered that there were natural levels of chlorine in the stratosphere, amounting to 0.6 parts per billion, which posed no threat to the ozone layer. However, because of increasing emissions from manmade CFCs, the chlorine levels began creeping up year by year: 0.8 parts per billion, 1.1 ... 1.4 ... 1.6 ... and nothing happened to the ozone layer. It was only when the amount of chlorine reached 2.0 parts per billion -- not a very large quantity -- that something dramatic occurred: this turned out to be a critical "triggering" level.

So unexpected was this sudden collapse of ozone, so inconsistent with any existing atmospheric theory, that the scientists at first found it unbelievable. The British team that discovered it in 1983 at their research station at Halley Bay in Antarctica spent nearly two years rechecking their data. They thought that something might be wrong with their measuring devices. They published the data in 1985, but there was still no clear link to CFCs and many theories were advanced to explain the ozone hole.

When the British data appeared, NASA went back to the drawing boards because we had satellites that were monitoring trace gases in the stratosphere and they had not come up with any loss of ozone over the Antarctic. These satellites generate millions and millions of pieces of information -- much too much for poor human beings to analyze. And so they are programmed not to bring forward data which are anomalies, which are beyond any possibility. The Antarctic ozone loss was far off the chart because it was not explainable by any theory. Thus the NASA scientists discovered that the data had been stored neatly away but the computer had not signalled it for the attention of the analysts because it was too incredible. It was not long after the British results that NASA was able to confirm that this extraordinary phenomenon was indeed happening.

I came across a similar threshold example in the forests of Poland, where I witnessed unbelievable destruction. The foresters told me that 12 years earlier, these barren slopes had been as thickly forested as nearby mountainsides where the prevailing winds were not depositing acid rain and heavy metals. By the time they saw the first signs of damage at the very tops of those great hundred-foot spruce and fir trees it was too late to stop the process. The subsequent decline was irreversible. In other words, there was no convenient early warning signal.

A Prudent Insurance Policy

As we look at the current positions of different countries, including our own, on the global warming issue, I am reminded of Pascal and his famous wager on the existence of God. This French philosopher asked himself what a reasonable person should believe and act upon given that there is no proof of God's existence. As a rationalist, Pascal focussed on the consequences of his beliefs and actions. If God doesn't exist, he reasoned, and I believe in Him anyway, there is no harm done; but if it is the other way around, there might be some very serious penalties to pay. Therefore, Pascal bet that God exists.

We are facing a similar dilemma with respect to some of these global environmental dangers where the evidence isn't all in. In the case of the ozone layer, the U.S. leadership policy and the eventual Montreal Protocol represented a prudent insurance policy in the face of the best, but still uncertain, information that science could provide. Soon enough we came to appreciate that imposing strong controls on CFCs was an action totally vindicated by the continued findings of science.

CONCLUSION: THE VIEW FROM OUTER SPACE

In concluding this brief survey of my personal experience with these issues, I'd like to recall a conversation I had a few years ago at a UNEP conference that brought together a Soviet cosmonaut and an American astronaut. The Russian was an air force major and the American was a scientist from the University of Washington. This was during the Cold War and there was still considerable friction on the political front. But in listening to these two people who had been in outer space, I was very impressed with the shared vision that they had of our planet. They spoke movingly of how they had looked back on Earth and saw it shimmering with life and light, alone and fragile in the black cosmos. It occurred to me how the traditional maps of diplomacy and the frontiers of geopolitics vanished from that perspective; what was evident to the astronaut and the cosmonaut was the interconnectedness of all the systems of our blue planet: the land, the water, and the atmosphere, and the life on it and in it.

This vision of Earth is very relevant to the international environmental negotiations now going on, dealing with climate change and similar dangers. In the realm of international relations there is always going to be resistance to change and there are always uncertainties -- political, scientific, economic, psychological uncertainties. Perhaps the greatest significance of the Montreal Protocol may be that it demonstrated that the international community of nations is capable of undertaking complicated cooperative actions in this real world of ambiguity and imperfect knowledge.

I hope, therefore, that the Montreal Protocol may serve as a paradigm for the new kind of global diplomacy that I mentioned earlier as being so necessary. It will be essential to understand the lessons of the ozone experience if we are to deal effectively with the far more complicated problem of global warming. In this new diplomacy, sovereign nations can, in partnership with industry and with citizens' groups, assume responsibility for the stewardship of this planet and for the security of the generations that will follow us.

QUESTIONS AND ANSWERS:

Advice to Scientists Studying Global Warming

Q. Do you have some special thoughts for the scientists in this Global Warming conference, and more generally for scientists working on issues that have significant social impact? What advice do you have for us when we get to the end of the conference where the question will be: where do we take it from here?

Amb. Benedick: My first thought would be: be true to the science. Look at it the way you have been trained to look at it. And try not to think about other implications. Let the economists think about the costs. By the way, I believe economics is far behind science in terms of the way they are looking at environmental issues. The basic tools of economic analysis are flawed. When economists employ present-value discounting to analyze future risks, they inevitably trivialize the costs to future generations of cleaning up the mess that we leave to them.

For the future, perhaps include at least a couple of diplomats and more non-scientists in your conference discussions, because I think that kind of interaction can be very fruitful. But that may come at a later stage.

Again: scientists should call it the way they see it. At this stage, you will look at the data with all of your training and your objectivity and you're not going to be able to come up with foolproof certainty. I would be surprised if you did so, but at least you can come up with ranges of probabilities. After all, if the economists were so good with predictions they would all be making money on the stock market ... maybe some of them are...

The Need for Domestic Reforms

Q. Since we in this country still have not come to grips with the single most difficult issue affecting global warming, namely, the interconnection of that problem with transportation, how are we going to assist third world countries?

How are we going to assist those countries in changing their development direction, including their transportation development, when we can't seem to do it?

Amb. Benedick: I think that the best way to assist the developing countries is to get our own house in order first. The National Academy of Sciences recently came out with an assessment of the climate change issue. They concluded that the United States could reduce its emissions of carbon dioxide by significant amounts at virtually no cost or even with some benefit, if we followed the right policies in the transportation and other sectors, particularly energy efficiency and conservation.

A major difference between the ozone issue and climate change is that there are so many contributing sources to the greenhouse effect. If you look at any one of those sources, you miss the whole picture. When I talk to people from General Motors they say "well, the U.S. automobile industry is only 2 to 3 percent of the problem." The Brazilians say "destruction of rain forests is only 6 to 7 percent of the problem." It's not like CFCs which were 70 percent of the ozone problem. If you add up all these 3 and 4 percents, they add up to about 140 percent ... the fact is that we have to start somewhere on climate change, even if the immediate impact does not seem decisive. There is no quick fix.

There are many reasons why our transportation sector needs reform, and global warming is only one of them. On your way to work on Monday you might want to ponder what some of the other reasons are, as you sit and watch the cars backed up on Key Bridge.

One thing that makes me very uncomfortable, even as a non-scientist, is the black and white comparisons that so often come out when some people exaggerate the difficulties of what can be done. The real world is actually shades of gray. When a government spokesman says that we can't effectively tackle the global climate problem because Americans don't want to do without their automobiles, he must be missing the point. It's not a question of either driving your car or doing something about carbon dioxide. There are all kinds of gradations in between: drive a different car, drive it less often.

Scientists can teach us a lot about avoiding the tendency to express things in extremes, because most good scientists do not make categorical statements. They recognize that nature and the real world are ambiguous. And it's often politicians who try to set these stark contrasts, which then lead them to the wrong conclusions.

Is the Ozone Problem Solved? What Ensures Compliance?

Q. First, with the Montreal Protocol to protect the ozone layer as it is now strengthened, have we solved the problem? Given that the molecules that attack the ozone layer could remain destructive for a hundred years or so, are we out of the dark? Second, what are the compliance mechanisms of the treaty? What is going to keep it in force and how are nations going to be held accountable?

Amb. Benedick: On the first of your questions: We're not yet home free. Absolutely not. Don Hodel was right: Wear sunscreen and broadrimmed hats and sunglasses. He was wrong, of course, when he said don't control the CFCs, because even with a phaseout, the effects of these long-lived substances are going to remain with us for many decades.

The protocol currently provides for an 85 percent reduction of CFCs by 1997 and a total phaseout by 2000. In practice, they will probably be phased out by 1997, given the pace of research into substitutes and the pressures of competition. They probably cannot be phased out much faster but that will not make much difference because the damage we have to live with for the next century was caused by earlier emissions.

There are other things we could do that would have more immediate positive effects on the ozone layer: specifically, getting rid of methyl chloroform. That chemical is responsible for about a sixth of the ozone depletion problem. Since it is shorter-lived than CFCs, a rapid phaseout of methyl chloroform brings a fast impact. You could significantly reduce the chlorine loading of the atmosphere within four or five years.

Methyl chloroform has up to now been treated too generously, even in the revised Montreal Protocol: it is to be phased out in 2005, but only slated for a 30 percent reduction by 1995. In my naivete, I don't see why we couldn't phase it out by 1995 -- the Germans plan a phaseout by 1995, and if they can do it, so can we. Substitutes do exist for methyl chloroform.

Still and all, you should not forget to wear your sunscreen. And all these Georgetown students in their twenties should really wear sunscreen because skin cancer takes 20 to 30 years to develop. Linc Gordon and I don't have to worry about it quite as much as the teenagers.

Anticipating the Unexpected

Q. Would you comment on the recent announcement of the doubling of the rate of the ozone layer depletion. Was that within the range of your uncertainties that you could easily accept in 1987 in Montreal? Was it a surprise? Was it known to you all along?

Amb. Benedick: It was not known to me all along, and it was not known to the scientists all along. It certainly was not within the range that we were thinking of when we signed the treaty in 1987. Only six months after the treaty was signed we had new evidence that indicated it would have to be strengthened, because the ozone layer was being depleted more rapidly than predicted. We had no evidence before then of ozone layer depletion apart from the Antarctic, which was an anomaly and was not linked with CFCs at the time the treaty was negotiated and signed.

This new upward revision of the rate of depletion of the ozone layer was unexpected, but in a way it was not a surprise. I recently spoke with Bob Watson and other scientists who were involved. It was not a surprise because we now understand that there are these unpredictable thresholds that I mentioned earlier, these non-linear reactions from Nature. And once you realize that there are billions of tons of chemicals going up in the atmosphere, and into the oceans, and into the ground water and the earth -- chemicals that do not belong there in those quantities -- we shouldn't be surprised anymore if we get unexpected reactions from that lighthouse out there blinking in the night.

Author Index

A

Ackerman, T. P., 1

B

Barkstrom, B. R., 55
Benedick, R. E., 292
Bierbaum, R. M., 237
Blake, D. R., 162
Broecker, W. S., 129

C

Cess, R. D., 46
Coppock, R., 222

F

Friedman, R. M., 237

H

Harrison, E. F., 55

K

Kasting, J. F., 175

L

Levenson, H., 237

M

Maul, G. A., 78

P

Price, L., 261

R

Ramanathan, V., 55
Randall, D. A., 24
Rapoport, R. D., 237
Rosenfeld, A. H., 261

S

Shukla, J., 113
Sundt, N., 237

T

Trexler, M. C., 201

W

Walker, J. C. G., 175

AIP Conference Proceedings

		L.C. Number	ISBN
No. 234	Amorphous Silicon Materials and Solar Cells (Denver, CO, 1991)	91-55575	088318-831-7
No. 235	Physics and Chemistry of MCT and Novel IR Detector Materials (San Francisco, CA, 1990)	91-55493	0-88318-931-3
No. 236	Vacuum Design of Synchrotron Light Sources (Argonne, IL, 1990)	91-55527	0-88318-873-2
No. 237	Kent M. Terwilliger Memorial Symposium (Ann Arbor, MI, 1989)	91-55576	0-88318-788-4
No. 238	Capture Gamma-Ray Spectroscopy (Pacific Grove, CA, 1990)	91-57923	0-88318-830-9
No. 239	Advances in Biomolecular Simulations (Obernai, France, 1991)	91-58106	0-88318-940-2
No. 240	Joint Soviet-American Workshop on the Physics of Semiconductor Lasers (Leningrad, USSR, 1991)	91-58537	0-88318-936-4
No. 241	Scanned Probe Microscopy (Santa Barbara, CA, 1991)	91-76758	0-88318-816-3
No. 242	Strong, Weak, and Electromagnetic Interactions in Nuclei, Atoms, and Astrophysics: A Workshop in Honor of Stewart D. Bloom's Retirement (Livermore, CA, 1991)	91-76876	0-88318-943-7
No. 243	Intersections Between Particle and Nuclear Physics (Tucson, AZ 1991)	91-77580	0-88318-950-X
No. 244	Radio Frequency Power in Plasmas (Charleston, SC 1991)	91-77853	0-88318-937-2
No. 245	Basic Space Science (Bangalore, India 1991)	91-78379	0-88318-951-8
No. 246	Proceedings of the Ninth Symposium on Space Nuclear Power Systems	91-58793	Casebound Pt. 1: 1-56396-021-4 Pt. 2: 1-56396-023-2 Pt. 3: 1-56396-025-7 Set: 1-56396-027-3 Paperback Pt. 1: 1-56396-020-6 Pt. 2: 1-56396-022-2 Pt. 3: 1-56396-024-9 Set: 1-56396-026-5
No. 247	Global Warming: Physics and Facts (Washington, DC 1991)	91-78423	0-88318-932-1
No. 248	Computer-Aided Statistical Physics (Taipei, Taiwan 1991)	91-78378	0-88318-942-9
No. 249			
No. 250	Towards a Unified Picture of Nuclear Dynamics (Nikko, Japan 1991)	92-70143	0-88318-951-8